Security of Energy Supply in Europe

LOYOLA DE PALACIO SERIES ON EUROPEAN ENERGY POLICY

Series Editor: Jean-Michel Glachant, *Loyola de Palacio Professor for European Energy Policy and Director, Florence School of Regulation, European University Institute, Italy, and Professor of Economics, Université Paris-Sud, France*

The *Loyola de Palacio Series on European Energy Policy* honours Loyola de Palacio (1950–2006), former Vice-President of the European Commission and EU Commissioner for Energy and Transport (1999–2004), a pioneer in the creation of an EU Energy Policy.

This series aims to promote energy policy research, develop academic knowledge and nurture the 'market for ideas' in the field of energy policy making. It will offer informed and up-to-date analysis on key European energy policy issues (from market building to security of supply; from climate change to a low carbon economy and society). It will engage in a fruitful dialogue between academics (including economists, lawyers, engineers and political scientists), practitioners and decision-makers. The series will complement the large range of activities performed at the Loyola de Palacio Chair currently held by Professor Jean-Michel Glachant (Loyola de Palacio Professor for European Energy Policy and Director, Florence School of Regulation, European University Institute, Italy, and Professor of Economics, Université Paris-Sud, France).

Security of Energy Supply in Europe
Natural Gas, Nuclear and Hydrogen

Edited by

François Lévêque
Professor of Law and Economics, École des Mines de Paris, France

Jean-Michel Glachant
Loyola de Palacio Professor for European Energy Policy and Director, Florence School of Regulation, European University Institute, Italy, and Professor of Economics, Université Paris-Sud, France

Julián Barquín
Professor, Institute for Technology Research, Universidad Pontificia Comillas, Madrid, Spain

Christian von Hirschhausen
Professor for Energy Economics and Public Sector Management, University of Technology Dresden, and Research Director, DIW Berlin (German Institute for Economic Research), Germany

Franziska Holz
Senior Researcher, DIW Berlin (German Institute for Economic Research), Germany

William J. Nuttall
Senior Lecturer in Technology, Judge Business School and Engineering Department and Assistant Director of the Electricity Policy Research Group, University of Cambridge, UK

LOYOLA DE PALACIO SERIES ON EUROPEAN ENERGY POLICY

Edward Elgar
Cheltenham, UK • Northampton, MA, USA

© François Lévêque, Jean-Michel Glachant, Julián Barquín, Christian von Hirschhausen, Franziska Holz and William J. Nuttall 2010

All rights reserved. No part of this publication may be reproduced, stored in a retrieval system or transmitted in any form or by any means, electronic, mechanical or photocopying, recording, or otherwise without the prior permission of the publisher.

Published by
Edward Elgar Publishing Limited
The Lypiatts
15 Lansdown Road
Cheltenham
Glos GL50 2JA
UK

Edward Elgar Publishing, Inc.
William Pratt House
9 Dewey Court
Northampton
Massachusetts 01060
USA

A catalogue record for this book
is available from the British Library

Library of Congress Control Number: 2009941262

Mixed Sources
Product group from well-managed forests and other controlled sources
www.fsc.org Cert no. SA-COC-1565
© 1996 Forest Stewardship Council

FSC

ISBN 978 1 84980 032 7

Printed and bound by MPG Books Group, UK

Contents

List of contributors	vii
Introduction	ix

PART I   NATURAL GAS

1 Supply security and natural gas   3
   Christian von Hirschhausen, Franziska Holz, Anne Neumann and Sophia Rüster
2 Seeking competition and supply security in natural gas: the US experience and European challenge   21
   Jeff Makholm
3 The new security environment for European gas: worsening geopolitics and increasing global competition for LNG   56
   Jonathan Stern
4 Natural gas and geopolitics   91
   David G. Victor

PART II   NUCLEAR POWER

5 European electricity supply security and nuclear power: an overview   109
   William J. Nuttall and David M. Newbery
6 Contractual and financing arrangements for new nuclear investment in liberalized markets: which efficient combination?   117
   Dominique Finon and Fabien Roques
7 Nuclear power and deregulated electricity markets: lessons from British Energy   155
   Simon Taylor
8 Nuclear energy in the enlarged European Union   166
   William J. Nuttall

PART III   HYDROGEN

9 Supply security and hydrogen   191
   Julián Barquín and Ignacio Pérez-Arriaga

10	Hydrogen from renewables *Dries Haeseldonckx and William D'haeseleer*	199
11	Build-up of a hydrogen infrastructure in Europe *Martin Wietschel, Philipp Seydel and Christoph Stiller*	221
12	The contributions of the hydrogen transition to the goals of the EU energy and climate policy *Anders Chr. Hansen*	248
13	R&D programs for hydrogen: US and EU *Steven Stoft and César Dopazo*	275

PART IV   CONCLUSIONS

14	EU energy security of supply: conclusions *Jean-Michel Glachant, François Lévêque and Pippo Ranci*	295

Name index 303
Subject index 305

Contributors

Julián Barquín is Professor at the Institute for Technology Research (IIT) at Universidad Pontificia Comillas, Madrid (Spain).

William D'haeseleer is Professor of Mechanical Engineering and Director of the KU Leuven Energy Institute (Belgium).

César Dopazo is Professor of the Mechanics of Fluids at Centro Politécnico Superior, University of Zaragoza (Spain).

Dominique Finon is Professor of Economics at Université Paris-Sud and Director, Laboratory of Networks and Energy Systems Economic Analysis (France).

Jean-Michel Glachant is Loyola de Palacio Professor for European Energy Policy and Director of the Florence School of Regulation, European University Institute (Italy), and Professor of Economics at the Université Paris-Sud (France).

Dries Haeseldonckx is a researcher at the Department of Mechanical Engineering, KU Leuven (Belgium).

Anders Chr. Hansen is Associate Professor at the Department of Environmental, Social and Spatial Change, Roskilde University (Denmark).

Christian von Hirschhausen is Professor for Energy Economics and Public Sector Management at the University of Technology Dresden, and Research Director at DIW Berlin (German Institute for Economic Research) (Germany).

Franziska Holz is Senior Researcher at DIW Berlin (German Institute for Economic Research) (Germany).

François Lévêque is Professor of Law and Economics at the École des Mines Paris / Paris Tech (France).

Jeff Makholm is Senior Vice President with National Economic Research Associates, Inc. (NERA), Boston (USA).

Anne Neumann is Senior Researcher at the German Institute for Economic Research (DIW Berlin), and was previously at the Chair for Energy Economics and Public Sector Management at the University of Technology Dresden (Germany).

David M. Newbery is Professor of Applied Economics at the Department of Economics and Director of the Electricity Policy Research Group, University of Cambridge (UK).

William J. Nuttall is Senior Lecturer in Technology at the Judge Business School and the Engineering Department and Assistant Director of the Electricity Policy Research Group, University of Cambridge (UK).

Ignacio Pérez-Arriaga is Professor of Electrical Engineering and Director of the Institute for Technological Research, Universidad Pontificia Comillas, Madrid (Spain).

Pippo Ranci is Professor at the European University Institute, Florence and former head of the Florence School of Regulation (Italy).

Fabien Roques is a research associate at the Electricity Policy Research Group, University of Cambridge (UK).

Sophia Rüster is a researcher at the Chair for Energy Economics and Public Sector Management at the University of Technology Dresden (Germany).

Philipp Seydel is a researcher at the Fraunhofer Institute for Systems and Innovation Research, Karlsruhe (Germany).

Jonathan Stern is Professor and Director of Gas Research at the Oxford Institute for Energy Studies (UK).

Christoph Stiller is a researcher at the Fraunhofer Institute for Systems and Innovation Research, Karlsruhe (Germany).

Steven Stoft is an energy economist and advisor for the California Energy Institute (USA).

Simon Taylor is Lecturer in Finance at the Judge Business School, University of Cambridge (UK).

David G. Victor is Professor at the School of International Relations and Pacific Studies, and Director of the Laboratory on International Law and Regulation, University of California, San Diego (USA).

Martin Wietschel is senior researcher at the Fraunhofer Institute for Systems and Innovation Research, Karlsruhe (Germany) and a lecturer at the University of Karlsruhe and ETH Zurich (Switzerland).

Introduction

1 THE ISSUE: SUPPLY SECURITY

Supply security is of utmost importance for the European Union (EU) and its member states, in economic, technical, and political terms. Secure energy supply is a cornerstone of the 'magic triangle' of energy policy, the two others being competition and sustainability. And in times of rising geopolitical conflicts, supply security has also increased in importance in the external relations of the EU.

This book presents the main ideas resulting from the CESSA project – CESSA stands for 'Coordinating Energy Security in Supply Activities', which gives the leitmotiv of the project. The project was funded by DG Research within the 6th Framework Programme, and it was also supported by DG TREN through information and access to decision makers. CESSA was coordinated by the Université Paris-Sud and the École des Mines de Paris/Paris Tech, with work packages attributed to the University of Cambridge, the Universidad Pontificia Comillas in Madrid, and the German Institute for Economic Research (DIW Berlin) in cooperation with the Chair of Energy Economics and Public Sector Management at the University of Technology (TU) Dresden. The Florence School of Regulation provided input to the project coordination and the conclusions. In addition, scholars from Stanford University and the Massachusetts Institute of Technology, among others, contributed to the work.

The overarching belief expressed in this book is that a sustainable energy policy, more competition, and better regulation will increase global welfare. The authors are convinced that ideas, theories and facts improve policies. There is also a consensus that the EU lacks an energy policy and national energy policies fail to match EU goals. Rather than engage in own theoretical research, the objective of CESSA was to work towards a consensus on critical issues in supply security of the EU. Five large conferences served as platform of the work (Berlin, May 2007; Cambridge, December 2007; Madrid, April 2008; Florence, June 2008; and Brussels, October 2008). Stakeholder meetings were regularly convened, and part of the discussion was carried out on the Energy Policy Blog where CESSA scholars regularly publish their research findings and policy recommendations (www.energypolicyblog.com).

2 STRUCTURE OF THE BOOK

The contributions in this book share the conviction that energy security is an important issue that should be addressed through economic analysis to yield policy-relevant conclusions. Although we believe in the necessity of solid economic theory, the book adopts a policy perspective. We refer to the underlying theoretical discussions in extensive references. This starts with the very definition of 'supply security': the reader will find 27 different approaches to supply security in this book. A single, universal definition of supply security does not exist.

The book consists of four parts. The first three are dedicated to specific energy sources: natural gas, nuclear fission, and hydrogen. These were chosen because they represent the 'original' energies, as compared to oil which dominated the supply security debate of the 1970s and 1980s. Each energy sector has different characteristics and specific needs for supply security actions. Supply security for the natural gas sector has some important geopolitical aspects. Nuclear power may not be part of the energy supply mix in the entire EU and faces great regulatory diversity between member states. Hydrogen is not yet a deployed technology and may not be part of the energy security measures in the short term, but it has a high long-term potential.

Parts I–III each begin with a summary of the CESSA work packages. They focus on the specific supply security issues in the sectors involved, the technical, economic and political aspects related thereto, and the policy implications. These chapters result from the 'policy briefs' that were written as a concise summary of the work packages, and destined for policy makers and other decision makers. In addition, each part contains three to four specific chapters.

Part IV summarizes the main findings of this book in 10 policy conclusions. Going beyond the sector-specific debates, we find that energy security of supply is a highly political matter. Thus, international relations, geography, and even infrastructure control are involved. However, economics has much to offer to improved policy making for security of supply. Markets are not sufficient to yield the optimal level of supply security, but enlarging markets is an important element. Just like other public goods, supply security needs 'good' regulation.

3 ACKNOWLEDGMENTS

The CESSA project was supported by the European Commission with funds from DG Research within the 6th Framework Programme, and by DG Energy and Transport with access to information and decision makers. The project benefited greatly from the lively discussions between scholars and stakeholders from the industry and regulatory and government agencies during our four conferences in 2007 and 2008. We hope that our recommendations not only reflect the diversity of points of view but also represent a consensus between all members of the CESSA community.

Last but not least, this book is the first volume of the 'Loyola de Palacio Series on European Energy Policy' edited by Professor Jean-Michel Glachant and published by Edward Elgar. We are proud to open this important series with an important European topic, and wish it every success!

PART I

Natural gas

1. Supply security and natural gas

Christian von Hirschhausen, Franziska Holz, Anne Neumann and Sophia Rüster[1]

> There is no inherent conflict between the liberalization of electricity and gas sectors that meet reasonable supply security goals as long as the appropriate market, industry structure, market design, and regulatory institutions are developed and implemented.
> Paul Joskow, Beesley Lecture, London, October 25, 2005, p. 2

> More transparency on prices and flows and more competitive internal markets could bring beneficial effects from international competition in the long-term, as well as improving gas security.
> IEA (2008): *Natural Gas Market Review 2008*, p. 15

1 INTRODUCTION

Energy security, and in particular the security of natural gas supplies, is currently the subject of intense discussion. In times of increasing competition for world natural gas supplies accompanied by increasing import dependency of many countries, the European Union (EU) has to position itself in the world natural gas market and develop a strategy for future energy policies. This chapter has two objectives: (i) to summarize the main issues about (European) supply security regarding natural gas, as discussed at the CESSA conferences; and (ii) to introduce the reader to the breadth of the research and policy debate that was carried out within the CESSA project, summarizing the important chapters, chosen among the nine working papers, that cover both regulatory and geopolitical aspects, including a North American and a European perspective.

The current discussion about security of natural gas supply is taking place amidst the most fundamental changes that the industry has seen for decades:

- The *restructuring* of the European natural gas industry is under way, trying to emulate positive experiences from the North American

example, without repeating the mistakes. The '3rd Package' on EU climate and energy policy, which was a regular 'friend' to the CESSA project, is intended to propel the European natural gas industry towards competition and competitiveness, as part of the 'Lisbon Agenda' to increase the competitiveness of the EU at large.
- Geopolitically speaking, natural gas *supply security* now occupies center stage with many countries and companies. This involves of course the 'usual suspects' of a cartelization, that is, the countries of the Arab Gulf and other potential members of the Gas Exporting Countries Forum (GASPEC). But it also includes Russia, the country with the largest natural gas reserves basin in the world, and which has shown on various occasions that – beyond its well-deserved economic interests – it considers natural gas as part of its foreign policy. As Jonathan Stern puts it in this volume (ch. 3) the 'dash for gas' takes place in an ever-worsening geopolitical context.
- Last but not least, *demand uncertainty* is larger than ever before, driven by both the quest for sustainable energy systems and a low-carbon world, and the unforecastable effects of the worst economic and financial crisis since 1929. Both factors, climate change and the financial crisis, have added an element of stochasticity to natural gas demand, and, hence, to investment into the sector, be it upstream production fields, midstream pipelines and liquefied natural gas (LNG) terminals, or downstream distribution infrastructure.

So, given this uncertain environment, why is our main message not to worry about natural gas supply? Based upon economic considerations, mixed with a grain of geopolitical neutrality, we believe that Europe is well prepared to meet the challenges of energy supply security, including natural gas, as long as it adheres to its reform-oriented agenda, and does not yield to the siren voices of national champions and vested national interests. In general, Europe benefits from a relatively well-diversified natural gas supply portfolio, whereas the strong dependence of East and South-East European countries still poses significant problems. However, as we argue in this introductory chapter, and all through Part II of this book, the future does not reside in national approaches, but in a coordinated, and regionally specific approach at the European level, with as little intervention as possible.

Along these lines, the next section of this chapter puts the issue of natural gas supply security in the economic perspective of energy supply security at large, and sets out the major controversies. In particular, we argue that supply security and competition are not antidotes, but can be organized to work together harmoniously. Section 3 discusses measures to improve natural gas supply security, both short term and long term. Diversification *and* market interconnectedness are two sides of the same coin. Section 4 sketches out different regional approaches to supply security that differ significantly from one another (the US model, continental Europe, and Asia). Section 5 then presents a survey of the empirical research on natural gas supply security carried out in CESSA: the papers include numerical partial equilibrium modeling as well as regulatory and institutional analyses, and a broad range of case study-based evidence. We also had to select three papers for this final volume (Chapters 2–4), and have chosen those with the highest specifics in terms of US–European comparison, and their geopolitical implications.[2] We conclude with policy recommendations in Section 6.

2 ISSUES IN NATURAL GAS SUPPLY SECURITY

2.1 Approaches to Supply Security

Natural gas energy security is an issue where regulation meets geology, and both meet geopolitics. Major consuming regions are often far from resource-rich areas; in addition, all countries with natural gas pipelines affect supply security via their regulatory policies. Whereas the traditional energy supply security discussion often focuses on electricity generation and transmission, there is also an important debate on natural gas supply security on both sides of the Atlantic as well as in Asian importing countries. Supply security has become a heavily discussed topic, due to rising natural gas demand worldwide, increasing import dependence in many countries, geopolitical conflicts, the globalization of formerly regional markets, and the need for a regulatory and policy response. Thus, between 2000 and 2005 the EU's natural gas demand increased on average by 2.1 percent per year whereas proven natural gas reserves remained nearly constant. Domestic production is at its plateau or even decreasing in several countries: thus, the UK is developing from a net exporter to a net importing country. The situation in the US was worrying until recently, with domestic conventional reserves less than 10 times current annual consumption, and imports from Canada diminishing.

There is no such thing as a 'one-size-fits-all' definition of energy supply security. At least two aspects have to be differentiated:

- the short-term effects, such as the physical supply of energy resources that may be threatened by supply disruption, caused by technical (shortfall of up-, down-, or midstream infrastructure), or political (exhaustion, temporary disruption). The physical supply may also be endangered by cartelization of upstream producers;

- the long-term effects, such as the adequacy of investments in (natural gas) infrastructure, such as terminals receiving LNG, transmission pipelines and storage facilities. Long-term issues are therefore related to an efficient energy supply at reasonable and stable prices.

In the rest of this section, we shall discuss the relation between supply security and the other two general objectives of energy policy, that is, competitiveness and sustainability.

2.2 Supply Security, Competition, and Sustainability

First, it is important to note that supply security does *not* imply the absence of high and volatile prices. The New England cold snap is the best example to illustrate this: although demand outstripped supply leading to very high prices, supplies were assured via a competitive process; once the cold snap was over, prices came back down again, approaching the competitive equilibrium.

Often the question is raised about the relation between supply security and the other two objectives of energy policy, that is, competitiveness and sustainability. There is a broad consensus that supply security is promoted by open and competitive markets that favor the exchange of information, the availability of resources and investments, leading to a diversified supply structure. The positive relation between supply security and competitive markets is strengthened by empirical evidence, for example, in the US (Hirschhausen, 2008), and the UK (UK Office of Fair Trading, 2007). This relation does *not* hold, of course, if there can be no downstream competition due to the dominant position of a vertically integrated company (the case of Bulgargaz). In this case, a more structured approach would be required.

How about supply security and sustainability? There may be a certain conflict between supply security and environmental objectives, in particular CO_2 mitigation, since these restrictions may limit the options of diversifying energy supplies. On the other hand, however, climate change policies may also help the transition towards a broader use of renewable energies. This would take pressure from natural gas as 'the' transition energy on the way to a hydrogen- or solar-dominated energy system.

3 MEASURES TO IMPROVE NATURAL GAS SUPPLY SECURITY

There are a large number of measures to increase the security of natural gas supply. However, one should not forget that supply security is not an objective in itself, but that each measure has costs and benefits. In this section, we discuss some general policy measures that we consider to be conducive to efficient supply security. Applications to specific regions (Europe, North America, Asia) follow in the next section.

3.1 Short-term Security Measures

Short-term supply security can be based on administrative measures, such as supply obligations in times of crises, and the market-based establishment of a liquid wholesale market that is capable of transmitting information about supply and demand very rapidly. Market-based instruments are generally more efficient to supply the right information, but some administrative measures may be needed for short-term security measures. This also requires the establishment of rules for cross-border support in cases of emergency. The financial compensation of such emergency support still needs to be explored in detail.

3.2 Increased Use of Storage

One advantage of natural gas is that, in contrast to electricity, it can be stored (EIA, 2004). Storage can be managed within the transmission grid ('pipeline storage') or in specifically designed storage facilities (salt cavern, aquifers, depleted oil and gas fields). Storage can therefore fulfill both reliability objectives (that can be shared between neighboring countries) and strategic objectives (delivery in the high-demand season, in Europe generally during the winter months). Traditionally, European producing countries used natural gas production swing as 'seasonal storage' – an economical solution as long as the fields are close to the market. However, swing supply from indigenous production is decreasing (for example, in the UK and the Netherlands). Industry argues that investments in seasonal storage capacities are necessary.

The more flexible storage sites are accessible, the higher will be the level of supply security, both in the short and the long terms. Vertically integrated storage facilities are generally not conducive to an efficient use of storage. Non-discriminatory access and regulated, efficient access prices of storage facilities are conducive to supply security. Also, merchant investment in storage should be favored, such as in the US, as it attracts investment that may not have been forthcoming in integrated companies.

3.3 Supply Diversification

Turning to the longer-term effects of natural gas supply security, diversification is certainly a strategy to limit the risks of import dependence. Supply security can be enhanced by a diversification of energy sources on the one hand, but also of suppliers and transport routes on the other. This may involve increasing import capacities (additional pipeline capacities as well as LNG import infrastructure) and the diversification of upstream sourcing.

In market economies, natural gas imports are the responsibility of business enterprises, and the state has no particular role to play therein. In the European Commission Green Paper 'A European Strategy for Sustainable, Competitive and Secure Energy' (EC, 2006, Part 3), it is suggested that the EU should improve 'the conditions for European companies seeking access to global resources'. It is not clear what this implies in practice, for example, what the limits of a 'European' energy company are, and how this state support to the business companies would work without discrimination. An efficient policy would be to create conditions conducive to business deals for the private sector, but to refrain from interventionist action.

3.4 Increased Market Interconnectedness

Adequate infrastructure investments are one key to ensure a resilient system and therefore secure natural gas supplies. This refers to both pipeline interconnections between neighboring countries, and investments along the vertical value-added chain. It is generally estimated that the downstream infrastructure investments in Europe as well as in the US are adequate; the challenge will lie in the initiation of sufficient upstream investments (that is, field development, addition of pipeline capacities and LNG liquefaction facilities and vessels).

Competitive markets favor the right timing of infrastructure investment. As long as prices are determined by supply and demand, competition will also incite the 'right' level of investment, indicated by scarcity rents in energy-only markets.

3.5 Increased Resilience through Information and Transparency

Competitive, transparent markets generate information about current and future demand and supply patterns, and thus increase the level of resilience in the system. There is a positive relation between information availability and supply security. Typically markets are an efficient way to promote the diffusion of information. Detailed information about future market conditions can also help potential investors to make economically rational decisions; with better information transparency, more-efficient investments in natural gas infrastructures are feasible. Competitive markets increase resilience since they support more-efficient reactions of producers and consumers to price movements. Liquidity of a market also leads to increased resilience, since the ability of the system to respond to local supply disruptions is higher.

A good example of how transparency and information diffusion can enhance supply security is secondary trading of transmission pipeline capacity in North America (US and Canada). Secondary trading not only improves the utilization of capacities, but also provides a way to secure forward deals in the natural gas market by eliminating the transportation risk. The US experience shows that secondary trading leads to liquid markets, a more-efficient use of the infrastructure, and a significantly increased supply security.

3.6 General Assessment: Open Markets Foster Supply Security

The arguments discussed above indicate that there is a positive correlation between the opening of markets for natural gas, and the level of supply security. In fact, there is an ongoing debate about the relationship between liberalization and investment incentives: industry representatives generally claim that industry restructuring in an unstable institutional environment places infrastructure investments that ensure supply security at risk. Market proponents, on the other hand, argue that a market-oriented approach is the best way of stimulating efficient investment, and thus enhance supply security. Both theoretical arguments and empirical experience (to be presented below) tend to be in favor of the latter opinion: supply security is enhanced by competitive markets. In fact, competition leads to a higher number of market actors, a diversification of infrastructures and new investment, and correct and efficient investment and price signals. A fully functioning internal energy market is an instrument to attain supply security and contributes to a better prediction of demand.

4 REGIONAL APPROACHES TO SUPPLY SECURITY

In this section, we summarize approaches to and results of supply security policies in three different regions: the US/UK, where early reliance on market mechanisms has produced a high level of transparency and supply security; continental Europe, where reforms were started only recently and supply security has so far been provided by monopolistic, often vertically integrated companies; and Asia, where the low-risk–high-price policy pursued by Japan and South Korea contrasts with the uncertainty facing natural gas supply and demand in India and China.

4.1 US and UK: Supply Security Provided by Restructured Natural Gas Markets

The US and the UK have followed similar routes towards a market-oriented natural gas industry, where supply security is generally assured by open and transparent markets (IEA, 1998). The US was the first country to 'restructure' (liberalize) its natural gas industry. FERC (Federal Energy Regulatory Commission) Orders 436 (1985) and 636 (1992) as well as other initiatives led to a well-functioning wholesale market for natural gas, at the level of both pipeline capacities and the market for the commodity natural gas (Makholm, 1996). The primary allocation of natural gas transmission pipeline capacity is still subject to (cost-based) regulation, while at the same time a highly liquid secondary market has been institutionalized that allows the resale of capacity (EIA, 2005). Natural gas hubs are very liquid trading points, characterized by a large number of participating producers and traders. Instead of oil indexation, supply contracts link future delivery prices to natural gas spot prices (for example, the Henry Hub). Stress situations (like hurricanes Katrina and Rita in 2005, or the cold snap in New England in 2005) have shown that the pipeline system is flexible, the actors are generally well informed, and this setting provides adequate supply security.

The UK approach has been similar, with some institutional differences. The natural gas industry has been liberalized with the Gas Act (1986) and the New Gas Act (1995); in 2000, British Gas was unbundled completely, with the National Grid becoming the transmission system operator, and Centrica taking over the commercial functions. The separated network monopolist is subject to a 'building block' regulation, including a price cap that takes into account the necessary return on capital. Trading takes place both bilaterally and on the virtual trading hub National Balancing Point (NBP). The churn ratio over 10 indicates the high liquidity of the market. Privatization and restructuring do not seem to have hindered investment in infrastructure: one newly built LNG import terminal (Grain LNG) started operation in 2005, and three more (Excelerate at Teeside, South Hook and Dragon LNG) have opened since. Furthermore, several pipeline projects have been realized (that is, Interconnector, BBL) and an expansion of natural gas storage capacities is planned. Like in the US, the natural gas market in the UK has delivered a high supply security and proven to be flexible in times of tight supply, such as in the winter of 2005/06, when cold weather was accompanied by a shortfall of the Rough storage site.

4.2 Continental Europe: Towards More Market-based Supply Security?

In continental Europe, supply security was for a long time assured by monopolistic suppliers, often vertically integrated, and in many cases state owned. Thus, the cost of the supply security was directly passed through to the customers, albeit at quite high costs. The EU Acceleration Directive 2003/55/EC introduced a formal separation between gas transmission and trading, but the effects of this unbundling have yet to materialize. The continental European natural gas market is still highly concentrated at the wholesale level; incumbents largely control imports, domestic production, and trading activities. Except for Zeebrugge, natural gas hubs are still relatively dry. Interconnection capacities are limited, incumbents are national and they often have a dominant position. While natural gas demand is increasing in many European countries, domestic production has reached a plateau in several producing regions, leading to an increasing supply–demand gap and hence, to increasing import dependency. This is one of the reasons why security of natural gas supply is back on the EU political agenda.

European consumers compete with the US for natural gas supplies in the Atlantic Basin; both import regions negotiate with exporters such as Trinidad & Tobago, Qatar or Nigeria. Furthermore, LNG enables the redirection of cargoes between the two regions if price differentials allow for arbitrage benefits.

A particular supply security situation is prevailing in Eastern Europe and the European countries of the former Soviet Union. Both regions are mainly landlocked and lack alternatives to the incumbent supplier, Russia. This has perpetuated a one-sided dependence on Russia, which is a source of unease to many East European countries, and which they are gradually trying to change. Thus, some Central European countries have sourced more expensive natural gas deliveries, for example, from Norway, in order to diversify. Coastal countries like Poland and Croatia are considering the construction of LNG regasification facilities. Russia has proven to be an unreliable supplier in this region, and it is clearly using natural gas exports as an instrument of its foreign policy. The fact that Russia has not yet signed the European Energy Charter also indicates that it is not seriously planning to accept market rules for its natural gas industry.

With respect to natural gas, Europe is in a relatively comfortable situation with respect to its supply sources. Europe is supplied by piped natural gas from the North Sea, Russia (in 2007: 148 bcm (billion cubic meters), or 35 percent of total imports), and North Africa (Algeria: 53 bcm in total); in addition, it can receive LNG from a very large number of suppliers from all over the world (Middle East: 8 bcm). Although Russia has an important share of the European natural gas imports, its role as the one-and-only strategic supplier to Europe is often exaggerated: model simulations show that the share of Russia in European natural gas imports will increase only slightly until 2030 (to 20–30 percent).

4.3 Supply Security in Asian Countries

Asia as a region is both a large producer of natural gas and also an increasing consumer, with some countries' markets expected to grow exponentially, particularly China and India. Major reserves are located in Australia, Malaysia, and Indonesia; transport is realized mainly in the form of LNG. Whereas certain countries have domestic natural gas reserves, others are fully dependent on LNG imports (for example, Japan and South Korea).

One can distinguish two very different strategies towards supply security: the 'traditional' approach followed by Japan and South Korea, and the 'emerging' approach characteristic for India and China. Natural gas markets in Japan and South Korea are mature. With 110 bcm per year installed regasification capacity, and 89 bcm imports in 2007, Japan is the world's largest LNG importer; South Korea with 40 bcm/a installed and 34 bcm imports in 2007 forms the second largest LNG importer. In these traditional natural gas importing countries, diversification of supplies and high prices paid as risk premia have led to a high level of supply security, and prices have been quite high. Recently, there have been indications that they would like to trade greater liberalization against the monopolistic supply situation.

In contrast, the emerging countries of China and India still have to define their approach to supply security, and, subsequently, their policy towards natural gas sector restructuring. Both countries are facing enormous uncertainty on both the supply and demand sides. A priori, the strong economic growth in these countries is associated with a rapidly increasing need for energy, also raising natural gas demand. However, both the extensive use of domestic coal and high natural gas prices may lead to substantially less demand than planned. Other open questions include the financing of the necessary infrastructure, the future CO_2-abatement policies, and the pricing policies imposed by the respective governments.

There is an interesting link between the different Asian markets on the one hand, and Europe and North America on the other, brought about by the recent globalization of natural gas markets via LNG (Jensen, 2004). Some Asian countries are likely to maintain high LNG imports (Japan and South Korea), or to increase their LNG imports. But on the supply side, it is unlikely that Malaysia and Indonesia, recently the main Asia-Pacific LNG suppliers, will be able to meet the Asian (expected) demand growth. Discussions about the trade-off between exports and domestic consumption are under way in certain export countries like Indonesia, limiting the rise of natural gas export volumes. Within the regional picture, Australia is likely to increase its role as an exporter, with not less than eight additional liquefaction plants currently under development. Furthermore, deliveries from the Middle East – a region able to export LNG to Europe and North America as well as Asia – will increase. Therefore, the future development of natural gas demand in Asia and the construction of import capacities will have a strong impact on the demand–supply balance of the 'world natural gas market' and, hence, on the EU as well.

5 OUTPUTS OF THE WORK PACKAGE ON 'NATURAL GAS'

This part of the book summarizes the major topics that dominate the supply security discussions. The CESSA project was in fact geared towards generating the broadest possible consensus among stakeholders (industry, governments, academics), an objective that was largely achieved, as evidenced by four lively conferences, nine working papers, and three special contributions that are reproduced in this volume.

5.1 The Conferences and Final Policy Event

The CESSA conferences aimed at the exchange of viewpoints between stakeholders:

- The first conference, entitled 'Economic Mechanisms Sustaining Robust Development of Natural Gas Investment and Infrastructure', was held in Berlin on June 1, 2007. Its objectives were to set out the issues for the subsequent project, and to oppose European and North American perspectives. There was a very broad overview of the relevant topics, ranging from pipeline infrastructure investments, strategic use of storage, geopolitical issues (with a focus on Eastern Europe) to the numerical modeling of international natural gas supplies.
- The second conference, held in Cambridge, UK, on December 13–15, 2007 was conceived as a 'sister' conference to that in Berlin,

involving more in-depth conceptual discussions, and also looking for intersectoral comparisons. Thus, besides exploring the relation between 'nuclear supply security' and 'natural gas supply security', we also introduced – politically incorrect in times of climate change – 'coal supply security'.
- The fourth conference, held in Florence, on June 14–16, 2008, was conceived as a cross-topical event focusing on 'supply security' at large.[3] With regard to natural gas, the highlights were the discussion between academics, policy makers, and industry about the conclusions to draw, in particular with respect to vertical integration and competition policy. Nonetheless, compromises were found.
- Last but not least, the final policy conference was held in Brussels, on October 2, 2008, to present the policy conclusions to European stakeholders. While most people thought the topic to be 'exhausted', given that Commissioner Andris Piebalgs had given a status report on the third energy and climate package, the financial crisis, which was about to wreak havoc in Europe and the world, told a different story!

5.2 Nine Working Papers

The task of the work package was to summarize the major trends regarding natural gas supply security. Nonetheless, the nine working papers that were produced as part of CESSA Work Package 3 cover a broad range of issues and span the academic debate. The papers (all available for download at cessa.eu.com) can be grouped into three broad categories:

- *Modeling international energy markets* Two papers provide a broad overview of modeling international natural gas trading. Ruud Egging, Steven Gabriel, Franziska Holz and Jifuang Zhuang (University of Maryland) produced 'A complementarity model for the European natural gas market' (Egging et al., 2007). This is in fact part of a larger exercise that has resulted in the World Gas Model (WGM), the largest model to date that can accommodate both fully competitive markets and strategic behavior by certain actors along the value-added chain (such as natural gas producers, LNG-facility and pipeline operators and downstream traders). The model shows that Europe has a relatively well-diversified portfolio of natural gas supplies, and that the share of Russia in future European gas supply will be important but not overwhelming. Yves Smeers (Université Catholique Louvain-la-Neuve) provides an account of 'Gas models and three difficult objectives' (Smeers, 2008). He highlights the

assumptions and limitations of the main models used in Europe to forecast natural gas supplies, and discusses the results of these models critically.
- *Geopolitics* was a topic that dominated throughout the CESSA project, and it will remain very important for a long time. In addition to the two chapters that were updated for this book by Jonathan Stern (ch. 3) and David Victor (ch. 4), Christoph Toenjes and Jacques de Jong (Clingendael Institute) contributed 'Perspectives on security of supply in European natural gas markets' (Toenjes and de Jong, 2007). While the Stern chapter takes a 'foreign policy' perspective, the paper by Toenjes and de Jong addresses supply security issues more from a company perspective, arguing in favor of strong political support for European natural gas companies in the international competition.
- *Regulatory issues* A third group of issues relates to regulating and restructuring the natural gas sector, to foster competitiveness *and* supply security. In addition to Jeff Makholm's paper 'Seeking competition and supply security in natural gas' which has been updated for this book (ch. 2), two papers were produced within the CESSA project. Farid Gasmi from the University of Toulouse discusses critical questions related to 'Investment in transport infrastructure, regulation, and gas-to-gas competition' (Gasmi, 2008). He argues that adequate regulation of the pipeline transport is necessary not only to ensure optimal network expansion but also for the investments to be financially viable. Christian von Hirschhausen has analyzed 'Infrastructure, regulation, investment and security of supply – a case study of the restructured US natural gas market'. While not everything is greener on the other side of the Atlantic, the US experience clearly shows the virtues of restructuring in terms of security of supply, and the contribution of gas-to-gas competition.

5.3 The Subsequent Chapters in Part I

The next three chapters provide a representative picture of the natural gas supply security issues. They cover two of the three broad topics, that is, regulation and geopolitics; the remaining important driver, sustainability, was less present in the written debate, but actively debated at the conferences.

In Chapter 2, Jeff Makholm, an internationally renowned expert and consultant on natural gas issues worldwide, provides a comprehensive comparison between the North American and European approach to (de-) regulating the natural gas sector, and, thus, to establishing rules to deal with supply security. He focuses on pipeline regulation since the availability of transport capacity in pipelines determines the degree of competitiveness of the natural gas market. In this field, the US experience of the last century can offer many insights for European policy makers. In the US, 'unbundling' of pipeline operation and natural gas sales had already been chosen in the 1930s to ensure competitiveness and supply security. With their character of a natural monopoly, pipelines have long been subject to regulation of their access tariffs. The federal US system required a clear separation and definition of state and federal competences with respect to regulatory rules and supervision. While there is now a situation of regulatory certainty in the US that enables network investments wherever necessary, there are still large uncertainties in Europe. However, Europe today has the chance to sustainably ensure its security of supply by establishing efficient regulation of its pipeline transport.

The following two chapters on geopolitical issues take very different approaches. In Chapter 3 Jonathan Stern, Director of the Gas Programme at the Oxford Institute for Energy Studies (OIES) explores 'The new securing environment for European gas' and, hence, focuses on the European perspective on its major suppliers. He also notes that, in addition to Russia and North Africa, the West African and Middle Eastern suppliers are increasingly important in the European import portfolio, in particular with regard to LNG deliveries. Stern sees a major short-term issue for European supply security in the transit of Russian gas to Europe. This chapter was written in early 2008 and the repeated tensions and even supply disruptions between Russia and Ukraine in early 2009 prove him right. In the longer term until 2020 or 2030, decreasing European gas production and potentially increasing demand for gas as a transition fuel to a cleaner energy system, will increase the European dependency on imports. Stern finds that several potential sources of additional pipeline imports are unlikely to deliver substantial amounts to Europe, such as the Caspian or the Middle East region. Institutional and political instability in these regions are a fundamental obstacle to pipeline investment. More hope lies on LNG imports from West Africa (Nigeria, Equatorial Guinea), but also from the Middle East and North Africa.

By contrast, in Chapter 4, David Victor, the Director of the Program on Energy and Sustainable Development (PESD) at Stanford University, analyzes 'Natural gas and geopolitics' in a longer-term perspective. Given the need for mature demand basins in Europe, Asia and North America to diversify their supply sources, Victor emphasizes the hindering effect of weak institutions in many gas-producing developing countries. In such an increasingly globalizing market, LNG and LNG players play a large role.

While LNG brings an element of flexibility to the market, the traditional long-term contracting behavior enforces contractual commitments but may also act as a deterrent to new market entrants. Citing a number of international case studies, Victor shows that the LNG market proves that private commercial players act more efficiently than states, especially as first movers with innovative commercial or technological processes.

6 POLICY RECOMMENDATIONS

In this chapter, we have set out the major issues concerning natural gas supply security, as discussed in Work Package 3 of the CESSA project and documented in a variety of oral and written contributions. The following policy recommendations are derived from these discussions; they have previously been presented and discussed at the 2nd CESSA conference (Cambridge) in December 2007, and were presented to the European stakeholders in the final policy conference in Brussels in October 2008.[4]

6.1 Supply Security and Geopolitics

1. *The Union itself should not intervene too directly in the supply choice of its natural gas importers* There is an emerging global market for natural gas and the EU currently has a relatively well-composed mix of imports. The increasing part of LNG in international natural gas trade will facilitate a further diversification, though at higher prices than previous long-term pipeline gas supplies. The well-founded objective of supply security should not lead to too heavy interventionism by member states of the EU in favor of its own incumbent companies.
2. *The energy dialog with Russia should be pursued, but placing too much emphasis on Russia (or any other natural gas supplier) does not seem to be justified* The role of Russia as a strategic supplier of natural gas to Europe is exaggerated. Although Russia enjoys an important share of the European natural gas imports, its share of European natural gas imports is unlikely to exceed 35–40 percent. With rising production costs and rising domestic demand, Russia's cost advantage will diminish, at least in the medium term. Since we know little about the use of natural gas in the long term, not too much emphasis should be placed on potential reserves in Russia that have yet to be explored.
3. *The diversification of natural gas supply to Central, Eastern, and South-Eastern Europe should be treated with particular attention* Recent years have clearly shown that supply security is a very local event,

in particular the supply crisis caused by Russia in January 2009. While Spain enjoyed ample gas supplies and warm homes, people in South-Eastern Europe were freezing almost to death. It is clear that natural gas supply security concerns are particularly strong in the new member countries of Central and Eastern Europe (for example, Poland, Hungary, and the Baltic countries) and in South-Eastern Europe (Romania and Bulgaria). These regions deserve particular consideration, for example, in the Trans-European Network (TEN) and the Priority Interconnection Plan (PIP).

6.2 Supply Security and Regulation

4. *Supply security within the EU could benefit from improvements in interconnectedness between physical and institutional markets, including the use-it-or-lose-it rule (UIOLI)* The EU is not making full use of the fact that a functioning internal market would enhance supply security considerably. Clearly a larger market favors supply security, but current transmission capacities are not used efficiently and hinder intra-European natural gas flows. Other technical and organizational obstacles also need to be overcome quickly to generate the benefits of a large market. The emergence of an integrated natural gas market in the US provides a useful benchmark.
5. *Ownership unbundling of production and trading activities from transport infrastructure should be implemented to favor the emergence of a competitive wholesale market* Both theoretical considerations and empirical evidence, for example, from the US and the UK, suggest that separation of infrastructure and the commercial part of the value-added chains is conducive to competition and supply security. Clearly an integrated natural gas company (transmission and trading) has incentives to discriminate network access of potential competitors. Some of the options currently discussed, such as 'weak' unbundling (administrative only) will be insufficient to achieve the incentives of a truly unbundled network operator.
6. *Where adequate, storage should be treated as an essential facility, investment and access should be regulated and a secondary market established* Storage has a critical role to play in a competitive natural gas market. Short-term supply security is significantly enhanced with ample storage capacities and liquid trading. The traditional role of storage as an inflexible reserve for annual gas flows of an integrated company is no longer adequate. A more commercial use of storage is conducive to both short- and long-term supply security. Merchant storage should be facilitated as well.

6.3 Supply Security and Climate Policy

7. *More efforts should be dedicated to establish 'clean gas'* The future of natural gas depends on its ability to establish itself as a 'clean' energy in a low-carbon world. Traditional wisdom of gas as a 'clean' source of energy is no longer valid in a 2° world of ambitious climate targets; rather, natural gas is a 'dirty' energy, emitting about 350g of CO_2 per kWh electricity produced. Thus, natural gas threatens to become a 'sunset industry' by 2030, caught in the middle between cheap 'clean' coal (carbon capture and storage: CCS), biogas, and other clean sources of energy.
8. *The deployment of CCS pilot units for natural gas, through financial incentives for early successful projects, should be accelerated* Given the medium-term CO_2 reduction objectives (60–80 percent), CCS is the only way to maintain fossil fuel electrification. Although attention is currently directed towards CCS for coal power plants, CCS for natural gas will soon become an important topic; although it is technically more complicated, it is feasible, and should be studied with the same intensity as CCS for coal or biogas.
9. *The feeding-in of (decentralized) biogas to the distribution networks should be facilitated* Biogas is considered to produce fewer environmental externalities than fossil fuel natural gas. Technically, blending and feeding in biogas is feasible, and there is a high potential to develop appropriate distribution networks.

7 CONCLUSIONS

Natural gas is a cornerstone of European energy supply security. It also has a large potential for the 'transition' towards a sustainable, largely renewable-based energy economy, and is therefore of utmost strategic importance. In this chapter, we have identified the major issues regarding supply security and natural gas, and provided a brief overview of the international state of the discussion. We have also introduced the differentiation between regulatory and geopolitical issues of security of supply, which will be covered in depth in the next three chapters.

NOTES

1. This chapter is the final summary chapter of Work Package 3 (Natural Gas) of the CESSA project. Earlier versions have been presented at the 2nd CESSA conference

(Cambridge, December 2007) and the 4th CESSA conference (Florence, June 2008). Thanks to Jean-Michel Glachant and François Levêque for initiating the CESSA project, to colleagues at TU Dresden and DIW Berlin for support to our team during the project and to the colleagues active in Work Package 3 and at the conferences for open, constructive discussions. The usual disclaimer applies.
2. Note that the working papers are available for download at cessa.eu.com.
3. The third conference, held in Madrid in April 2008, was entirely dedicated to hydrogen.
4. The authors of this chapter assume the responsibility for these conclusions.

REFERENCES

Egging, R., S.A. Gabriel, F. Holz and J. Zhuang (2007): 'A complementarity model for the European natural gas market', CESSA Working Paper 10.
Energy Information Agency (EIA) (2004): *The Basics of Underground Natural Gas Storage*, Washington, DC: US Department of Energy.
Energy Information Agency (EIA) (2005): *Natural Gas Transportation Infrastructure*, Washington, DC: US Department of Energy.
European Commission (EC) (2006), Green Paper: 'A European Strategy for Sustainable, Competitive and Secure Energy', Com (2006) 105, Brussels.
Gasmi, F. (2008): 'Investment in transport infrastructure, regulation, and gas-to-gas competition', CESSA Working Paper 17.
Hirschhausen, C. von (2008): 'Infrastructure, regulation, investment and security of supply: a case of the restructured US natural gas market – is supply security at risk?', *Utilities Policy*, Vol. 16, No. 1, pp. 1–10.
International Energy Agency (1998): *Natural Gas Pricing in Competitive Markets*, Paris: OECD.
International Energy Agency (2008): *Natural Gas Market Review 2008*, Paris: OECD.
Jensen, J. (2004): *The Development of a Global LNG Market*, Oxford: Oxford Institute for Energy Studies.
Joskow, P. (2005): 'Supply security in competitive electricity and natural gas markets', Cambridge, MA, Paper prepared for the Beesley Lecture in London, October 25.
Makholm, J. (2006): 'The theory of relationship-specific investments, long-term contracts and gas pipeline development in the United States', Paper presented at the ENERDAY, University of Technology, Dresden, April 21.
Smeers, Y. (2008): Gas models and three difficult objectives', CESSA Working Paper 15.
Toenjes, C. and J. de Jong (2007): 'Perspectives on security of supply in European natural gas markets', CESSA Working Paper 13.
UK Office of Fair Trading and UK Office of Gas and Electricity Markets (2007): Note provided to the Roundtable on Energy Security and Competition Policy, held in Paris, February 21.

2. Seeking competition and supply security in natural gas: the US experience and European challenge
Jeff Makholm[1]

1 INTRODUCTION

Many economists and policy makers acknowledge that liberalization in gas and electricity markets is consistent with energy supply security – but only with the important caveat that industry structure and underlying institutions in those markets support genuine competition. The structural questions for natural gas seem easy to assess. Gas is a natural resource (not a manufactured product like electricity), with well-defined production and import sources and major consumers, all interconnected by a highly predictable pipeline network. The institutions that create conditions necessary for efficient and competitive gas markets, however, are not so easy to assess.

The contrast between the American and European gas systems is a case in point. In the twenty-first century, both display continent-sized pipeline networks connecting various major sources of supply with large gas distribution, power generation and industrial gas customers. And yet, America displays a freely competitive gas market typified by vigorous spot trading at many hubs nationwide, including a robust forward/futures trading market. That market has dispensed with the long-term gas contracts typical through the 1980s. The American market also exhibits an unregulated, 'Coasian' (what we also call 'contractualized') market in gas transport to any part of the existing network – the entirety of which remains, paradoxically, subject to cost-of-service regulation by the Federal Energy Regulatory Commission (FERC).[2] Parallel to the growth of an independent, contractualized transport market, the job of the FERC has shifted away from just rate setting to include overseeing the property rights and information flows that make the contractualized gas transport market work.

The institutional foundation for gas trade on the European pipeline network is radically different. Even though there is some gas trading at a few points on the network (all constituting less than 1 percent of gas consumption), the European gas trade remains dominated by long-term contracts and vertically integrated pipelines in the various member states. It has nothing remotely like the American spot market in gas or the unregulated, contractualized market in pipeline capacity.

What does this spell for security of supply? In the American gas market, security comes in the form of a large array of sellers and buyers transacting in highly liquid and competitive markets for both gas and gas transport. The American gas market did not always have supply security in gas – in living memory it saw gas shortages that cost Americans billions of dollars yearly in social costs (Pierce, 1988). But the market has learned to deal successfully with extreme winter weather, California energy crises and natural disasters.[3] Europe, in contrast, worries about the potential dominance of Russia, Algeria or Norway as gas suppliers, despite the fact that none has much more than a 20 percent share of Europe's gas sales, either now or in the foreseeable future.

The answer to gas security lies in the pipelines – those inanimate, sunk, steel, low-technology assets. The treatment of the pipelines defines the possibilities for the creation for gas security in Europe based on market liberalization. And with respect to those pipelines, the answers lie in the analysis of the institutions.

2 THE CENTRAL ROLE OF INSTITUTIONS

In order to explain a particular industry's market structure, economists in the field of industrial organization traditionally examine the cost structure of firms. For major gas pipelines this style of economic analysis does not go far enough. The most basic economic analysis would appear to paint pipelines as almost classic natural monopolies, but the pipeline business is much more complex and difficult to categorize. To most policy makers, the structure of pipeline markets, for both oil and gas, remains something of a mystery.

There are two reasons for the mystery. First, the world's major pipelines make rather lousy natural monopolies.[4] On first examination pipelines do exhibit one traditional indicator of a natural monopoly, namely declining unit costs (larger pipelines will have lower unit costs than smaller ones). It is less clear if declining unit costs (or natural monopoly) plays a significant role in the structure of actual pipeline markets. The second problem concerns economists' preoccupation with analyzing the pipeline industry in isolation. Pipelines themselves have utterly no value on their own – they are part of a larger, tightly interconnected supply chain that transports fuel from production wells and gathering systems to distributors (and from there to households, power generators and commercial establishments).

The world's actual gas pipeline markets are shaped by how pipelines *transact* with those suppliers and customers to whom they are physically connected. The risk of stranding invested capital, or 'hold-up', is extreme for gas pipelines. For privately owned pipelines, the transaction requires a meticulous form of long-term contract to motivate investor-supplied capital. Those in the capital markets who finance major investor-owned pipelines (in reality a small and specialized corner of the larger capital markets) know well the mutuality of contractual obligations required to commit major blocks of capital to sunk costs and immobile assets. Without such a contract, there are only three alternative approaches to building a pipeline: vertically integrate into production and/or end-use, regulate the pipe as a large-scale public utility monopoly, or have the government build and operate gas pipelines. For jurisdictions wishing to tap competitive rivalry in the creation, expansion and use of gas pipeline networks, as an element in providing for long-term, market-oriented security of supply, none of the three choices is particularly attractive.

3 THE TWO GAS PIPELINE STRUCTURES

The United States and Europe have fully formed gas pipeline networks that supply between 20 and 30 percent of the energy needs in each market, with a higher growth projected for Europe (Figure 2.1). In both markets, the sources of gas are relatively distant from the major market centers, requiring the extensive pipeline transport system. This section describes the development and characteristics of the US and European markets for natural gas.

3.1 American Gas Pipelines

The American gas pipeline networks connect various gas basins and market areas in the country. There are a few important points, from Canada and from a handful of existing LNG terminals (with many more on the drawing board).

Growth of the American gas pipeline network
Gas pipelines in the United States move gas from a few major fields, the principal ones in the Gulf of Mexico, to the market areas in the upper Midwest, the Northeast and the West Coast. The system developed in three stages (Makholm, 2006):

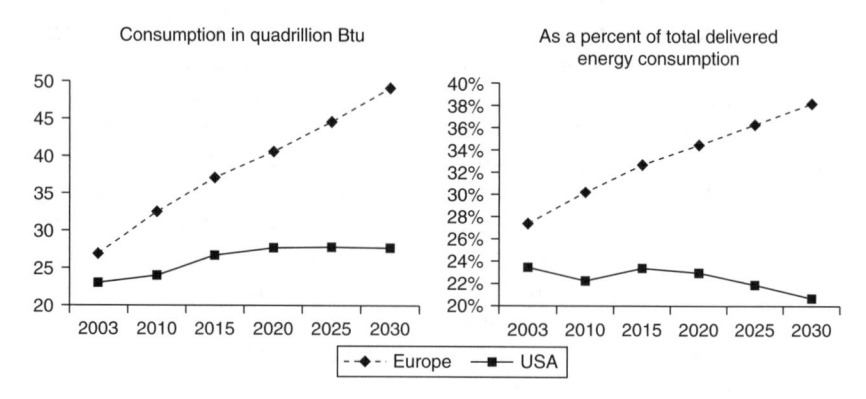

Consumption in quadrillion Btu / As a percent of total delivered energy consumption

Source: US Energy Information Administration, *International Energy Outlook 2006*, June 2006 (DOE/EIA-0484, 2006).

Figure 2.1   Delivered natural gas consumption, Europe and US, 2003–30

1. *The unregulated, vertically integrated era (1889–1937)*   Gas pipelines existed generally as entities integrated either with gas producers or gas distributors. When gas pipelines crossed state lines, they left the jurisdiction of the various state regulators. There was no federal licensing or rate regulation. Vertical integration was forcibly broken with federal legislation in 1935.
2. *The era of delivered gas (1938–83)*   Authority over entry (licensing) and rates was assigned in 1938 to a federal commission (the FERC).[5] Vertically separate gas pipeline companies sold delivered gas to distributors purchased under long-term contracts in the gas fields.[6]
3. *The era of contract-based gas pipeline transport (1985–present)*   By 2000, after 15 years of development, an unregulated market for gas pipeline capacity exists in its own right. Gas pipelines companies are not permitted to own the gas they transport in their trunk pipelines. The FERC still licenses new pipeline capacity projects and 'primary' pipeline prices, according to the 1938 legislation, but it does not regulate 'secondary' capacity prices.

Throughout all of these eras, consumption of natural gas in the United States grew rapidly, as did pipeline construction. In the first period, pipelines were financed through vertically integrated firms, and an extensive pipeline network appeared throughout the East Coast and from the major gas basins in Kansas/Oklahoma to the upper Midwest. Gas distributors and petroleum-producing companies owned both the interstate pipelines and the gas in them. In the second period, the American life insurance industry developed a new lending method that would accept federal Commission regulation as security in long-term pipeline loans (Hooley, 1968). Gas pipeline companies owned large blocks of gas-producing properties, but federal accounting treatment for gas costs encouraged the purchase of gas from third parties. Nevertheless, most of the gas flowing through the interstate gas pipelines was purchased in the producing area and owned by the pipeline companies as it flowed through the major trunk pipelines.

During the third era, conflict developed regarding unfair competition between pipeline-owned gas and that owned by third parties. As a result, the Commission imposed rules in 1992 to divorce the ownership of gas within the pipeline. Thereafter, all the gas flowing in the major trunk gas pipelines was owned by third parties; mostly the gas distributors and power-generating companies (and some gas marketers as the gas market became more liquid and competitive).

Independence of gas and oil markets
One of the distinguishing features of the modern gas market in America is its independence from the oil market. Gas and oil trade on independent exchanges, based on the demand and supply of each at the various locations. Due to the liquidity of the trading at the hubs, there are generally no gas contracts indexed to oil, or vice versa – a feature of less liquid gas markets. Figure 2.2 shows prices at the largest trading hub, the Henry Hub. Prices at the various other trading hubs are often significantly different, reflecting the availability of firm transport between them.

The US pipeline market handles (and learns from) stress
Liquid and competitive markets should adjust to stress and exogenous shocks to the supply and demand for the underlying product. After each shock to the market, those operating in the market should be able to anticipate a response to any similar event in the future and the market disruption should dissipate more readily. Natural gas markets should be no different. Three recent market stresses highlight the response the US market for natural gas has had to the following events:

- a localized, weather-related spike in demand in the Chicago area during the winter of 1995–96,
- a confluence of supply and institutional constraints leading to massive electricity supply shortages during the California energy crisis of 2000–01, and
- the large supply disruption resulting from a natural disaster in an area of natural gas production during the hurricane season of 2005.

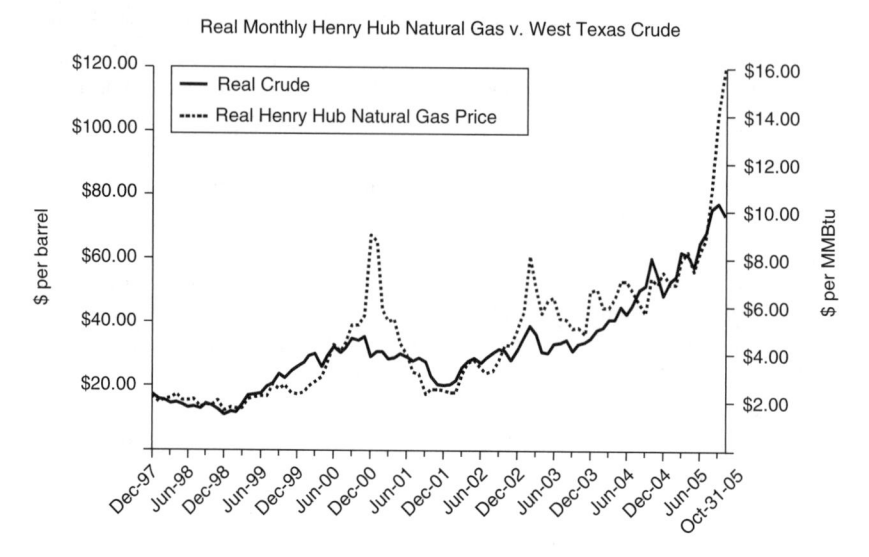

Real Monthly Henry Hub Natural Gas v. West Texas Crude

Note: Prices indexed to CPI, Base Year = January 2000.

Figure 2.2   Gas and oil prices, December 1998–October 2005

The first example of a stress on the natural gas system occurred when the beginning of the heating season of 1995–96 began with below normal temperatures. This resulted in large natural gas storage withdrawals that could not be readily replaced with storage injections because the low temperatures and thus high natural gas demand persisted for an extended period of time. When temperatures again dropped dramatically across the Midwestern US, there was not enough available gas in storage to meet the spiking demand. Accordingly, the local price of natural gas spiked. Figure 2.3 displays the price differential (a.k.a. the price basis) for the Chicago city gate pricing point relative to the Henry Hub pricing point located near much of the country's natural gas supply in Louisiana. Indeed, in the case of cold weather in Chicago, Figure 2.3 shows how the market learned to deal with the relatively new contractualized regime. The cold snap in 1997 was much like the one in 1996, but market and traders had learned from the year before, and the temporary rise in basis differentials was only one-fifth as high.[7]

The second stress on the US natural gas markets occurred during the highly publicized California energy crisis of 2000–01. Supply constraints, among other factors, resulted in widespread electricity shortages across the western US. Accordingly, the price of natural gas spiked because of the increased value of electricity generated in natural gas burning power plants. Figure 2.4 shows the basis price for the Southern Californian Gas

Seeking competition and supply security in natural gas 27

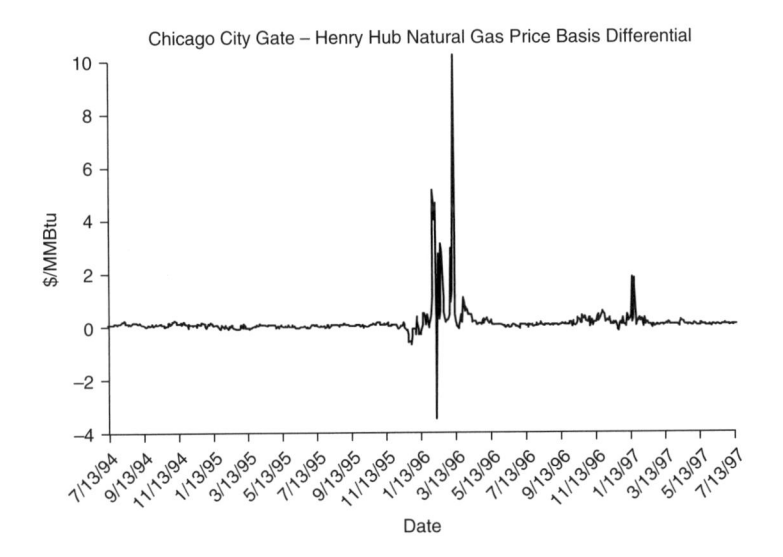

Source: Natural Gas Intelligence Press.

Figure 2.3   Chicago cold snap of winter 1996

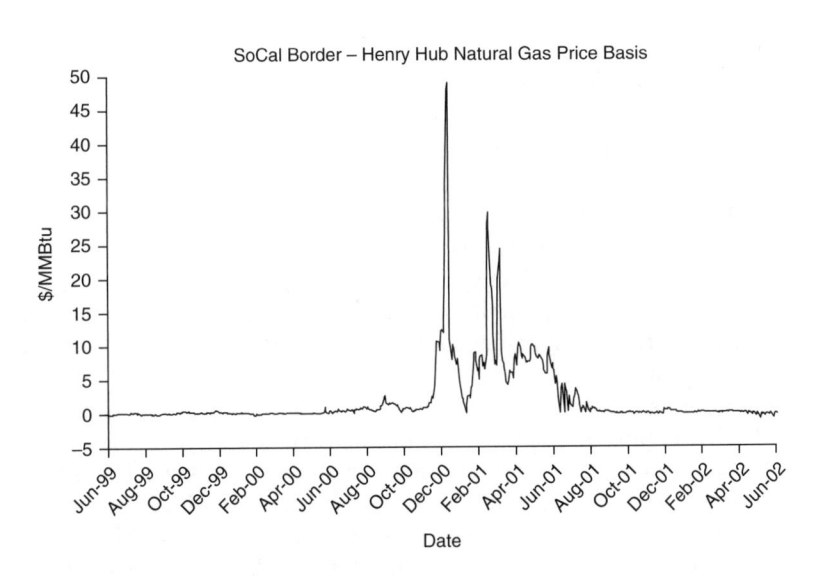

Source: Natural Gas Intelligence Press.

Figure 2.4   California energy crisis of 2000–01

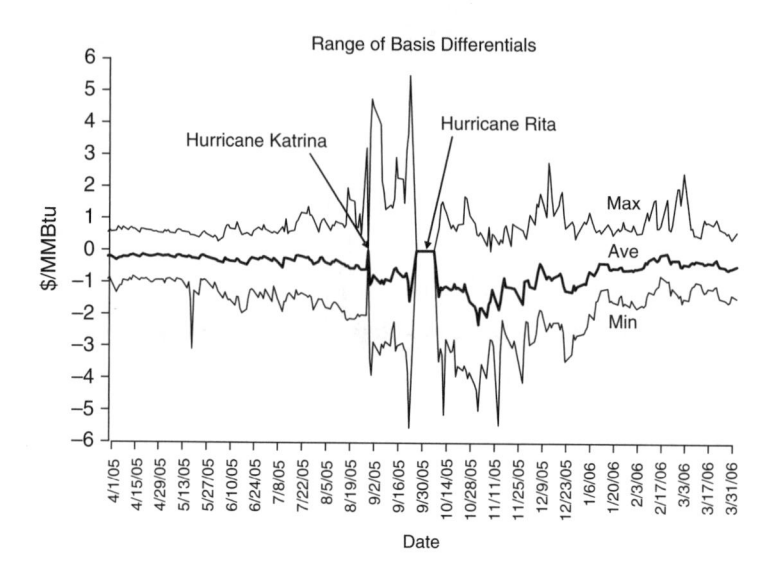

Figure 2.5 The hurricane season of 2005

Comp. (SoCal) border natural gas pricing point relative to the Henry Hub price.[8]

The third and most recent stress on the US natural gas system occurred in the summer of 2005, during the hurricane season in the Gulf of Mexico. Figure 2.5 shows the range and average of the 84 basis differentials relative to the price at Henry Hub in Louisiana between April 2005 and April 2006. During this period of already tightening energy supplies, two hurricanes disrupted a large portion of the US natural gas supply and production. In addition to completely shutting down the Henry Hub for a day and a week, respectively, hurricanes Katrina and Rita led to different and larger than normal supply–demand imbalances across the country, and thus larger basis spreads.[9]

These events illustrate how the flexible, well-informed and contractualized US gas pipeline market facilitates supply security. In each of the three cases above, the market responded to an exogenous shock to supply and/or demand, the spot price moved according to the local supply and demand for natural gas, and the market was able to clear. In order for this to occur, adequate pipeline capacity must be available as well as able to respond to changing market conditions.

3.2 Gas Pipelines in Europe

Gas transmission pipelines in Europe developed first to connect various smaller gas fields to market in the early part of the twentieth century, such as in Poland and Romania. But the first significant gas pipeline development began in the 1950s and 1960s to connect internal fields to local consumers. For example, two gas pipelines in the center-south of Italy were constructed to take gas discovered in the province of Chieti to Terni and Rome. Similarly, the first international pipelines in Europe were built to supply the Netherlands (and later to Belgium, France, Germany and Italy) with gas from the Groningen field upon its discovery in 1959. Since this discovery, natural gas pipeline deliveries in Western Europe grew at an average rate of 10.8 percent (Zhao, 2000).

Until the 1970s, imports from the Soviet Union came via small pipes to Poland (dating back to 1949) and also by a small pipeline from Ukraine into former Czechoslovakia and Austria. The period between the 1970s and the 1990s was one of rapid growth for trans-European pipeline networks. In the early 1970s two pipeline systems began to deliver gas from Western Siberia, namely Transgas and the Orenburg pipelines. Transgas included the Trans Austria Gas Pipeline 'TAG', which delivers gas to pipelines to Czechoslovakia, Austria and Italy (built in 1974), and the MEGAL pipelines to Austria (1974), Germany (1976) and France (1979). The Orenburg pipelines delivered gas to Hungary, Romania and Bulgaria (1975). Also in that period, the Trans Europa Naturgas Pipeline (TENP) began operations in 1974. The TENP is a 968 km long natural gas pipeline, which runs from the German–Netherlands border near Aachen to the German–Swiss border near Schwörstadt, where it is connected with the Transitgas Pipeline. It carries North Sea natural gas from the Netherlands to Italy and Switzerland. Also from the North Sea, Norpipe has delivered natural gas to Germany since 1977. Algerian gas was connected to Europe through Italy via the Transmed pipeline (1983) and Spain (1996). Also in 1996, the Interconnector linked UK gas to continental Europe. Table 2.1 summarizes the flows of natural gas throughout the European continent and the data on the gas pipeline infrastructure currently in place.

4 THE INSTITUTIONAL DIVIDE IN THE US AND EUROPEAN GAS PIPELINE NETWORKS

Transacting is uniquely difficult for gas pipelines, perhaps more so than for any other major business. The heavy cost and risk of contracting caused virtually all major pipelines outside America to be built by governments.

Table 2.1 Gas pipelines in Europe

Name (ownership)	Length (km)	Principal origin	Principal destination	Diameter (inches)	Date of operation
Transgas (RWE)	3,763	Russia	Europe	32, 36, 48	1968
Transitgas Pipeline (Transitgas AG)	291	Netherlands and Germany	Italy	10, 12, 16	1972
Trans Austria Gas Pipeline (ENI, OMV)	380	Slovak–Austrian border	Austrian–Italian	34, 36	1974
Trans Europa Naturgas Pipeline[1]	968	German–Netherl. border	German–Swiss	35, 37	1974
Norpipe (Gassco)	440	North Sea	Germany	36	1977
Vesterled (Gassco)	361	North Sea	UK	32	1978
Megal (Ruhrgas and Gaz de France)	1,070	Russia	Germany, France	36, 48	1980
Trans-Mediterranean Pipeline[2]	2,475	Algeria	Italy	48, 42, 26, 20	1983
STEGAL (Wingas)	320	Czech Republic	Germany	31	1992
MIDAL (Wingas)	702	North Sea	Germany	31, 35, 39	1993
Zeepipe (Gassled Partners)	814	North Sea	Belgium	40	1993
Rehden–Hamburg gas Pipeline[3]	132	Rehden	Hamburg	31	1994
NOGAT Pipeline System[4]	257	Dutch continental shelf	Netherlands	24–36	1994
Netra[5]	341	Coast of the North Sea	Eastern Germany	48	Mid-1990s

Europipe I (Gassco)	670	North Sea	Germany	40	1995
HAG Pipeline (OMV)	45	Austria	Hungary	28	1996
Maghreb–Europe Gas Pipeline[6]	1,450	Algeria	Spain, Portugal	48, 8, 22	1996
JAGAL (Wingas)	111	Yamal–Europe pipeline	Wingas grid	47, 55	1997
Yamal–Europe Natural gas Pipeline[7]	4,196	Russia	Germany	56	1997
Interconnector[8]	235	UK	Belgium	40	1998
Wedal (Wingas)	320	MIDAL pipeline	Belgium	31	1998
Franpipe[9]	840	North Sea	France	42	1998
Europipe II (Gassco)	658	Norway	Germany	42	1999
Green Stream[10]	520	Libya	Italy	32	2004
Blue Stream (Gazprom and ENI)	1,213	Russia	Turkey	55	2005
Iran–Armenia Natural Gas Pipeline	140	Iran	Armenia	27	2006
BBL Pipeline[11]	235	Netherlands	UK	36	2006
South Caucasus Pipeline[12]	690	Caspian Sea	Turkey	42	2006
Turkey–Greece Pipel. (BOTAŞ, DEPA)	296	Turkey	Greece	36	2007
South Wales Gas Pipel. (Nat. Grid UK)	316	Milford Haven	Pembrokeshire	48	2007
Langeled Pipeline[13]	1,200	Norway	UK	42, 44	2007

Table 2.1 (continued)

Name (ownership)	Length (km)	Principal origin	Principal destination	Diameter (inches)	Date of operation
Nabucco Pipeline[14]	3,300	Turkey	Austria	56	2014*
Medgaz[15]	747	Algeria	Spain	24–48	2010*
Balticconnector[16]	80–120	Finland	Estonia		2010*
Nord Stream (Gazprom, BASF, E.ON)	1,113	Russia	Germany	48–56	2012*

Notes:
* expected.
1. E.ON Ruhrgas and Eni.
2. Sonatrach, Sotugat, ENI.
3. Wingas E.ON Hanse.
4. 45% Energie Beheer Nederland B.V. and 23% by Total S.A.
5. E.ON Ruhrgas, BEB Erdgas und Erdöl, Norsk Hydro and Statoil Deutschland.
6. Sonatrach, the Moroccan State, Enagas, Transgas.
7. PGNiG, Gazprom (both 48% of shares) and Gas-Trading S.A.
8. BG Group (25%), E.ON Ruhrgas (23.59%), Distrigas (16.41%), ConocoPhillips (10%), Others (25%).
9. Gassled partners (65%) and Gaz de France (35%).
10. Agip North, Africa BV, NOC- Tripoli.
11. N.V. Nederlandse Gasunie 60%. E.ON Ruhrgas and Fluxys both 20%.
12. BP (UK) 25.5%, Statoil (Norway) 25.5%, others 49%.
13. The Ormen Lange licensees, ConocoPhillips and Gassco.
14. OMV, MOL, Transgaz, Bulgargaz, BOTAŞ each 20%.
15. CEPSA 20%, Sonatrach 20%, BP 12%, Endesa 12%, Gaz de France 12%, Iberdrola 12%, Total 12%.
16. Gasum, Eesti Gaas of Estonia, Latvijas Gāze of Latvia and Gazprom of Russia.

Only within the United States were pipelines built by investor–owners based on the idea that the individual pipeline projects would pay for themselves. The worldwide privatization wave of the late twentieth century placed a number of those pipeline systems in the hands of new investor–owners and government regulators, a change that has generated a morass of new, highly dissimilar, and often unsuccessful regulatory arrangements for gas pipelines. During this period, the rest of the world viewed the institutions that govern gas pipelines in the United States as mysterious, archaic, inefficient, and both impossible and undesirable to emulate. The United States viewed other global pipeline regulatory efforts as largely built on sand – with few of the institutions that support credible regulation and creditworthy pipelines. The transactions-based analysis of the new institutional economics helps to sort through the mutual failure of understanding.

There are a number of elements of the gas pipeline market that differ importantly from the market in Europe and are critical to support contractualization. I present them generally in the order that the institutions arose in the American gas market:

1. Gas pipelines are private carriers for their customers, without common carriage or third party access (TPA) obligations (1906).
2. Gas pipelines are divorced, by unusually strong legislation, from gas distributors (1935).
3. Gas pipelines are subject to a single regulatory authority that licenses all new capacity and sets regulated pipeline prices based on its own new and meticulous accounting rules (1938).
4. Gas pipelines tariff cases follow specific administrative procedures with the value of their property that is constitutionally protected from expropriation by the regulator (1944–46).
5. Gas pipelines 'volunteer' to transform gas delivery contracts to pipeline transport contracts after an unscripted series of events in the domestic and world energy markets puts their finances in jeopardy (1986).
6. Gas pipelines ordered to ship only gas owned by others in their trunk lines, an application of the age-old 'Commodities Clause' for US rail transporters (1992).
7. Gas pipelines forced to cede to contract shippers the control of property rights inherent in the value, in excess of cost, of the capacity on their trunk lines (2000).

It is impossible to overstate the importance of each of these events in the development of the institutions support competition on American gas pipelines. Each was accompanied by vigorous legislative debate, Supreme Court action, regulatory litigation or complex regulatory rulemaking procedures. In each case there was vigorous and essential conflict between private interests, whose business profitability, or ability to serve their own customers economically, depended on the outcome of the proceedings.

The problem with understanding the institutions that support the competitive American gas markets, and particularly the role of competitive pipelines in it, is that the institutional foundations are so old. To a large extent, the institutional memory is shaky and many of the economists who were important in its development are long dead (for example, E. Troxel, W. Splawn, J. Dirlam, J. Bonbright, and E. Rostow). But the institutional history of the industry is so critical to its present competitiveness that a brief review of these landmarks is essential to describe how the market came to be.

It may not be completely fair to perform a straight comparison of the institutional foundations for gas markets in the US and Europe according to the list above. Unfair or not, it is instructive to do so. The comparison serves to illustrate the great divergence from the necessary institutional foundations, namely a contractualized network and a competitive gas market, that would confront supply security fears and prevent any supplier, especially ones like Russia or Norway, that holds as little as one-fifth of the market, from profitably withholding supplies or constraining the market.

4.1 Private Carriers, without TPA Obligations

From the first natural gas pipeline constructed in 1872 – from white pine, stretching 25 miles from West Bloomfield to Rochester, New York – the most defining characteristics of American pipelines (either in oil or gas) is that they have all been financed by investor–owners under the assumption that each pipeline individually would pay for itself (Castaneda, 1999). No piece of legislation, or new regulation, was ever introduced for any pipelines in America without having to deal with the interests of those investor–owners.

A case in point happened in the year 1906. It was then that Congress, fed up with the rail and pipeline abuses of the Standard Oil Company, decided to try to constrain the market dominance of that company by regulating oil pipelines. Congress responded to the ineffectiveness of state regulation of railroad and pipelines with a bill to give the Interstate Commerce Commission (ICC) expanded ratemaking authority over those elements of interstate commerce. The legislation was called the Hepburn Amendment to the Interstate Commerce Act of 1887. Pages of Senate debate concerned whether gas pipelines (which were not owned by Standard Oil) ought to be included in the legislation along with oil pipelines (which were generally owned by Standard Oil). The most vocal and effective proponent for excluding gas pipelines from the jurisdiction of the ICC was Senator Joseph Foraker of Ohio. Foraker argued on the floor of the Senate, again and again, that gas pipelines were not common carriers and that trying to regulate them as such would kill the business, which depended on the ability of the pipeline to be committed to a particular enterprise. At the critical point of the Senate debate in 1906, Foraker argued:

> [N]obody is interested in that [gas pipeline] enterprise, except only the people who are building the line with the idea of bringing the gas to Cincinnati, to do a great public service, and they have had trouble enough to set the enterprise on foot. They are just now in the midst of their trouble, trying to raise the money. They have not yet been able to raise it all. If it should go out, after they have raised the money to build the line, that any man can take possession of it to bring gas there for his own purposes, and that the line is to be under the charge of the Interstate Commerce Commission, I think it will be the end of the enterprise.[10]

That was enough for the Senate – it excluded gas pipelines from the Hepburn Amendment, which is still the defining statute for regulating American oil pipelines. That exclusion ensured a 32-year reprieve from regulation for the gas pipeline industry. During those 32 years, American regulators made great advances in the principles of regulatory accounting and licensing. By the time it came finally to impose federal regulation on the gas pipeline industry, Congress would avoid the common carriage (or TPA) model entirely and turn to utility regulation.

It is impossible to predict how the industry would have developed had Senator Foraker not pushed to exempt gas pipelines from the ratemaking and routing jurisdiction of the ICC, which over the following 70 years made a complex mess of regulating oil pipelines. In any case, it is no understatement to call Senator Foraker the father of the modern American competitive gas pipeline markets. He might have been amused by the title.[11]

The US never had TPA for gas pipelines, as such. In contrast, Europe starts from the requirement of TPA for all of the national networks. The Second EU Gas Directive, adopted in June 2003 to be effective in July 2004, required regulated third party network access on gas pipelines in the European Union. The directive (EC, 2003) states:

> Member states shall ensure the implementation of a system of third party access to the transmission and distribution system . . . applicable to all eligible customers . . . applied objectively and without discrimination between customers.

Further, the Second Gas Directive states:

> Further measures should be taken in order to ensure transparent and non discriminatory tariffs for access to transportation. Those tariffs should be applicable to all users on a non discriminatory basis.

Mindful of the different legal basis for most of Europe and the common law in America, this provision has similarities to the following passage in the Interstate Commerce Act in the US:

> Every common carrier subject to the provisions of this act shall, according to their respective powers, afford all reasonable, proper and equal facilities for the receiving, forwarding and delivering of ... property ... and shall not discriminate in their rates and charges.[12]

A common carrier, or TPA provider, holds itself ready to serve the general public to the limit of the facilities that the carrier is prepared to offer. By contrast, the private carrier transports only a narrowly defined clientele – it discriminates in favor of those with contracts, affording a secondary service to those without.[13] Gas pipelines in Europe are directed, by Article 18 of the Gas Directive, to be TPA providers. Gas pipelines in the US escaped that responsibility in 1906. When Congress passed legislation in 1938, it rejected common carriage (after having considered it) and made pipeline private carriers, who would not be obliged to carry any gas for any customer without a multi-year pipeline company contract.

This is a very large difference between the two regimes. Contract carriage would not work in the US, and the contracts for capacity would be of little value, except that capacity rights are scarce commodities and only obtainable in the primary market through long-term, contractual commitments.[14] It has been the value of these capacity rights which has allowed US pipelines to be financed on a stand-alone basis. Common carriage, or TPA, was created for oil pipelines in the US, and it caused their vertical integration and joint venture structure throughout the twentieth century, and the lack of a contractualized capacity market in the twenty-first century.

To be sure, the entire European gas pipeline network is not bound by TPA rules. A distinction exists between the national networks for which no contractualization exists in the US sense, and the large international supply pipelines (called 'interconnectors') that transport gas from producing countries that do have long-term contracts (EC, 2003). No common carrier rights apply to these basic supply routes, as the pipelines were developed through contracts between the pipeline owners and shippers, and any requirement of open access would impair capacity availability for those shippers. Any new shipper would need to have capacity built on request and enter into a bilateral contract with the pipeline company on negotiated terms. The elements of carriage by contract evident on these basic European supply lines do not embrace the range of requirements (information requirements, tradability, segmentations, and so on) that accompany contractualization in the US. A major difference is that the exemption from TPA granted in Article 22 remains within the discretion of the Commission, whereas the US has no TPA whatsoever on its interstate pipeline network.

4.2 Gas Pipelines Divorced from Distributors

This subsection deals with the split between transportation pipelines and local gas distributors. This split in the US did not require that the transportation pipelines divorce themselves from the ownership of the gas in the pipelines (referred to as the 'Commodities Clause' in legislative circles pertaining to transport regulation).[15] The decisive federal regulation of gas pipelines in the 1930s was intertwined with the simultaneous federal investigation of the manifest abuses of multi-state utility holding companies. The holding company structure adopted by electric and gas utilities during the 1920s and 1930s enabled a number of financial abuses that state regulators could not effectively control.[16] Many of the types of abuses recognized at that time can be solved by modern accounting regulations and meticulous scrutiny of affiliate transactions in experienced regulatory jurisdictions. But in the 1930s, however, American regulatory methods were not equipped to handle them.

In 1928, the Senate asked the Federal Trade Commission (FTC) to conduct an investigation of the public utility holding companies. The FTC produced a comprehensive and massive report in 1934 and 1935, which ultimately comprised 96 volumes. The report showed that over half the gas produced and more than three-fourths of the interstate gas pipeline mileage in America was controlled by 11 vertically integrated holding companies. The FTC called the abuses of the unregulated, integrated gas pipeline companies a 'positive evil'.[17]

Congress dealt with the abusive market behavior of the holding companies by passing the Public Utility Act in 1935, Title I of which was called the Public Utility Holding Company Act.[18] The act gave the Securities and Exchange Commission (SEC) jurisdiction over public utility securities. As part of their new jurisdiction, the SEC was given great powers to simplify the holding company structures of gas and electric utilities. Troxel (1937, pp. 21–2) wrote:

The Holding Company Act was a severe law. It was the most stringent, corrective legislation that ever was enacted against an American industry. Yet forceful actions were needed to straighten out the corporate organization and control of the electric and gas industries. The remedy was suited to the patient. Many holding companies were dissolved, partially liquidated, or reorganized. Unified, technically related systems replaced conglomerate financial arrangements that served the interests of financiers. Investment banker control of electric and gas industries was eliminated or modified; and the engineers became more important.

Like the Hepburn Amendment before it, the Holding Company Act was a legislative assault on the existing structure of private American businesses. In both cases, however, abusive and acquisitive practices by those businesses, and the resulting public outcry, overcame Congress' normal aversion to dealing with the complex internal structure of American corporations.

Make no mistake about it: the Holding Company Act of 1935 was a gigantic step in the direction of ultimate supply security for the American gas market, even though it would take many more years, and many wrong turns, to realize it. Splitting gas distributors from gas pipeline companies created two powerful and sophisticated constituencies that worked to shape gas regulation going forward. Congress never forced such a structural split on the American *oil* pipeline system, despite almost continuous calls to do so in Congress over many decades, from the 1930s through the 1970s (Johnson, 1967). The oil pipeline industry thus continued to develop with a high degree of vertical integration, with more than three-fourths of the industry's pipeline capacity owned by the 18 major vertically integrated oil companies, mostly interlocked with joint ventures, by the 1980s.[19] The US Department of Justice once lamented this interlocking of shipper ownership on that pipeline system, and the resulting lack of strong contending constituencies:

> This level of customer/supplier dominance in the ownership of a regulated industry is perhaps unique to the petroleum pipeline industry [in the United States]. In other regulated industries, the clash between those regulated and their immediate customers provides the necessary tension to achieve effective and even-handed regulatory scrutiny. Here, however, the absence of adverseness requires the regulator to take affirmative steps to regulate effectively.[20]

It is no overstatement to say that the Holding Company Act was the *sine qua non* in the future regulatory battles over liberalization and eventual contractualization of the gas pipeline market in the US.

Europe has never had the chance to confront the problems of investor-owned, unregulated and vertically integrated utility holding companies.

Its policies regarding the tie between pipeline companies and gas distributors are thus rather milder. The Second Gas Directive deals with vertically integrated gas companies as follows:

> Where the transmission system operator is part of a vertically integrated undertaking, it shall be independent at least in terms of its legal form, organization and decision making from other activities not relating to transmission. These rules shall not create an obligation to separate the ownership of assets of the transmission system from the vertically integrated undertaking.[21]

Despite the unbundling required by this directive, the European Commission had strong words to say regarding the 'vertical foreclosure' still inherent in the gas pipeline network (EC, 2007, p. 7):

> The current level of unbundling of network and supply interests has negative repercussions on market functioning and on incentives to invest in networks. This constitutes a major obstacle to new entry and also threatens security of supply.

The connection between gas transporters and distributions, within the same ownership, was a corporate form utterly rejected in the Public Utility Holding Company Act of 1935, and an 'uncommonly powerful' regulator, in the form of the SEC was charged with breaking them up. For reasons that those in Europe may consider manifestly obvious, the Directive never hints that public policy should include the structural separation of gas pipelines from gas distributors. The 'clash between those regulated and their immediate customers provided [that] the necessary tension to achieve effective and even-handed regulatory scrutiny' divides the European and US gas markets.

4.3 Gas Pipelines Subject to a Single Regulatory Authority

Throughout the 1930s, Congress wrestled with how to regulate gas pipelines. It had become clear in the courts that crossing state lines exempted gas pipelines from state regulation, and no federal body had any jurisdiction over the business at all. The Supreme Court has said as much when it struck down an order issued by the Kansas Corporation Commission that fixed city gate rates charged by the Cities Service system, one of the largest multi-state holding companies. The Court stated:

> The transportation, sale and delivery constitute an unbroken chain, fundamentally interstate from beginning to end, and of such continuity as to amount to an established course of business. The paramount interest is not local but national – admitting of and requiring uniformity of regulation. Such

uniformity, even though it be the uniformity of governmental non-action, may be highly necessary to preserve quality of opportunity and treatment among the various communities and states concerned.[22]

There was clearly a hole in regulatory authority that Congress had to fill. In drafting legislation to do so during the 1930s, however, Congress had to deal with two powerful, battling constituencies. The first was the pipeline industry itself, which had strong representation in Congress. The second was the Cities Alliance, a group of 100 Midwestern city and town governments, which was organized in the mid-1930s to lobby for gas pipeline regulation (Sanders, 1981). In order to pass gas pipeline legislation, Congress had to satisfy both of these constituencies.

The bill that accomplished this delicate political balance was the Natural Gas Act of 1938.[23] The pipeline interests conceded that some form of federal ratemaking authority was inevitable, but it wanted to shield pipeline companies from competitive pressures. Cities Alliance wanted to cap the price of gas delivered to the state-regulated gas distributors, but also wanted gas pipelines to be forced to compete with one another.[24] What emerged from Congress was a utility-style regulatory statute that capped rates but required Commission licensing for any new line to a region already served by an existing line.[25] The Cities objected to the licensing provision, which could limit pipeline competition. But the bill's sponsor disagreed, saying: '[t]hat is what regulation is, monopoly controlled in the public interest'.[26]

The United States has a multiplicity of state and federal regulatory authorities. But no issue is more important than the dividing line between them. Jurisdiction is not shared – it is meticulously divided on the basis of the requirements of the US Constitution.[27] Such a division in regulatory responsibility does not exist within the EU. The Second Directive states:

> Member States shall designate one or more competent bodies with the function of regulatory authorities. They shall, through the application of this article, at least be responsible for ensuring non-discrimination, effective competition and the efficient functioning of the market, monitoring in particular [allocation of capacity, congestion mechanisms, publication of information, unbundling of accounts for transmission and distribution, the level of transparency and competition, etc.]

This provision has also provoked comment in the Competition Report of January 2007 (EC, 2007):

> To ensure the implementation of the regulatory framework in this respect, the Second Gas Directive requires the creation of national energy regulators

[note omitted]. . . . Market integration is also hampered by limitations in the competences of national regulators. In the absence of any single cross-border regulator, national regulators must cooperate with each other in monitoring the management and allocation of interconnection capacity. . . . Moreover, the matter in which Community rules have been implemented varies between Member States, and may in some cases even give rise to regulatory vacuum – especially in cross border situations. In addition to the requirements under Community law, there is also a considerable scope for Member States to apply their own specific national rules.

This type of overlapping of jurisdiction is abhorrent to regulators in the US. The existence of a single interstate pipeline regulatory authority, the FERC, simplifies greatly where parties will 'clash' to pursue regulatory remedies in their interests. It also manifestly simplifies the creation of a single 'code' for shipping gas throughout the network.[28]

4.4 Pipeline Property and the Regulation/Administration of Rates

In the US, Supreme Court decisions define the legal limitations on regulators' ability to take action on charges that may damage the value of utility investors' property. The best-known case is that of *Federal Power Commission v. Hope Natural Gas*, in which the Supreme Court set a standard for determining 'just and reasonable' returns, a standard that has stood the test of time.[29] Even today, normal utility tariff reviews, as well as substantial changes in regulatory rules, reference this particular judicial precedent. For the purposes of the future contractualized gas pipeline market, the *Hope* decision was critical. It sharply limited investor or shipper uncertainty regarding the ability of regulators to act in a manner that would damage the value of the assets that investors would devote to regulated enterprises.

The other pillar of certainty associated with American rate regulation is the Administrative Procedures Act of 1946. During the 1930s, considerable scholarly analysis was devoted to determining the legality of utility regulation's growing impact on the value of regulated property. At the time, regulators had the power to augment or shrink the value of their investors' property in their jurisdictions. Accordingly, legal scholars and the courts questioned whether utility regulators were acting within the confines of authority actually granted by legislatures. Existing regulatory statues gave discretion to regulatory commissions that were not extended by specific legislative mandate and seemed to violate the US Constitution's prohibition of the taking of property without due process.

Congress addressed these issues by passing the Administrative

Procedures Act of 1946, which laid out meticulous procedures to be followed by all regulatory commissions that would assure Constitutional due process (Moynihan, 1998). It also specified timing limits, the need to act upon evidence, the ability of witnesses presenting that evidence to be cross-examined, and many other aspects of the work of regulators. The Administrative Procedures Act imparted much greater fairness, predictability, and transparency than had theretofore been the case in American regulation.

These two limitations on the discretion of regulators was no academic exercise. They were fundamental to the further financing of the investor-owned gas pipeline business in the US. Prohibiting vertical integration after 1935 effectively closed off vertical sources of equity funding for gas pipelines. With these two limitations, and other features of FERC regulation, investors could know that pipeline loans would be reliably repaid over the 30–50 year lives of major pipelines. Seeing this, the American life insurance industry created new loan instruments specifically for gas pipeline financing. With those new loan instruments life insurers underwrote the industry. During the gas industry's sixfold expansion of interstate gas shipments between 1946 and 1959, approximately 78 percent of natural gas pipeline bonds were held by life insurance companies. The remaining 22 percent of bonds were funded by 'trustee investments' such as private pension funds and personal trusts that looked to the life insurance industry for guidance (Hooley, 1968, pp. 13, 45). Without the strong restrictions on the discretionary power of the FERC over private property, such long-term financing, at low interest rates, would not have been forthcoming.

In Europe, the administration of regulated rates is a more recent, and less exacting and consistent, affair. The Second Gas Directive calls for 'published tariffs, applicable to all eligible customers',[30] but does not further describe the ratemaking formula or rules on the level of permissible revenues. Compared to the restrictive rules on what can constitute an element of regulated pipeline tariffs in the US, the rules for gas pipelines in Europe, as a regional pipeline market, are not so well defined. The licensing powers of the FERC allow it to define a cost of service for new pipelines relating predictably to its book investment cost, a cost that will predictably be returned to investors over the life of the new pipelines in regulated rates, through the standard FERC cost-based ratemaking formulas. That kind of predictable licensing, accounting and regulated rate administration does not exist for new pipelines in Europe. As such, it is much more likely that new pipelines will be built by vertically integrated firms, or joint ventures of such firms, in order to spread the investments risk.

4.5 Structural Separation for US Gas Pipelines (the 'Commodities Clause')

The 'Commodities Clause', which would forbid common carriers to own the product they shipped was imposed as part of the Hepburn Act in the US in 1906.[31] For the 30 years that passed between 1950 and 1980, there was deliberate and seemingly endless litigation between American gas producers and distributors, and later between gas pipelines and distributors, over the delivered price of gas on the nation's pipeline network. The tug of war between producing interests, who favored complete gas price deregulation, and consuming interests, who favored limiting the shift of economic rents that deregulation would cause, never ceased. Economists were stalemated (Kahn, 1960). The difficulty lay in the fact that the institutions employed to regulate the pipelines were created to regulate naturally monopolistic utilities, not inherently competitive and relationship-specific inland transport companies. Indeed, during the battle over gas prices the proposal to have pipeline companies exit the business of buying and selling gas never appears to have come up at all.

What prompted institutional change in that direction was an accidental side-effect of the struggle to deregulate gas prices. Both the FERC and Congress failed to anticipate the destructive economic incentives that their partial gas deregulatory policies would place on pipeline companies. The menu of old and new gas prices combined with the method of regulating pipeline re-sale rates sent pipeline companies on an expensive gas-buying spree during the late 1970s and early 1980s that ultimately crippled their finances. It was then, when it was given the power to extract concessions from a weakened industry, that the FERC extracted the 'voluntary' institutional concessions from pipeline companies that created the contractualized gas pipeline market of the twenty-first century.

The FERC found that open access was not enough to foster competitive gas markets if pipelines owned the gas that they shipped. It was at this point, in 1992, that the FERC required that pipelines transfer title to their own gas supplies by the time the gas entered the main trunk pipelines. In this way, all of the gas in their trunk lines was owned by others, and no gas supplier could claim an operational advantage over any other (as the pipelines had theretofore successfully been able to do). In essence, the FERC imposed the Commodities Clause that Congress had declined to apply to oil pipelines in 1906 or to gas pipelines in 1938.

As far as the Second Directive and more recent publications are concerned, there is a considerable amount of discussion about fair access to networks, but almost no discussion on the competitive problems that may arise when pipeline network owners ship their own gas. The lack of structural separation creates problems, as the DG Competition Report pointed out (EC, 2007, p. 58):

> The Commission has also gathered indications that one TSO [Transmission System Operator] grants its affiliated supply company substantive rebates for the transportation fees as compared to non-affiliated network users. In doing so, the TSO directly supports the competitive position of the related supply company. This appears to be an overall business strategy carried out by some integrated companies despite the formal Chinese Walls created by the Second Gas Directive. The introduction of ownership unbundling would make this competitive advantage of the affiliated suppliers disappear, given that the transport tariffs would follow market principles and thus tend to be the same for all suppliers. If the ownership link is broken the incentives facing the network operator will change. It will seek to optimise its network business as opposed to acting in the overall interest of the vertically integrated group.

There are a number of cases of discriminatory behavior cited in that report, all seemingly stemming from the lack either of ownership separation or of separation between transport and gas sales. Within many gas companies, trading names, brands and logos are still shared, and there is no application of what is known in the US as the Commodities Clause, which would prevent transport pipelines from owning the gas shipped in their trunk pipelines.

4.6 Provision of Information on Pipeline Transport

A critical element of the competitive pipeline market in the US is the free and transparent flow of information. In its far-sighted Order No. 637 in 2000, the FERC dealt with this issue squarely:

> The Commission finds that the disclosure of detailed transactional information is necessary to provide shippers with the price transparency they need to make informed decisions, and the ability to monitor transactions for undue discrimination and preference. Shippers need to know the price paid for capacity over a particular path to enable them to decide, for instance, how much to offer for the specific capacity they seek. . . . The disclosure of all transactional information without the shipper's name will be inadequate for other shippers to determine whether they are similarly situated to the transacting shipper for purposes of revealing undue discrimination or preference. . . . Finally, to be meaningful, for decision making purposes, the transactional information must be reported at the time of the actual transaction.[32]

Basically, while the FERC acknowledged that some shippers thought that its information reporting requirement may cause some burdens, and also that it may 'give shippers knowledge of their competitors 'general marketing strategy',[33] it was more than swayed by the need for the market to be fully informed to operate efficiently and to uncover undue discrimination or market manipulation when it would appear. Thus the FERC chose to require the most comprehensive and immediate provision of all information on the identities and quantities, locations, and so on of all shippers.

For the FERC there are no 'trade secrets' with respect to the use of the regulated pipeline network – it is an open book. If it erred, the FERC erred on the side of transparency and full disclosure of data. In contrast, the Second Gas Directive says the following:

> Without prejudice to Article 16 or any other legal duty to disclose information, each transmission, storage and/or LNG system operator shall preserve the confidentiality of commercially sensitive information obtained in the course of carrying out its business, and shall prevent information about its own activities which may be commercially advantageous from being disclosed in a discriminatory manner.[34]

The greatest contrast between this provision and that of the FERC's is that the latter admits to no 'commercially sensitive' information on the use of the network that outweighs the need for the market to be fully informed, in order to prevent undue discrimination and market manipulation. The DG report echoes a frustration about the provision of information on the European gas network (EC, 2007, p. 90):

> The Sector Inquiry confirms that gas wholesale operators have contrasting views on the question whether the amount of information available on network capacity is sufficient. Incumbents are usually satisfied, whereas most new entrants find that information is lacking, suggesting that vertically integrated incumbents have privileged access to information.

The DG Competition Report says (ibid., p. 90):

> It may be a concern that excessive transparency could facilitate collusion between the major market players, particularly on an oligopolistic market. A balance must certainly be found as to what data is published and how it is published, in order to improve transparency without enabling collusion.

There is a considerable contrast between the market information required in the US and Europe. For the US gas pipeline market, the issues are considered virtually black and white by the FERC – users of the regulated pipeline network have no right to secrecy (Olson, 2005). Such a point of view is entirely consistent with the FERC's firm, decades-old control over regulatory accounting. In Europe, where regulatory accounting is yet an unsettled and controversial issue, the issue of transparency and information provision on the pipeline network is far from resolved.[35] As the DG Competition Report said 'the large number of different pipeline systems and the high number of operators controlling the capacity on these routes [note omitted] render the access conditions to these transit pipelines opaque' (EC, 2007, p. 73).

4.7 Crystallizing Property Rights in Gas Pipeline Capacity: Regulating Rents

In 2000, in order to cement the property rights that pipeline firm shippers could exercise with their capacity contracts, and to facilitate the market in which pipeline capacity could trade, the FERC implemented the following five changes into its regulations:[36]

1. Removal of the price cap on secondary pipeline capacity sales.
2. Requiring pipeline companies to permit shippers to 'segment' capacity for their own use or release. Segmenting broke up capacity into separate segments in a complete chain, to facilitate using some and releasing others.
3. Limiting imbalance management and penalty provisions only to those needed to protect system reliability.
4. Consolidating and enforcing pipeline reporting requirements to improve price transparency and more effectively monitor the exercise of market power.
5. Requiring 'incremental pricing' for all new pipeline transport capacity.[37]

Without realizing it, and certainly without referencing Professor Coase, the Commission's changes settled the ability of contract shippers to trade and profit from marketing their capacity. It was a key step in creating durable property rights in shippers' contracts for cost-based interstate pipeline capacity. In his 1960 article, Coase argued that if property rights are clearly specified, parties have an incentive to negotiate a mutually beneficial trade. Coasian markets have been created in pollution rights, carbon allowances, radio bandwidth, and other commodities through the creation and clear specification of property rights (Kwerel and Rosston, 2000; Ellerman et al., 2003). This kind of market for gas pipeline capacity now exists in the United States – and only there.

The new era has made new pipeline construction a straightforward affair, rather than a political or public relations battle between interests contending for regulatory favor. The FERC has jurisdiction over pipeline licensing under the Natural Gas Act, under a 'public interest' standard.[38] Pipeline developers once engaged in two forms of competition to meet the

FERC's public interest requirement: (i) they worked to get distribution companies to favor them; and (ii) they tried to outmaneuver rival pipeline projects to demonstrate the economic necessity of their own project. If a particular pipeline company or consortium won that test, they received the Commission's approbation in the form of a 'certificate of public convenience and necessity', which was required to build.[39] The new market cut through this regulatory tangle when it successfully dealt with the issue of pricing new capacity on an incremental basis.

In 2000, again after some wrong turns, the FERC mandated that pipeline companies price new or expanded services on an incremental basis. Setting incremental pricing as the default rule for new pipeline construction put an end to the fight over traditional certification. If a pipeline company approached the Commission with a new proposal, it was no longer required to demonstrate connected gas supplies. With a portfolio of letters of intent from committed incremental shippers, the Commission had virtual prima facie evidence of economic need. The typical time period needed to plan and construct a typical large-scale pipeline expansion project dropped from five years to two.[40]

In the US, property rights for independent shippers are founded on the long-term contracts under which the entirety of the network since 1944 was built.[41] Except for the exemption under the 2003 Gas Directive for 'interconnectors', the European gas network was not built with such long-term shipper contracts. As such, the ready-made basis for contractualizing the pipeline capacity does not exist in Europe. That being the case, the remaining elements of contractualization (incremental pricing, segmenting of capacity, the market for secondary trades, and so on) are generally inapplicable.

5 SUMMARY

It is not a ridiculous overstatement to call the regulation of American gas pipelines the Stradivarius of inland transport regulation schemes pertaining to pipelines. The use of the network is freely competitive with a robust secondary market in capacity rights. New capacity is independent and competitively constructed and relatively easy to license and finance on the basis of long-term capacity contracts with shippers. Those pipeline users in areas served only by one or two pipeline companies are protected from price gouging or denial of availability by traditional cost-based regulation. While those rate cases were once hugely contentious, they are now largely perfunctory. Long-term gas commodity contracts – once typical – have evaporated on the network, with most gas traded through predictable and reliable physical hubs and established markets. The pipeline market has shown great robustness to bad weather and environmental disasters, clearing the transport market through the price mechanism in each case. Even with the administrative and market meltdown of the California market in 2000–01, the gas market cleared, facilitated by a market-sensitive and flexible pipeline network.

But lest any users, owners or regulators of that American pipeline network try to claim too much credit, one only has to look at the similarly vast American crude oil and oil products pipeline network to see a broken-down country fiddle of an inland transport market by comparison. That latter network has no capacity rights, no such independent and competitive market for new pipeline capacity, obscure tariffs and access restrictions and grotesquely lengthy tariff cases.

The difference between the American Stradivarius and the broken-down country fiddle, on pipeline networks that otherwise look the same, is a function of the development and growth of economic institutions over the past 101 years. Largely because of Ohio's Senator Joseph Foraker, the unsung hero of the modern gas market in the US, the two pipeline networks took different evolutionary institutional paths in 1906. Today, the two networks are institutionally so completely unlike one another that very few Americans in the legal, economic or regulatory spheres have experience with both. That fact is odd in and of itself. Perhaps even odder is that the final institutional development of contractualization evident in the twenty-first century American gas pipeline market was largely unscripted.

What does the US experience hold for security of supply on the gas pipeline network in Europe? More than anything else, the US experience points to the critical nature of a number of regulatory, contractual and ownership institutions that are new or untested in Europe. This chapter has briefly touched on seven of the critical institutional building blocks for the contractualized gas pipeline market in the US. In only a few cases have similar institutions developed in Europe (often for reasons that will appear quite obvious to Europeans).

What are the choices for competition and security of supply on the European gas pipeline network? The contractualization route in Europe would require the definition and distribution of property rights and the imposition of meticulous accounting regulations currently foreign to European regulators. It would require the synchronization of pipeline capacity across national borders within Europe and a single regulatory jurisdiction for the major trunk gas pipeline companies that requires vastly enhanced market information requirements. It would also perhaps require the structural separation of pipeline companies (critical in the US, virtually impossible to foresee in Europe) and the imposition of the Commodities Clause. To say that this is a challenging administrative, legal and political job would be a crass understatement. And to be sure, any Americans who would suggest otherwise should simply be reminded of that broken-down, 101-year-old fiddle yonder in their oil pipeline market.

The other choices? The other choices look more like command and control. One choice could be the UK's system that treats the network like a large integrated vessel, with notional trading points – sort of like a large natural gas tank into which suppliers inject gas and from which distributors, power generators and others withdraw it. It could be a remedy for an uncompetitive gas commodity market, but it would overlook the achievement of competitive efficiency on the pipeline network itself. One of the great advantages of contractualization in the US was that the market identified regions of unexpected excess capacity, and highlighted the demand for capacity in others. Pursuing the UK's system would ultimately require the network to be an administered – not a competitive – affair, much like electricity transmission in the US.

Another similar choice is to continue to treat the great national pipeline systems like national utility monopolies, with reciprocal agreements with others for reliable and well-informed transborder shipment rules. Again, this removes many competitive pressures from the pipelines themselves, but it may be a reasonable option nonetheless, given the evident challenge of creating an institutional basis for Europe-wide pipeline contractualization.

It would not be fair to conclude without emphasizing again the great difference in regulatory certainty between the US and Europe. The growth of federal regulatory institutions for the interstate gas network has been evolving for more than a century in the US. The body of legislation, case law (under the legal principles of common law precedent) and collaborative work of regulators,[42] all developed through the dogged efforts of adverse interests working through the administrative and civil law courts to pursue and retain their rights, makes the FERC regulation of gas pipelines a highly predictable affair. Pipeline developers in the US can take that predictability to the bank – literally – and the bankable nature of an independent, contractualized network makes supply security a reality. It takes time, adversarial effort and judicial review to develop such predictable institutions, and in this area Europe has only just begun.

Ultimately, security of gas supply in Europe will come through a more flexible and transparent transport system, increased flexibility in the current supply contracts,[43] and a lower cost for the use of the network itself. Either competition or efficient regulation could advance this goal, for which the number of viable sources of gas rises and the ability of any supply to constrain the market shrinks. Which path will Europe follow? As far as the inland gas transport violin in Europe is concerned, the luthier is still at work.

NOTES

1. Helpful comments by the participants at the 1st CESSA conference in Berlin, May 31, 2007 as well by Fabrizio Hernandez and Wayne Olson are gratefully acknowledged.
2. Coase (1960) argued that if property rights are clearly specified, parties have an incentive to negotiate a mutually beneficial trade. Coase also recognized that transaction costs matter. The initial allocation of property rights matters because of the transaction costs associated with reallocating those rights via the market.
3. Hurricanes Katrina and Rita disrupted the gas supplies from the Gulf of Mexico in 2005, and the latter left the principal American gas trading hub – the Henry Hub – under water.
4. Kahn (1971, pp. 153–4) put his finger on the weakness of the natural monopoly idea with respect to pipelines: 'As far as the actual carriage of gas is concerned, economies of scale could not possibly require a single chosen instrument for the entire national market. Pipelines travel from one point to another; in consequence there is ample room for a large number of criss-crossing lines, with ample resultant possibilities of competition both in areas between lines and their points of junction. The main potentials of scale are to be found in employing pipe of the maximum diameter available. . . . [note omitted] But these economies taper off sharply once the largest possible pipe available is used and even more sharply when the limits of further expanding capacity in the manner indicated are reached'.

    Nelson (1966, p. 3) put the whole 'natural' concept brilliantly into context: 'One of the most unfortunate phrases ever introduced into law or economics was the phrase "natural monopoly." Every monopoly is a product of public policy. No present monopoly, public or private, can be traced back through history in a pure form. . . . Roads? The "King's Highway" was usually more an easement than a facility until well into the eighteenth century, except where the admittedly monopoly-minded Romans had done their work; the highway was lifted from its literal morass only by private turnpike companies, sometimes on a quasi-competitive basis. . . . So "natural monopolies" in fact originated in response to a belief that some goal, or goals, of public policy would be advanced by encouraging or permitting a monopoly to be formed, and discouraging or forbidding future competition with this monopoly'.
5. Section 7 of the 1938 Natural Gas Act provides for Commission licensing authority. Such licensing does not apply to the extensive oil pipeline network in the United States, under that controlling legislation (the 1906 Hepburn Amendment to the 1887 Interstate Commerce Act). We know of no such central repository of European gas pipeline expansion projects as exists for US pipelines.
6. Some gas was transported, under individual licenses, to industrial customers.
7. Trapmann and Todaro (1997) concluded: 'Competition is increasing in US gas markets. The overall nature of the market outcome – prices and volumes – depends on the interaction of the entire set of participating entities: producers, consumers, and infrastructure operators (e.g., storage, transportation, and hubs). The system seemed to perform better in 1996–97 than in the prior heating season. Although prices were higher, the system avoided the extreme price spikes that occurred in some localities (e.g., Chicago) during the 1995–96 season. The 1996–97 price pattern reflects the improved interconnectedness of the system, which supports effective competition between regions of the Lower 48 [states]. Storage utilization during the past heating season may be questioned in light of subsequent events, but the strategy does not

appear to be unreasonable. The early reliance on storage gas in 1995–96 left lower-than preferred levels of gas as inventory, which became a critical factor when the severe temperatures persisted in major consuming locations. On the other hand, the lesser reliance on storage gas in early 1996–97 greatly contributed to increased prices for marketed production'.

8. The EIA's Natural Gas Weekly report (EIA, 2000) describes the events of December 2000 as follows: 'Variability was the order of the day for spot prices at the Henry Hub last week. After beginning the week with spot prices reaching double-digit highs of $10.17 per million British thermal units (MMBtu), prices dropped sharply to $7.52 per MMBtu by Thursday, then gained over $0.30 to end Friday at $7.83 . . . The past week was highlighted by unprecedented prices in the large California market where contributing energy problems include: long-delayed maintenance at several nuclear facilities in California, reduced generating capacity, low availability of hydroelectric power, unseasonably cool temperatures, below average natural gas stock levels, and reduced transmission capacity to southern California. This resulted in midpoint prices at California's PG&E and SoCal citygates of $44.00 and $59.42 per MMBtu, respectively, on Monday with prices reaching a high of $72.00 for a period of time on SoCal. In response to this situation, the US Department of Energy imposed its authority to require independent electricity generators to continue to operate and the Federal Energy Regulatory Commission re-imposed rate controls on electricity generation. These governmental actions along with moderating temperatures appear to have calmed the markets for the time being. The prices at the California border on Friday ranged from $10.85 in the north to $17.06 in the southern parts of the state'.

9. The EIA's Natural Gas Weekly report (EIA, 2005) describes the events of hurricanes Katrina and Rita and their impact on the natural gas market as they unfolded, including the following: 'For the week covered by this report (September 21–28), prices have declined largely owing to Hurricane Rita weakening as it reached major gas supply areas in the Gulf of Mexico. Nonetheless, the combination of Hurricanes Katrina and Rita has disrupted natural gas supplies and continued to prop up prices at near-record highs around the nation . . . The Henry Hub is not operating, but large price decreases prevailed in other Louisiana spot markets and East Texas. The average price in the two trading regions yesterday was $12.60 per MMBtu, a decline of $1.67 on the week. Prices also declined in major consuming regions in the Northeast and Midwest . . . Prices decreased significantly in the Rockies and the West Coast as well, albeit slightly less so than in the East'.

10. Congressional Record – Senate, 59th Congress, 1st Session (May 4, 1906), p. 6371.

11. Senator Foraker, who was then in his second term, was an ironic champion for modern, competitive gas pipeline markets. Elected to the Senate in 1896, he was the only Senate Republican (Theodore Roosevelt's party) who voted against the Hepburn Bill, a position that may have been related to payment he received from the Standard Oil Company for legal advice he provided during his first term. When news of this involvement became public in 1908, exposing a seeming conflict of interest, Foraker was forced to retire from Congress.

12. Interstate Commerce Act of 1887, Section 3.

13. The DG Competition Report of January 2007 (EC, 2007b) discusses the issue of 'contractual congestion' where access to pipelines is denied on the basis that all capacity is already reserved. (Section II.3.3). Such 'contractual congestion' is of course an essential element of US gas pipelines, for the property rights in capacity are what motivates gas distributors, power generators and others to commit to long-term payments (which is the support for pipeline financing). Such 'contractual congestion' is not considered a problem in the US, however, for two reasons: (i) a market exists with sufficient information to make any attempt to monopolize a pipeline route utterly and instantly discoverable and obvious; and (ii) the pipeline company profits through marketing 'interruptible' pipeline capacity (which is essentially firm for 30 days) to the extent that it has the ability to do so given firm contract holders' use of their lines.

14. In the secondary market, freed of direct responsibility to the pipeline company to underwrite the pipeline venture, firm capacity contracts come in any variety of time periods, configurations and prices. No matter, the primary capacity holder is obligated to the pipeline company through the underlying long-term, multi-year contracts that support pipeline financing.
15. I deal with the split between pipeline transport and gas ownership in Subsection 4.5, below.
16. Those abuses included, among other things, the writing up subsidiary property values and charging excessive service fees through affiliates. Phillips (1993) provides an excellent discussion of those abuses perpetrated by US utility holding companies.
17. Final Report of the Federal Trade Commission to the Senate of the United States pursuant to S. Res. 83, 70th Cong., 1 Sess. (1935), p. 615.
18. 49 Stat. 803 (1935). The Act was only repealed in 2005 by the Energy Policy Act (EPACT) of 2005 (Section 1263), which replaced it with the Public Utility Holding Company Act of 2005, which provides for federal access to books and records of holding companies and their affiliates.
19. About 60 percent of pipeline shipments were on pipelines owned jointly by groups of oil companies constituting their major shippers, primarily the major integrated oil companies. See: Anderson, R.E., and Rapp, R.T., *Competition in Oil Pipeline Markets: A Structural Analysis*, National Economic Research Associates Inc. (NERA), April 1983, p. 2.
20. In the matter of valuation of Common Carrier Pipelines, Docket No. RM-78-2, Statement of the Department of Justice (Donald A. Kaplan, Chief, Energy Section, Antitrust Division), October 23, 1978, p. 9.
21. EU Gas Directive, Article 9. The Directive goes on to say (preamble paragraph 10): 'It is important however to distinguish between such legal separation and ownership unbundling. . . . However, a non-discriminatory decision-making process should be ensured through organisational measures regarding the independence of the decision-makers responsible'.
22. *Barrett v. Kansas National Gas Co.*, 265 US 298, P.U.R. 1924 E78. Troxel (1937) presents a very good discussion of all of these cases.
23. Intra-state gas pipelines, which did not cross state lines, continue to be regulated by the various states.
24. American pluralistic politics – the desire to satisfy the greatest number of contending interest groups – has a lot to do with the birth and structure of regulations applied to American industries. Professor Theodore Lowi is a key contributor to the body of political science on this subject. For a classic description of how the pluralistic exchange of benefits shapes particular Congressional legislation, see Lowi (1973).
25. The licensing provision was extended by Congress in a subsequent amendment to all new gas pipeline capacity additions, whether in new or existing pipeline markets, in the 1940s.
26. Sanders (1981, pp. 41–2). Before serving in Congress, Representative Clarence Lea had been a member of the California Public Utilities Commission, which would explain his confident recitation of the purpose of utility regulation and his sponsorship of such a utility-like statute.
27. Article I, Section 8, Clause 3 of the United States Constitution, known as the 'Commerce Clause', empowers the United States Congress 'To regulate Commerce with foreign Nations, and among the several States'.
28. A split of American regulatory jurisdictions exists for electricity transmission, and it greatly complicates the job of promulgating a single set of rules that could transform that business into a reliable inland transport network in its own right.
29. *Federal Power Commission v. Hope Natural Gas*, 320 U.S. 591 (1944).
30. Article 18, Section 1.
31. Oil pipelines were exempted from the clause after considerable Senate debate, mostly because of the highly vertically integrated nature of the US oil business at the time.

Many bills proposing to enforce the Commodities Clause for oil pipelines were debated in Congress over the decades, but none was ever passed. The clause was not a part of the Natural Gas Act (as gas pipelines were regulated as utilities in 1938), but in essence it was accepted by gas pipelines as part of concessions extracted from those pipelines with FERC Order No. 436 in 1986.

32. Order No. 637, pp. 184–5.
33. Ibid., p. 183.
34. Second Gas Directive, Article 10.1.
35. The gulf between European and American conceptions of accounting regulation is apparent in the various efforts in European regulatory jurisdictions to pursue 'cost benchmarking' as a tool for creating tariffs for regulated companies instead of the accounting costs of the particular company in question. Such cost benchmarking is fundamentally antithetical to the American regulatory accounting rules and administrative procedures developed in the 1930s and 1940s, which emphasize objectivity, transparency and the protection of private property under the US Constitution.
36. The FERC said in its own summary (90 FERC 61,109, CFR Parts 154, 161, 250, and 254 (Order No. 637), February 9, 2000): 'In this rule, the Commission is revising its current regulatory framework to improve the efficiency of the market and provide captive customers with the opportunity to reduce their cost of holding long-term pipeline capacity while continuing to protect against the exercise of market power'.
37. Incremental pricing requires all new capacity projects to support their own regulated cost of service, preventing incumbent pipeline companies from subsidizing new capacity by drawing on the economic rents (between cost of service and market value) in existing contracted capacity. See *Policy Statement on Determination of Need*, 1902-AB86, FERC Docket No. PL-3-000.
38. The FERC does not have such authority over oil pipelines under the Hepburn Amendment, a key problem in its regulation of that pipeline sector in America.
39. For instance, the 'Northeast U.S. Pipeline Expansion Projects' proposal from the late 1980s involved expanding gas pipeline service to the northeastern US. The FERC received dozens of applications from competing transport companies; by March 1988, the Commission had made a preliminary determination that 20 of the projects appeared to be entitled to consideration in a hearing. In June 1988, the Commission limited consideration to nine discrete proposals, requiring 13 others to be consolidated into a single investigation. By November, most of the parties settled with each other, agreeing to form three new proposals, the first of which, the Iroquois Gas Transmission System, L.P., was eventually awarded a certificate by the Commission in 1990. Iroquois entered service on January 28, 1992, more than five years after the original project inquiries reached the Commission. See: FERC, *Iroquois Gas Transmission System L.P.*, Docket No. CP89-634-000 et al., Order Making Preliminary Determination and Establishing Procedures, 52 FERC 61,091, at 61,344–61,345 (July 30, 1990).
40. For example, it took only two years for Kern River Gas Transmission Company to plan and construct its $1.27 billion 2003 Expansion Project to add 906 million cubic feet per day (approximately 26 million cubic meters per day) of gas pipeline capacity between the Rockies and Southern California. The project was announced on March 22, 2001 and completed on May 1, 2003.
41. From 1936 to 1944, during which time the Natural Gas Act was passed into law, no major gas pipelines were constructed in the US.
42. I refer to the National Association of Regulatory and Utility Commissioners (NARUC) which creates well-researched guidelines for regulators, such as for the critical definition and role of depreciation as a component of regulated rates.
43. These contracts in Europe are largely long term and linked by formula to oil, which is the best evidence of a lack of a robust independent market in gas. The length of this chapter prevents a review of those commodity contracting methods.

REFERENCES

Castaneda, D.J. (1999): *Invisible Fuel, Manufactured and Natural Gas in America, 1800–2000*, Twayne Publishers, New York.
Coase, R.L. (1960): 'The problem of social cost, *Journal of Law and Economics*, Vol. 3, pp. 1–44.
Ellerman, A.D., P.L. Joskow and D. Harrison, Jr. (2003): *Emissions Trading in the US: Experience, Lessons and Considerations for Greenhouse Gases*, Pew Center on Global Climate Change, Arlington, VA.
Energy Information Agency (EIA) (2000): Natural Gas Weekly Update, December 18, available at: http://www.eia.doe.gov/pub/oil_gas/natural_gas/data_publications/natural_gas_weekly_market_update/historical/2000/2000_12_18/pdf/ngweek.pdf (accessed May 16, 2007).
Energy Information Agency (EIA) (2005): Natural Gas Weekly Update, September 29, available at: http://tonto.eia.doe.gov/oog/info/ngw/historical/2005/09_29/ngupdate.asp (accessed May 16, 2007).
European Commission (EC) (2003): Directive 2003/55/EC concerning common rules for the internal market in natural gas and repealing Directive 98/30/EC, Brussels, June.
European Commission (2007): DG Competition Report on Energy Sector Inquiry, SEC (2006), 1724, Brussels: European Commission.
Hooley, R.W. (1968): *Financing the Natural Gas Industry*, AMS Press, New York.
Johnson, A.M. (1967): *Petroleum Pipelines and Public Policy*, Harvard University Press, Cambridge, MA.
Kahn, A.E. (1960): 'Economic issues in regulating the field price of natural gas', *American Economic Review*, Papers and Proceedings, Vol. 50, No. 2, pp. 506–17.
Kahn, A.E. (1971): *The Economics of Regulation: Principles and Institutions*, Vol. II, John Wiley & Sons, New York.
Kwerel, E.R. and G.L. Rosston (2000): 'An insider's view of FCC spectrum auctions', *Journal of Regulatory Economics*, Vol. 7, No. 3, pp. 253–89.
Lowi, T.J. (1973): 'How the farmers get what they want', *Legislative Politics U.S.A.*, 3rd edn, ed. T.J. Lowi and R.B. Ripley, Little, Brown, Boston, MA, pp. 184–91.
Makholm, J.D. (2006): 'The theory of regulation-specific investments, long-term contracts and gas pipeline development in the United States', Paper presented at the ENERDAY, Dresden University of Technology, April 21.
Moynihan, D. (1998): *Secrecy: The American Experience*, Yale University Press, New Haven, CT.
Nelson, J.R. (1966): 'The role of competition in regulated industries', *The Antitrust Bulletin*, Vol. 11, pp. 1–36.
Olson, W.P. (2005): 'Secrecy and utility regulation', *The Electricity Journal*, Vol. 18, No. 4, pp. 48–52.
Phillips, C.F. (1993): 'The regulation of public utilities', Public Utilities Reports, Inc., Arlington, VA.
Pierce, R.J. (1988): 'Reconstituting the natural gas industry from wellhead to burnertip', *Energy Law Journal*, Vol. 9, No. 1, pp. 1–57.
Sanders, M.E. (1981): *The Regulation of Natural Gas: Policy and Politics, 1938–1978*, Temple University Press, Philadelphia, PA.

Trapmann, W. and J. Todaro (1997): 'Natural gas residential pricing developments during the 1996–97 winter', Energy Information Administration, *Natural Gas Monthly*, August, available at: http://tonto.eia.doe.gov/FTPROOT/features/trapman.pdf (accessed May 16, 2007).

Troxel, C.E. (1937): 'II. Regulation of interstate movements of natural gas', *Journal of Land and Public Utility Economics*, Vol. 3, No. 1, pp. 21–30.

Zhao, J. (2000): *Diffusion, Costs and Learning in the Development of International Gas Transmission Lines*, International Institute for Applied Systems Analysis, Laxenburg (Austria).

3. The new security environment for European gas: worsening geopolitics and increasing global competition for LNG

Jonathan Stern

1 INTRODUCTION: A NEW SECURITY ENVIRONMENT

Security of European gas supply became a very topical subject following the cuts in Russian supplies to Ukraine in the first days of 2006 which had the consequence of briefly restricting the availability of supplies to some European countries. Much of the subsequent discourse has been concerned with 'the arithmetic of gas security' expressed as current and projected national or collective dependence of European countries on non-OECD suppliers (or groups of suppliers) over the next 15–25 years. Increasing dependence is directly correlated with growing insecurity, defined as the likelihood that gas exporting countries will cut off, or threaten to cut off, supplies to importing countries in support of their commercial and political (foreign policy) demands. The European Union (EU) has responded to the prospect of growing import dependence with the publication, since 2000, of two Green Papers (EU, 2000, 2006a) plus a security of supply directive (EU, 2004) and an energy policy document (EU, 2007).

Even if these projections of future dependence are believed to be correct, concerns about the resulting commercial and political leverage form only a small part of a security environment. It also includes a cluster of short- and long-term issues among which are resource availability, technical breakdown and accident, terrorist attack, political instability, and lack of timely investment, as well as disagreements in relation to existing and future supplies and prices, transit and facilities.

The central proposition of this chapter is that, in both the short and the longer terms, a 'new security environment' for European gas supplies is evolving. So degraded has the term 'security' become, in relation both to gas and to energy in general, that it is essential to define the geographical focus, the precise problems and the timeframes which are being considered. While the chapter focuses on Europe, it also takes into account progress towards a globalizing liquefied natural gas (LNG) market in the Atlantic and Pacific Basins, including the potential requirements of China and India. The principal issues discussed in the chapter are:

- the worsening political and geopolitical relationships between key gas exporting and importing governments – particularly between Russia and EU countries; and
- the increasingly competitive market for LNG supplies.

These problems are viewed in two timeframes – the next 1–2 years and the period up to 2020.

Some of the developments which shape this environment have been evolving since 2000; others have occurred only since 2005. The issue of whether they can all be termed 'new' is therefore questionable. However, the conclusion of the analysis is that collectively these trends suggest both a short- and a longer-term supply outlook for the European gas market which is significantly different from the one which has generally been assumed and projected.

Most traditional gas projections are based on some combination of reserve availability and of economic and commercial incentives to bring the reserves to markets. This chapter makes two broad and bold assumptions:

- that sufficient reserves have been established within an economic radius of European markets to meet any conceivable level of gas demand over at least the next three (and probably more) decades.
- that, at 2006 prices, all of these reserves would be commercially viable when delivered to European gas markets.[1]

In the late 1990s and early 2000s, the liberalization of EU gas markets was considered to be a major potential security of supply problem (Stern, 2002). This issue remains important, particularly in relation to incentives for the timely provision of peak supplies and storage facilities. But the new – less favorable – security outlook is fundamentally due to something else: a worsening geopolitical environment in both the short and longer terms.

2   DECLINING PRODUCTION AND RISING IMPORT DEPENDENCE

An important factor in the longer-term natural gas supply is the clear trend towards declining European gas production and resource discovery. While this trend is not 'new', and indeed has been foreseen for many years, the 2000s have produced increasing evidence that it is really occurring. UK gas production is projected to decline steeply to the point where the country may be 40 percent dependent on imports in the early 2010s, rising to as much as 80 percent by 2020 (DTI, 2005). Dutch production may be maintained at current levels until 2010–15 with output from the long-established Groningen field compensating for declines in the smaller fields. An overall production cap of 425 billion cubic meters (bcm), imposed by the government for the 10 years from 2006 to 2015, places limits on annual Dutch production increases. Thereafter, both Groningen and the small fields will experience accelerating decline (CPB, 2006). Elsewhere in continental Europe, most countries will experience a gradual decline in production. The only exception to the trend of declining gas production in OECD Europe is Norway, whose production and exports will increase strongly up to 2010; thereafter they are projected to level off.[2]

On present knowledge of European gas resources, indigenous gas production will not increase beyond 2010. How fast it will decline is a matter of debate, but in the absence of substantial additional discoveries increasing import dependence – identified by both the EU and the International Energy Agency (IEA) – is incontrovertible, but again not new. The European Union projects that gas imports will increase to 80 percent of EU demand by 2030 (EU, 2006a) while the IEA predicts that OECD European dependence on gas imports will increase to 65 percent by the same date (IEA, 2004a). Both these sources project relatively high levels of demand compared with work published by the Oxford Institute for Energy Studies (OIES); this is principally because the latter assumes that power generation will account for only 54 percent of incremental demand, compared with 63 percent in the IEA study.[3]

As already noted, sufficient reserves exist in a range of countries within economic reach of European gas markets – Russia, North Africa, Middle East, Caspian, and a number of intercontinental LNG suppliers – to bring sufficient gas supplies to Europe to meet the projected levels of demand. But such imports, far from being seen as the solution to European gas security, are almost universally seen as 'the problem.' A question addressed later in this chapter is whether the historical record supports the contention that increasingly import dependence should be automatically considered to be equivalent to decreasing supply security.

The main argument advanced in this chapter is that despite the political and public fixation on economic and political vulnerability arising from import dependence, this is not the principal security threat to European gas supplies in either the short or the longer term. Nevertheless, the debate on the security of European gas supply has focused overwhelmingly on supplies from external countries, particularly Russia.

3 RUSSIAN GAS SUPPLIES AFTER THE 2006 UKRAINE CRISIS

Despite the large number of commentators who discovered the subject of security of Russian gas supplies on January 1, 2006, it is not 'new'.[4] What has changed in the 2000s is that Russian gas supplies are delivered to increasingly pan-European destinations and in much larger volumes. In 2006, Gazprom, the dominant Russian gas company, exported 162 bcm gas to 22 European countries via its export subsidiaries, principally Gazexport.[5] Russia is the largest single supplier of gas to Europe, providing around 25 percent of European gas demand. However, dependence on Russian gas is not uniform throughout Europe: some Central and East European countries are totally dependent and there is significant dependence in North-West Europe. But the Iberian Peninsula imports no Russian gas, and the UK (Europe's largest gas market) has so far imported only relatively small quantities.[6]

Irrespective of national positions, the crisis on January 1–4, 2006 when Russia cut gas supplies to Ukraine, with the consequence that Ukrainian consumers diverted substantial quantities of gas in transit through their country to Europe, produced a huge negative reaction from governments and commentators on both sides of the Atlantic.[7] A year later, there was a similar crisis in relations with Belarus which was primarily about gas prices but principally affected transit of Russian oil to Europe via Belarus (Yafimava and Stern, 2007). Gazprom's imposition of steep increases in gas prices on CIS (Commonwealth of Independent States) importing countries since 2005 has been interpreted both within and outside those countries as politically motivated, despite the continuing gap between those prices and the corresponding EU import price. Many CIS governments (as well as some in Central and Eastern Europe) appear to believe that, if they could only obtain access to non-Russian supplies of pipeline gas and LNG, they would be able to import such supplies on more favorable terms.[8]

During February and March 2006, there was a period of exceptionally cold weather in both Russia and many parts of Europe. Moscow experienced its coldest winter for more than 60 years; temperatures well below minus 30 degrees Celsius for more than a week raised gas demand in Russia and much of Central/Eastern Europe to extremely high levels. This placed a huge strain on Russian gas and power networks which coped extremely well. During this period, there were again diversions of Russian gas in transit to European countries through Ukraine. These diversions – mostly not disputed by the Ukrainian government – prevented Gazprom from being able to meet the very high demand requirements of some European customers. Buyers in Poland, Hungary, Italy and Austria reported that deliveries were between 10 and 35 percent below requested volumes on a substantial number of days in January and February.[9]

The overwhelming conclusion of the political and public commentary throughout Europe during this period was that, by this action, the Russian government was exerting political pressure on the Ukrainian government and president in order to reassert its influence on a country attempting to make a decisive move towards the EU and the North Atlantic Treaty Organization (NATO) and away from Russian political influence.[10] The lack of any public official European censure of Ukraine for taking gas supplies to which it was not entitled, clearly demonstrated where European politicians believed the blame lay for this episode.[11]

Irrespective of contractual obligations and rights (prices, payments, obligations to supply and entitlements to take gas) these early 2006 episodes, and ongoing problems and uncertainties in the Russian–Ukrainian relationship, have raised serious doubts in the minds of European politicians as to whether Russian gas can be considered reliable.[12] There were suggestions that the Russian government was by this action 'sending a signal' to Europe that it had the power to cut off gas supplies should it choose to do so and that, should European countries act in ways which it did not like, it might choose to do so. This is based on an increasingly popular view of Russian foreign policy which holds that the Putin administration sees energy trade as an important means – perhaps the principal means at Russia's disposal – of projecting its political power and influence internationally.[13]

This growing perception of the undesirability of importing increasing quantities of Russian gas was not addressed by the March 2006 EU Green Paper on energy security, which envisaged a deepening of the existing energy partnership with Russia and argued that the G8 should intensify efforts to secure Russian ratification of the Energy Charter Treaty and its Transit Protocol (EU, 2006a). But these suggestions were not new and the failure of the European Commission to play any significant role during or after the events of January 1–4, 2006, using the institutions of the EU–Russia Energy Dialogue and the EU–Ukraine summits, did not inspire confidence in its role in any future crisis management.[14]

These events were followed by strongly adverse reaction to the following two sentences in a Gazprom press release of April 18, 2006 (Gazprom, 2006b):

> [O]ne cannot forget that we are actively developing new markets such as North America and China . . .
>
> It is necessary to note that attempts to limit Gazprom's activity in European markets and to politicize gas supply issues, which are in fact solely economic, will not lead to good results.

These produced front-page banner headlines in the *Financial Times*:[15] 'Gazprom in threat to supplies: EU told not to thwart international ambitions; Group says it may divert sales to other markets.' This reaction ignored the fact that Gazprom has no current capability to divert European supplies to North America or Asia and – in the most optimistic of all possible scenarios – will not have such capability for a decade.

This commentary also almost completely ignored other passages in the press release which read:

> Alexey Miller noted at the meeting that Gazprom was and is the main supplier of natural gas to Europe. We understand our responsibility and henceforth will remain the guarantor of energy security for the European consumers. All the contracts signed to supply gas will be implemented. There are no doubts at all . . .
>
> Gazprom is interested in developing mutually beneficial energy cooperation with partners in Europe. A good example is the North European pipeline project. We have signed new contracts to supply gas and, for the first time, have started working jointly with German companies along the entire chain from production, transmission and up to gas sales to the consumer. This enhances cooperation reliability for all project participants and even broader – for all consumers of the Russian gas in Europe.

The reaction to the April 18 press release was followed, in early May, by US Vice President Dick Cheney's speech to a conference of East European leaders in Lithuania. He noted in relation to Russia:[16]

> No legitimate interest is served when oil and gas become tools of intimidation or blackmail, either by supply manipulation or attempts to monopolize transportation.

[T]he IEA then made a direct connection between Gazprom's export monopoly and security:

[T]he IEA is worried about the increasingly monopolistic status of state-controlled Gazprom. Europeans cannot import gas from Russia unless Gazprom agrees. This restriction undermines European energy security'.[17]

Since early 2006, similar events have been avoided. However, there is continuing nervousness in Europe about the possibility of interruptions of Russian gas supplies flowing to Europe via Ukraine (and other transit countries), particularly given political changes in Ukraine over the past two years. Security uncertainties in Ukraine and Belarus have accelerated the development of the Nordstream Pipeline from north-west Russia through the Baltic Sea to northern Germany. The first string of this pipeline, in which the German companies E.ON and Wintershall, and the Dutch company Gasunie have agreed to take a 49 percent equity share, is due to be completed in 2010 with the second string to be built soon thereafter, adding a further 55 bcm to Russian gas export capacity to Europe.[18] This would increase nameplate Russian export capacity from around 230 bcm in 2006 to 285 bcm by the early 2010s.[19] But because of the deterioration of the Ukrainian network, total Russian export capacity to Europe in 2006 probably does not exceed 185 bcm. Should the lack of adequate investment in the Ukrainian network continue, by the early 2010s – even with the construction of two Nordstream pipelines – this figure will probably not exceed 215 bcm/year.

From the European side the two new Nordstream pipelines are already proving controversial in relation to the increased dependence on Russian gas that they will create for North-West Europe. These two pipelines will reduce dependence on Ukrainian transit routes, at least until such time as total Russian exports require all available transport capacity to be utilized. However, if Russian–Ukrainian gas relations fail to show sustained improvement, the majority of Nordstream pipeline capacity will be used to replace Russian exports via Ukraine, rather than for incremental exports. The same reasoning may be applied to the South Stream Gas Pipeline across the Black Sea providing a route to South-Eastern Europe via Bulgaria, possibly as far north as Hungary.[20]

It is a major contention of this chapter, that following the events of early 2006, a political limit to Russian gas supplies to Europe is in sight. The judgment that a limit may be imposed on Russian gas supplies follows from the European political reaction to the events of early 2006 (which would be reinforced by any repetition of these events). This type of political reaction would not be based on any analytical appraisal of European dependence on Russian gas, or the likely consequences of a supply disruption. It is rather, as Skinner (2006) has noted, related to a psychological notion of security which, despite being purely subjective, is just as important – arguably more important – for policy formation than analysis of likely scenarios.

Perhaps surprisingly, the possibility of a limit being placed on Russian gas imports by European governments is not inconsistent with Russian export aspirations. There is little sign that either Gazprom or the Russian government has ambitions to increase exports significantly above 200 bcm/year. The Russian Energy Strategy sees total exports, including those to CIS and Europe, rising from 194 bcm in 2000 to 250–265 bcm in 2010, and 273–281 bcm in 2020, suggesting very moderate increases in the second decade of the century.[21]

There are several reasons for limited Russian export aspirations of which the most important are the limits to Gazprom's production horizons post-2010, due to the need to invest in a new generation of fields on the Yamal Peninsula. Lead-times for the development of these fields mean that they cannot now be producing large volumes (that is, 100 bcm/year) prior to 2015 (Stern, 2005). For a number of commentators, the IEA most prominent among them, this carries the implication that:'Gazprom could face a gradually increasing supply shortfall against its existing [European] contracts beginning in the next few years if timely investment in new fields is not made' (IEA, 2006, p. 33).

Other reasons for not increasing exports to Europe include Gazprom's desire to diversify gas exports to North American and Asian markets, both of which will involve large-scale investments in pipelines to East Asia and LNG projects in the Russian Far East (Sakhalin) and the Barents Sea. The gas which will be sold to these markets will remain largely undeveloped unless export projects go ahead.[22] With no Russian gas currently being sold in either market (aside from occasional spot cargoes purchased from other producers), there is less political sensitivity in relation to gas import dependence. There is also increasing evidence that Russian commentators believe that it would be desirable to reduce Gazprom's (and Russia's) financial dependence on European gas exports.[23]

But probably the key long-term uncertainty for Russian gas exports to Europe is the development of domestic demand which is subject to major uncertainties, though analysis of this is not helped by the lack of any consistent and convincing historical data. Gazprom's supplies to Russian customers increased just over 2 percent during the 2001–05 period to 307 bcm in 2005.[24] Total gas delivered to Russian customers increased by 7 percent during the same period and by more than 2 percent per year since 2002 (Gazprom, 2006a, p. 41). In 2006, the increase in gas demand was around 4 percent – twice the level of the previous five years – a development which created serious concern in Moscow and was probably the main reason for a longer-term commitment to increased domestic prices. But in 2007, gas demand barely increased and the 2005–07 period may serve as a useful statistical exercise in understanding the difference in gas demand during a very cold, followed by a very warm, year.

In Stern (2005), I have set out scenarios which show how price developments in the domestic and European markets will impact on both investment in new supply and the attractiveness of sales to the different markets. This has been highlighted by the November 2006 commitment by the Russian government to increase prices for industrial and power customers to European netback levels by 2011. Substantially higher domestic prices will create downward pressure on domestic gas demand which will allow Gazprom to either increase exports or reduce production. But in the power sector, rapidly increasing demand, and the long lead-times necessary for building either coal or nuclear plants – even if that were deemed to be the correct policy – may create short-term gas demand which is relatively insensitive to price.[25]

A key conclusion of this chapter is therefore that even by 2020, Europe should not count on having at its disposal more than 200 bcm per year of Russian gas, and should not count on any increase in Russian supplies thereafter. Any suggestion that Gazprom might be concerned about meeting its long-term export commitments would probably be signaled by curtailing short-term gas supplies – in countries such as the UK – and reluctance to renew long-term contracts when they expire. However, during 2005–06 the majority of Gazprom's long-term contracts with its largest European buyers were extended by 20–30 years.[26] Although exports in 2007 declined by around 10 bcm due to warm weather, as noted above, the company has major plans to expand its European sales, particularly in the UK.[27]

4 MIDDLE EAST, NORTH AFRICAN AND WEST AFRICAN GAS TRADE: HUGE POTENTIAL, DIFFICULT POLITICS

In any discussion of global natural gas reserves, the Middle East and North Africa (MENA) are among the leading countries (see Table 3.1). Although Russian reserves are larger than any single MENA country, many of the latter countries have reserve to production ratios exceeding 100 years, suggesting ample potential for exports.[28] For these reasons as well as the geographical proximity of particularly North African countries to Europe, MENA countries have always been seen as a huge potential import resource for European gas markets. This potential was highlighted in the International Energy Agency's *World Energy Outlook 2005* on this

Table 3.1 Middle East and North African* gas export projections 2003–30 (bcm)

	To Europe			Total exports			
	2003	2010	2030	2003	2010	2020	2030
Middle East	2	35	117	34	102	185	244
North Africa	61	83	170	63	86	143	200
Total	63	118	287	97	188	327	444
Major exporters:**							
Qatar				19	78	126	152
Algeria				64	76	114	144
Iran				–	5	31	57
Egypt				–	10	19	28
Libya				1	2	13	34
Iraq				–	1	7	17
Total				84	172	310	432

Notes:
*   In addition to the countries listed, MENA includes: UAE, Kuwait and Saudi Arabia.
** Figures are for 'net trade'.

Source: IEA (2005, pp. 178–9, 560, 564, 568, 580, 592, 596, 600, 604).

region. Table 3.1 shows the IEA's projections of MENA gas exports to 2030 both to Europe and in total. This is an extremely positive outlook for European gas supplies but may be overoptimistic in a number of respects.

The first of these is that MENA exports are projected to increase roughly fourfold within a period of less than 30 years. In absolute terms, this would require an increase of nearly 350 bcm/year, of which the majority (over 250 bcm) would need to come from the Middle East. But the table shows that in 2003, Middle East gas exports had reached only 34 bcm, a figure reached 25 years after the start of LNG exports.[29] Likewise North African projections foresee exports from that region increasing more than threefold to 200 bcm/year over the next 25 years, when around 40 years were required for exports to reach the 2003 level of 63 bcm.[30]

These levels of gas exports could certainly be sustained by known proven reserves (let alone what may be discovered in these countries over the next two decades), although a significant number of new fields will need to be developed.[31] New LNG and pipeline projects, both under construction and in advanced stages of planning, would support the projections to 2010. Cost reductions in LNG (and to a lesser extent pipeline) projects during the 20 years up to 2004 meant that the economics of any project under discussion were positive. Since 2004, cost increases of up to 50 percent, due to the rise in raw material prices and competition for engineering, construction and contractors' services, were more than offset by price increases. For these reasons, the availability of investment funds has so far not proved to be a significant constraint; but cost overruns, particularly for large projects, and lack of available services have led to delays in implementing projects.

A more serious doubt is whether such a huge rate of increase in exports, sustained over a 25-year period, is realistic from the institutional, political and geopolitical point of view. In a number of countries, particularly Iran but also Algeria, increases in domestic consumption of gas (either directly or for reinjection in oil fields) may curtail availability for export (Hallouche, 2007). The rather conservative projections for Libya in Table 3.1 were made prior to the opening up of exploration acreage in that country which promises to dramatically increase the existing 1 bcm of LNG export capacity and 8 bcm of export capacity via the Green Stream pipeline to Italy.

A second reason why the IEA may be overoptimistic arises from the projection in the table that exports to Europe as a percentage of total MENA gas exports remain at 60–65 percent throughout the period. The share of Middle East exports delivered to European markets is projected to increase to more than one-third by 2010, and to nearly one-half by 2030. The table also suggests that Europe will retain the overwhelming majority of North African exports – 85 percent in 2030. Somewhat surprisingly, the US market is projected to take less than 20 percent of MENA exports by 2030, of which more than half will be from North Africa. Out of a total of 270 bcm of MENA LNG exports in 2030, the IEA believes that Europe will capture a minimum of 113 bcm or 42 percent, and perhaps up to 50 percent.[32] This suggests that Europe largely 'wins the battle' for global LNG supplies with the US and the Pacific Basin for both Middle East and North African LNG. This is a very optimistic projection for Europe and, given developments in the North American and Pacific markets (see below), there must be a question about whether it is realistic.

The third reason to question the IEA projections arises from the fact that the six countries shown in the bottom half of Table 3.1 account for more than 90 percent of projected MENA gas exports in the 2010–30 period; two countries – Algeria and Qatar – account for 70–90 percent of total exports.[33] Should any political or geopolitical problems prevent these two countries from developing exports as anticipated in the table the consequences for European gas supplies and the Atlantic Basin (and global) LNG market will be significant. Saudi Arabia, the other major country with significant gas reserves, has shown no interest in exports, preferring to use gas domestically and export oil.[34]

In the 2000s, West Africa has emerged as an important LNG exporting region, with Nigeria as the major supplier and Equatorial Guinea and Angola likely to start deliveries over the next few years. After more than 30 years of discussion and disappointment, the Nigeria LNG (NLNG) project began exporting in 1999. Within a decade of starting these exports, NLNG will have six trains in operation delivering nearly 30 bcm/year of supplies to the Atlantic Basin. Two more NLNG trains are planned which would add a further 22 bcm of export capacity. In addition, two more projects are awaiting investment decisions which could see up to 47 bcm of additional LNG export capacity, bringing total export capacity to nearly 100 bcm/year. This is in the same range as Qatar and Algeria and would make the country one of the world's leading gas and LNG exporters. In addition, Equatorial Guinea and Angola may add up to another 12 bcm of exports per year. West African gas export potential currently appears somewhat less than either North Africa or the Middle East, but additional discoveries could significantly expand current expectations.

5 A WORSENING GEOPOLITICAL ENVIRONMENT FOR EUROPEAN GAS SUPPLIES

Just as there is a common assumption that the principal threats to European gas security are externally focused, so there is a common assumption that, within that external focus, the policies of exporting countries and/or probable political events within exporting countries will be the principal threats to European gas security. Thus, in respect of both Russia and the Middle East, much European commentary is focused on the general political and economic policies of governments – as well as narrower oil and gas policy frameworks, which are believed to 'threaten' European (and possibly wider OECD) gas security.

Part of this stronger recent sensitivity towards exporting countries is the product of a new assertiveness of oil and gas producing and exporting countries in the wake of the post-2003 increase in prices, and of a widespread perception that such price levels will be at least a medium-term phenomenon (Mitchell, 2006). This new assertiveness – often termed 'resource nationalism' – has created significant commercial challenges to both international oil and gas companies and OECD government policies in countries as geographically diverse as Venezuela, Bolivia, Russia and Iran, combined with a desire to challenge the political and geopolitical status quo which they see as imposed by US and EU governments.

Increasing producer/exporter assertiveness is resulting in reduced access to resources for international oil and gas companies (IOGCs), and demands by host governments and national energy companies for increasing shares of the rent from joint activities with IOGCs. In addition, OECD companies are facing increased competition for energy exploration and development opportunities particularly with Chinese and Indian companies. Overlaying these general commercial developments are trends which have specific and potentially serious consequences for European gas supplies:

- increasing bilateral and geopolitical tensions between Russia and both the US and European governments;
- continued deterioration of political stability in the Middle East region as well as increasing tensions between potential gas exporting countries, particularly Iran, and US and European governments; and
- uncertainty about political stability in African LNG exporting countries, especially Nigeria.

Geopolitical scenarios, such as the Clingendael Institute's 'Regions and Empires' (Clingendael, 2004), and Shell International's 'Low Trust Globalisation' (Shell, 2005), have produced comprehensive storylines that are strongly negative for oil and gas trade. Correlje and Van der Linde (2006, p. 538) have observed that under 'Regions and Empires' there is likely to be 'a slowly emerging [gas] supply gap, as a result of lagging investments as a consequence of ideological and religious contrasts, particularly with regard to the North African suppliers, the potential supplies in the Persian gulf and the Caspian Sea region'.

5.1 Russia and CIS Countries

Much has been said already about the geopolitics of Russian gas supplies but in important respects, European and US reactions to the 2006 Ukrainian crisis reflected a significant deterioration of Russian political relationships with those governments. The disillusion of OECD governments with what they perceive as the Putin administration's weak commitment to democracy and economic reform has been exacerbated by the new confidence and assertiveness of the Russian government and companies in projecting their oil and gas interests internationally. OECD objections have been met by Russian accusations of hysteria and double standards (in relation to judging democratic and economic reform credentials of different states) combined with a growing feeling in Moscow that the fundamental concern of most OECD governments is related to Russia's growing economic and political strength, following a protracted period of weakness during the 1990s.

This is a specific problem in relation to CIS countries where the past two years have seen governments elected in Ukraine, Moldova and Georgia which have sought to distance themselves from Russian influence and developed aspirations (however distant and unrealistic) of becoming members of NATO and the EU. Meanwhile in Central Asia and the Caspian, US policy is aimed at removing oil and gas export flows from Russian influence by creating a new export corridor via Turkey. Needless to say, such aspirations run directly counter to Russian interests which are continued control over Central Asian resources. None of these tensions seems likely to be quickly resolved.

5.2 Middle East and Caspian Region

Over the next 2–3 decades, problems may arise within the important gas exporting countries, or between these countries and OECD importers. Qatar is a small state in terms of both population and geographical area. In a relatively short time it will become the world's second largest gas exporter (after Russia) and the world's largest LNG exporter. The scale of the industrial facilities needed to develop such a large LNG export capacity (plus some of the world's largest gas to liquids facilities) has the potential to create internal political strains. At present there is a moratorium on new gas development as the country assesses the consequences of the scale of development to which it is already committed.

In the mid-2000s, the Iranian political relationship with the international community (and especially the United States) has become increasingly difficult. The Ahmedinejad regime has had regular verbal confrontations with OECD countries on issues ranging from nuclear power development to the existence of Israel. Should a reference to the UN Security Council, in relation to nuclear materials, lead to international sanctions being imposed on Iran, then most (if not all) substantial international investments in Iranian gas projects would become impossible.

Reinforcing current events is a pattern of political and economic development in Iran during the past 25 years which has consistently prevented the country from fulfilling what seemed likely to be a leading role in international pipeline gas and LNG trade. In 2005, Iran exported less gas than it had prior to the Iranian revolution of 1980. Moreover, with imports from Turkmenistan which increased in 2006–07, more than offsetting exports to Turkey, Iran became a net importer of gas in the 2000s – an unthinkable position for a country with the second largest gas reserves in the world (after Russia and just ahead of Qatar).

Iran has a 30-year history of LNG and pipeline export project failure with a range of buyers. There was a reminder of this history in 2006 with the apparent repudiation of the price clause in the Iran–India LNG project – signed before the most recent increase in oil and therefore gas prices – for which deliveries have not yet started.[35] While such action may be understandable, it will not encourage potential investors in, or customers of, Iran to have confidence that the country can be relied upon to honor long-term gas export contracts. Because the economic value of Iranian gas, reinjected into oil fields to promote increased oil production and exports is currently several times greater than that of gas exports, the incentive to conclude gas export contracts has been significantly reduced.[36]

There are therefore good reasons to question whether Iran will become a substantial gas exporter to Europe over the next 25 years.[37] Export volumes projected in Table 3.1 are relatively modest in relation to total MENA exports, and would be very unlikely to include pipeline gas exports dedicated to European markets, as opposed to LNG exports from Iran for which Europe would be in competition with Pacific and (assuming a resolution of current political problems) North American markets.

These problems with Iran would be less significant for Middle East gas trade if the security situation in neighboring Iraq were not so serious. Iraqi gas reserves are relatively modest by Middle East standards but the country's potential is believed to be significant and its proximity to Turkey – plus the existence of a previous gas trade contract between the countries – means that Iraq could become a significant source of European supplies. But the current security situation and outlook mean that secure and stable large-scale gas exports from Iraq seem a very distant prospect.[38]

Thus with the exception of Qatar, the prospects for Middle East gas exports – and, in particular, pipeline exports to Europe – are relatively poor for at least the next decade and probably much longer. The best hope is for a dedicated pipeline via Turkey carrying supplies from Caspian countries – Azerbaijan, Turkmenistan and Kazakhstan. Of these countries, only Azerbaijan has shown an inclination to commit substantial volumes to the European market and, as already noted, it is uncertain whether some could be considered secure suppliers. But diverse sources of supply flowing through a single pipeline would decrease the importance of any individual supply source. This appears to be the concept underpinning the Nabucco pipeline currently being promoted by a number of Central and South-East European utilities and the European Commission.[39] Such pipelines from the Middle East/Caspian region are strongly endorsed by the US, EU and south-eastern European governments to promote diversification away from Russian gas supplies and transport routes. However, two points should be recalled in relation to pipeline gas projects from the Middle East and Caspian region:

- they are not a new idea; there have been regular initiatives to create such projects for at least the past 30 years without success;
- it is not clear – given the number of borders which they will need to cross and the potential for problems within and between countries along the route – whether such pipeline routes can be considered more reliable than existing and new supplies from and through Russia which they are intended to displace.[40]

5.3 North and West Africa

During the mid-2000s, the Algerian political situation has been relatively calm in comparison to the decade of the 1990s. Confidence can be drawn from the fact that exports have increased substantially over the past 15 years, a period during which Algeria experienced internal upheavals and conflicts akin to a civil war. Libya has recently returned to acceptance within the international community after a long period of isolation due to international trade sanctions. The return of international energy companies to Libya is therefore a recent phenomenon and there is uncertainty about how soon gas can be developed, which partly accounts for the relatively cautious projections in Table 3.1.

The importance of North Africa for future European gas supply, however, goes beyond the purely numerical aspect of projected volumes. North Africa is likely to be the only supply source which will increase the volume of pipeline gas dedicated to Europe. There are not only possibilities of expanding the existing pipelines – the Enrico Mattei (Trans-Mediterranean), Pedro Duran Farrell (GME) and Green Stream lines – but also of building new ones such as the proposed Medgaz line to Spain and the Galsi line to Sardinia and Italy.

The political situation in West Africa is problematic. In Nigeria, the most important LNG exporting country, petroleum-related political unrest has increased in the mid- to late-2000s, when local communities protested against the lack of benefits conferred upon them by central government in return for what they see as the destruction of their environment by energy companies. In Equatorial Guinea, where LNG exports will commence in 2007, the governance of the ruling regime has prompted serious transparency and human rights concerns, while Angola has only recently emerged from a 27-year civil war.

6 THE GAS EXPORTING COUNTRIES FORUM: AN OPEC FOR GAS?

It remains uncertain whether the creation in 2001 of the Gas Exporting Countries Forum (GECF) was an event of no importance or the start of an 'OPEC for gas' (Hallouche, 2006). Since its creation, the GECF has been a rather chaotic organization without stable membership, well-defined membership rules, mission or objectives. For external observers, this situation has not been helped by a lack of information about the Forum and its activities, with a rather sparse website in existence only since 2007.[41]

Many believe that the creation of the Forum can be attributed to the need felt by producers to respond to European liberalization and the application of competition rules to the natural gas sector.[42] These EU initiatives, which were not arrived at in consultation with producers, provided a rationale for the latter to create their own organization. Since its creation, by far the most active members of the Forum have been Iran, Algeria and Qatar, with Trinidad and Venezuela becoming more active since 2004. Plans for a Venezuelan presidency in 2006 collapsed and the Forum failed to meet that year.

The Forum has been notable for its relative lack of active pipeline gas exporters: Canada and the Netherlands are completely absent; Norway is only an observer; Russia attended all of the meetings but took very little active part.[43] This changed in 2007 with a very high-level Russian delegation attending the meeting in Doha, and the next meeting of the Forum in Moscow in April 2008. Algeria and Libya, which are pipeline as well as LNG exporters, as well as Iran, a pipeline exporter, are exceptions to the general trend with Algeria and Iran being among the more active members. No meetings of the Forum have been held in the Pacific, and Australia, an important LNG exporter to that region, has not been involved. The Forum therefore appears to be biased towards LNG exporters and, in terms of active members, heavily biased towards Atlantic rather than Pacific Basin LNG trade.

Key Forum members have strenuously denied any intention of becoming a 'gas OPEC' in the sense of a price-setting or volume-controlling organization, and the only attempt to agree a common position on gas pricing ended in failure. Only two countries have spoken out unambiguously in favor of price-setting ambitions: Iran which is a net gas importer, and Venezuela which does not yet export gas. Indeed there seem to be significant tensions among the members around issues of sharing commercially sensitive information and of collaboration on gas sales. At present, the GECF shows little prospect of metamorphosing into anything akin to a gas OPEC; it would need to develop considerably greater institutional capacity and cohesion for this to become a reality. In a longer-term perspective of one or two decades, the possibility of some type of price-setting organization should not be ruled out. The most likely characteristics of such an organization would be the following:

- initially at least, it is more likely to be focused on exports of LNG rather than pipeline gas, possibly because of the greater flexibility and arbitrage possibilities;
- it is more likely to develop with a regional focus – Europe or the Atlantic Basin – rather than as a global cartel; and
- it is more likely to develop quickly in the context of a crisis for exporters (for instance, if prices sank to levels which threaten the profitability of new projects), rather than in the price environment of the post-2003 period.

The biggest threat could come from an agreement between the pre-eminent LNG exporters to the Atlantic Basin (Qatar, Algeria, Nigeria and Egypt) which, by acting together in a tight LNG market, could exert significant market power over importers. The sensitivity of importers to any such possibility was demonstrated by the Italian reaction to a press release following the visit of a Gazprom delegation to Algeria (Gazprom, 2006c): 'the parties reviewed possibilities of jointly implementing "full cycle" projects encompassing hydrocarbon exploration, production, transmission, processing and marketing in Algeria, Russia and third countries'. The references to 'marketing' and 'third countries' were immediately interpreted in terms of gas price collusion to the detriment of Italy, causing an appeal by that government to the European Commission (International Gas Report, 2006).[44] In 2007, the Memorandum of Understanding signed between Gazprom and Sonatrach expired and was not renewed.

7 SECURITY AND IMPORT DEPENDENCE: EMPIRICAL OBSERVATIONS FROM THE PAST 25 YEARS

The traditional inclination among politicians and the media in OECD countries is to regard energy supplies which are produced domestically as 'secure', and supplies which are imported as 'insecure'. This dates at least as far back as the 1973 Arab oil embargo, which was a formative experience for the current generation of senior politicians and decision makers in terms of energy security. A survey of gas security incidents since 1980 classified three types of incidents: source, transit and facility (Stern, 2002).

During the 1980–2001 period, there were one or two source incidents and some transit incidents relating to Russian gas supplies through Ukraine, but no significant facility incidents.[45] There was one incident which could be labeled as 'terrorism' in 1997 when an explosion on the Trans-Mediterranean Pipeline cut the flow of Algerian gas to Italy.[46]

Since 2001, three serious facility incidents have affected European gas supplies: the liquids contamination of the Interconnector UK pipeline in 2002, the fire at the Algerian Skikda liquefaction plant in 2004 and the fire at the UK's Rough storage facility in 2006. During this period the only other European incidents that caused significant supply shortfalls were the 24-hour interruption of Russian gas supplies to Belarus in February 2004 and the January/February 2006 Ukraine crisis. The 2004 pipeline explosion in Belgium, which killed 16 and injured 120 people, is not included here since, having occurred in July, it appears to have caused no significant supply disruption.[47]

Summarizing the security incidents which have occurred over the past 25 years in Europe: there have not been very many; and those that have occurred have been divided between the three main causes (source, transit and facility) but facility incidents appear to have increased over recent years. In particular, as far as the UK is concerned, the risk of facility incidents became increasingly problematic in the mid-2000s due to the tightness of the supply/demand balance and the lack of storage capacity (Stern, 2004). Despite references by the EU to problems of importing gas from 'regions threatened by insecurity', it is difficult to think of any historical incident involving political instability which has prevented gas from being delivered to Europe.[48]

There is no evidence from Europe (or anywhere else in the world) that imported gas supplies have been – or are necessarily likely to be – less secure than supplies of domestically produced gas. Indeed history suggests that all serious security incidents – those in which customers have lost gas supplies for a considerable period of time – have stemmed from failure of indigenous supplies or facilities. While there is no guarantee that the future will be the same as the past, no empirical experience would lead to the conclusion that a country with substantial dependence on imported gas supplies is necessarily less secure, in other words, more prone to disruption, than one which is self-sufficient. Increased security, whether for domestically produced gas or imports, requires increased diversity of sources, of transportation and transit routes, and of facilities such as pipelines, LNG terminals, processing plants and storages. Clearly the higher the percentage of gas in a country's energy demand, the greater is the importance of diversity as protection against security incidents.

Exporting countries have a very strong incentive to maintain continuous and secure deliveries due to the revenues which they earn and the importance of those revenues to corporate and national budgets. For most non-OECD gas exporting companies and countries, earnings from gas export revenues are not only very significant in absolute terms, but also as a proportion of their total revenues. Even for a company as large as Gazprom, gas export revenues in 2005 were around 55 percent of the company's total receivables and around 17 percent of total Russian foreign trade earnings outside CIS countries.[49] This is a long-term stream of earnings that would not lightly be put in jeopardy by an exporting company or government and which could not easily or quickly be replaced by any other commodity. LNG suppliers have a greater range of export options than pipeline exporters and could choose to supply, or not to supply, certain markets for political as well as commercial reasons. But unless there is a significant global shortage of LNG, or a concerted boycott of a particular country by a group of exporters, it is not likely that an individual importing country will be completely unable to access LNG supplies. Equally likely, if not more so, is a refusal of importing countries to trade with certain LNG exporters for political reasons.[50]

8 SECURITY INVESTMENTS IN LIBERALIZED MARKETS

Two dimensions of European gas security which are only just beginning to receive the attention which they deserve are the potential problems which can be caused by infrastructure breakdown, and the question of how to ensure the availability of adequate gas storage in liberalized markets. This chapter is not the place to explore these issues in any detail, but it is important to note that the fire at the Rough storage site in February 2006 – arguably Europe's most important gas security incident of that year – deprived the UK of access to around 80 percent of its stored gas for several months. Had the incident happened any earlier or later in the winter, the consequences might have been substantially more serious than the price spikes which the market experienced in the few weeks before temperatures rose and demand declined.[51]

The huge investments in both new supplies and new storage which are under way in the UK certainly contradict the views expressed in the late 1990s and early 2000s that multibillion dollar investments would be impossible to finance in a highly liberalized market. For a discussion of such views, see Stern (2002). However, these projects will arrive several years after the market needed them and, even when all of the storage capacity which UK investors are currently seeking to build is complete, they will be substantially less than other major markets in Europe. A useful comparison could be made with Italy where a combination of shortages of Russian gas (due to the problems with Ukraine mentioned above) and very cold weather in the winter of 2005–06 forced the use of strategic storage. The Italian government considered that the country had a narrow escape when only 3.9 bcm of gas remained in strategic storage on March 22, 2006 (Garriba, 2006). This volume, however, was roughly equal to total annual storage capacity in the UK – a much larger gas market than Italy – at the same date. The case of the UK raises important issues about the ability of liberalized gas markets to deliver timely, market-based security investments (Clingendael, 2006).[52]

9 INCREASING GLOBAL COMPETITION FOR LNG SUPPLIES

9.1 The Emerging LNG Market in the Atlantic Basin

Since 2000, the LNG market in the Atlantic Basin has been transformed from a relatively limited and rigid set of bilateral trades into an increasingly liquid market with a much larger number of players. There are a number of reasons for this transformation:

- substantial cost reduction in all phases of the LNG chain up to 2004, although this trend has subsequently been decisively reversed;
- the transformation of the US and UK from surplus markets with low prices to shortage markets with high prices;
- the slow pace of liberalized access to pipeline networks in continental Europe which makes LNG a more attractive transportation option; and
- greater emphasis on diversification of gas supplies to promote security, particularly in Southern Europe and the UK.

The key issue for the evolution of the Atlantic Basin LNG market has been the transformation of the North American gas market. This market (comprising the USA, Canada and Mexico) is roughly 30 percent larger than that of Europe and is experiencing a similar trend in relation to indigenous production.[53] The major difference is that while Europe developed a range of imported supplies over the past 30 years, North America has remained almost completely self-sufficient, aside from marginal quantities of imported LNG. Around 2000, the North American gas market changed as both US and Canadian production began to decline. Since 2001, natural gas prices – which had been around $2/mmBtu for 15 years prior to that date – have been in the range of $4–10/MMBtu with much greater volatility. Significant additional resources remain to be developed in North America and the late 2000s has seen US production decline reversed by the development of unconventional (especially shale) gas reserves. The largest-known undeveloped conventional gas fields are in the Canadian Arctic and Alaska. Mackenzie Valley production can make a contribution in Canada equivalent to a baseload LNG terminal but much of it may be devoted to developing Canadian tar sands. A pipeline from Alaska, for which costs have risen to $25 billion (2006) could provide around 60 bcm/year of additional gas supplies, but the scale of that project combined with corporate, regulatory and logistical complexities means that it cannot be fully operational until 2018 at the earliest (Martin, 2006). For Mexico, the issues of gas development are less related to resources, and more to a constitution which prevents foreign investment for their development.

Thus North American countries (and particularly the United States) have begun a major drive to import LNG supplies which has seen a profusion of proposals for new regasification terminals, and an expansion of existing terminals. In July 2006, there were five existing and 45 proposed receiving terminals in North America, all but six of which were in the United States, and all but eight on the east and Gulf coasts.[54] Of the proposed terminals, 23 had received federal regulatory approval and, of these, seven were either under construction or in the advanced stages of planning. The capacity of the existing terminals and those under construction would exceed 140 bcm/year; terminals which have received federal regulatory approval would add a further 110 bcm/year of capacity. The US Energy Information Administration expects US LNG imports to rise from less than 18 bcm in 2005 to more than 80 bcm in 2015 and to 125 bcm by 2030. This suggests either that many terminals which are currently anticipated will not be built, or that a significant amount of excess import capacity will be created over the next two decades; others have made significantly higher import projections (Martin, 2006).

In 2005, Trinidad was the major source of LNG for the US, with significant quantities from Algeria and Egypt and additional small deliveries from other African and Middle East countries.[55] Projects under construction clearly show that Qatar, Egypt and Nigeria will become much more significant suppliers of LNG to the US. Although LNG projects to the USA from South American countries, such as Venezuela, Bolivia and Peru, have been promoted, current politics both within those countries, and between them and the United States, make many of these developments impossible.[56] Another major potential supplier of LNG to the US is Russia, with both the Shtokman and Baltic LNG projects currently targeted at North America. However, a combination of rising costs, lengthening lead-times and worsening US–Russian political relations may create problems for these projects. In general, Trinidad aside, it seems most likely that incremental LNG to the US over the next 10 years is most likely to come from Middle East (Qatar) and African countries (Egypt, Nigeria and possibly Libya) which would suggest much greater competition with Europe than was indicated above.

The question of how North American gas supply and demand will unfold in a price environment of $5–10/MMBtu, which is projected to continue over the next few years, is highly uncertain.[57] If unconventional gas production continues to develop strongly and demand fails to recover, prices may continue in this range and it is not certain – from the evidence of 2007–08 – that this will be high enough to attract substantial LNG cargoes from other markets. It still seems likely that North America will be a strong competitor for Atlantic Basin and global LNG supply in the 2010s but prices may need to be substantially higher than these levels, given developments in major Pacific Basin markets of Japan and Korea. The traditional major LNG supplier Indonesia is struggling to meet current contractual commitments and is not likely to renew most of the 16 bcm/year of existing export contracts which expire around 2010–11, the Pacific Basin may be facing a short- to medium-term supply shortage, especially in the winter months, and spot prices of $15–20/MMBtu have been common in the late 2000s.

The price that LNG can command in North America, Europe and the Pacific Basin at any point in time will be an extremely important determinant of where some of the available LNG supplies will be landed. As both the Atlantic and the Pacific Basins become increasingly liquid LNG marketplaces, this entails both positive and negative security consequences for European importers. On the positive side, a more liquid marketplace will mean that cargoes will always be available if an importer is willing to pay a sufficiently high price. On the negative side, cargoes which importers would previously have been considered firmly 'contracted' to be landed at national terminals, may be drawn away by higher prices at a different location.[58]

Extreme temperatures and other weather events have become major determinants of short-term trade flows and prices on both sides of the Atlantic – but especially in North America – and will affect short-term production and demand unpredictably.[59] Gas prices on both sides of the Atlantic have appeared to fluctuate in a band where the floor is set in the summer by the heavy (residual) fuel oil price and the ceiling in the winter by the gasoil price. For Europe, this band is roughly determined by the indexation of long-term gas contracts. In the US, it appears to be set by interfuel competition which dictates that if the summer price of gas (at the margin to meet the air conditioning load) falls below that of heavy fuel oil, gas demand, and therefore prices, will increase; by contrast in the winter if gas prices rise above those of gasoil for heating, large consumers with fuel flexibility switch to gasoil causing gas demand and prices to fall (Foss, 2007). These processes mean that Atlantic Basin gas prices have very similar dynamics, albeit somewhat delayed in Europe due to the operation of contracts, which would suggest that no significant price differential is likely beyond that of short-term supply and demand fluctuations caused by weather-related events.[60] The frequency and extent of those fluctuations, and whether they produce similar or opposite price movements, will determine the extent of LNG arbitrage opportunities and the future development of short-term trade. But with LNG becoming an increasingly global business, since 2006 the Atlantic Basin has lost an increasing share of incremental cargoes to the Pacific Basin with every sign that the trend will continue.

9.2 Competition for LNG from China and India

As the 2000s have unfolded, it has become clear that developing countries, particularly China and India, are having an increasingly significant impact on global energy demand. In neither country is gas yet an important fuel, providing less than 3 percent of Chinese, and 8 percent of Indian, primary energy demand. But with coal-dominated energy balances, serious urban pollution problems and limited indigenous gas resources, both countries have a significant need for imported energy, particularly gas. During the early 2000s, both countries developed plans for very substantial gas imports with LNG apparently the dominant import mode under consideration.

In 2008, China had two operating LNG terminals, one under construction and up to a further nine in various stages of planning. However, the significant increase in oil and therefore LNG prices since 2003 has undermined the Chinese program. The majority of the regasified LNG cannot be used in power plants where the competing fuel is coal, and it is not possible for LNG to compete in the power sector at the prices seen in the Pacific market since 2004.[61] Since late 2005, aspiring Chinese LNG importers have lost out to their Japanese and Korean competitors and, under pressure from suppliers, were forced to agree to a significant upward revision of prices under current contracts (Gas Matters, 2006b). Since then, Chinese companies have continued to press ahead with new terminals and appear to have come to terms with paying competitive prices for LNG, mainly for use in the residential, commercial and industrial sectors.

Whether, at gas prices equivalent to more than $100/bbl oil, imports can be increased to 70 bcm/year which some projections suggest by 2015 is uncertain (EIA, 2006b). Chinese gas import diversification towards pipeline gas from Central Asia, is not likely to ease the price problem, given the distance which the gas will need to travel. Continuing inability to agree on large-scale pipeline gas imports from Russia – easily the most attractive source of imported gas for China – will restrict the role of the fuel in China's energy balance over the next decade (Fridley, 2008).

Indian LNG importers, although they are better located in respect of Middle East supplies, are in a similar position to their Chinese counterparts, that is, unable to compete on price with the richer Pacific competitors and therefore struggling to obtain either long-term LNG supplies or spot cargoes. In this commercial environment, the three existing LNG receiving terminals may be underutilized for some years as soon as substantial new domestic gas from the Krishna Godavari Basin is fully onstream in 2009 (Joshi and Jung, 2008).

The position of India and China is similar in another respect in that both countries have opportunities to import substantial volumes of pipeline gas: China from Central Asia, Eastern Siberia and the Russian Far East; India from the Gulf and Central Asia. The location and magnitude of the Russian resources in Eastern Siberia and the Far East means that the natural market will be China and there will be only limited competition for these resources from other importers and also that for East Siberian gas the most efficient means of transport will be via pipeline rather than as LNG.

Agreement was reached between the Turkmen and Chinese governments for a pipeline carrying 30 bcm/year for 30 years starting in 2009.[62] This will require a pipeline of around 2,000 km through Uzbekistan and Kazakhstan, just to reach the Chinese border; a gas pipeline from Kazakhstan parallel to the oil pipeline between the countries will be around 1,000 km. Both pipelines will connect to the second West–East pipeline taking gas distances of up to 4,000 km to customers in China. Despite the highly unattractive commercial aspects of these arrangements, they show the determination of the Chinese government to secure pipeline gas supplies from countries amenable to Chinese influence.[63]

India has the opportunity to import gas from the Gulf both by pipeline and as LNG. Plans for both types of projects from Iran are already well advanced, but the pipeline project requires transit across Pakistan. There is also the long-discussed TAP pipeline project from Turkmenistan to India via Pakistan and Afghanistan. India also has the possibility of importing gas from the east, with both Bangladesh and Myanmar as possible sources of supply.[64] But there are significant political problems in relation to all

Indian pipeline gas imports: first, dependence on transit through Pakistan to the west or Bangladesh to the east, raises serious issues of security given the at best uneasy, and at worst hostile, bilateral relationships between India and its neighbors; second, there are problems with Iran as a supplier given US opposition to large-scale energy projects involving that country (Joshi and Jung, 2008).

Thus for India and China, a strong case can be made that their natural suppliers are Gulf and Central Asian countries and Eastern Siberia/Russian Far East, respectively, and that the most advantageous mode of transportation is pipeline gas. It makes less economic sense for these relatively poor countries to attempt to compete with much richer OECD importers for LNG, when they could import pipeline gas supplies for which – certainly in the case of East Siberian gas and arguably for other sources – they have few competitors. Should they choose to compete for LNG supplies with Europe (and the US) then this will be for Middle East supplies, principally from Qatar. This picture suggests that neither China nor India is likely to become, or seek to become, a serious competitor to Europe for gas supplies up to 2020. However, to the extent that the political and geopolitical problems suggested in this chapter create limits on Russian and Middle Eastern gas exports to Europe and North America, China and India may become more attractive markets for those exporters.

10 THE NEW SECURITY ENVIRONMENT: SHORT- AND LONGER-TERM CONSIDERATIONS

10.1 Short Term

For the next few years, the European gas security discourse will be dominated by the problems between Russia and the countries which transit its gas to Europe, principally Ukraine but also Belarus. The current problems in the gas relationship, particularly between Russia and transit countries, are commercially and politically complicated and will take time to resolve. During this period there will be nervousness about maintaining Russian exports to Europe, especially during winter months. Established institutions, such as the EU–Russia Energy Dialogue and the Energy Charter Treaty, should play a role in helping to resolve these problems but due to the downward spiral in Russian relations with the EU, particularly after the August 2008 Georgian conflict, this seems unlikely.

Over the same time period at least as much, and probably more, attention should be devoted to dealing with the risk that end users could be deprived of supply due to a combination of infrastructure failure and insufficient storage to meet extreme weather conditions. Ensuring adequate supplies to meet peak demand, and preventing domestic infrastructure failure, particularly in countries such as the UK which have limited storage capacity and deliverability relative to the size of their markets, will be of paramount importance.

10.2 Longer Term

Over the next 10–15 years, European gas supply availability will be adversely affected by a combination of three factors: first, ongoing indigenous resource depletion; second, political and geopolitical problems between Russia and CIS countries, within the Middle East/Caspian region and between these regions and EU countries; and third, the globalizing market for LNG in the Atlantic and Pacific Basins.

For the period up to 2020, this chapter has advanced a series of propositions about the development of European gas supply:

- European gas production will not increase significantly after 2010 and is likely to fall; this decline is likely to accelerate after 2015.
- Russian gas exports to Europe will plateau at around 200 bcm/year over the next decade and will not rise thereafter. This limit will result from a combination of several factors: European unwillingness to become more dependent in either volume or percentage terms on Russian gas, due to a deterioration of the political climate between Moscow and European capitals and between Moscow and Washington, DC; Gazprom reluctance, and perhaps inability, to increase exports above this level due to a combination of shortage of available gas in the 2010s, a desire to diversify exports away from Europe towards Asia and North America; decreasing commercial attractiveness of European sales compared with Russian domestic sales.
- Large-scale (50–100 bcm/year) exports of Middle East and Caspian gas to Europe *by pipeline* are extremely unlikely given the institutional/political/geopolitical outlook. Several Middle East and Caspian exporters could combine supplies through one or more pipelines, but this will be a complex task with no guarantee of success prior to 2020.
- The best prospects for substantial additional pipeline gas dedicated to the European market will be from North African countries. But these producers have domestic gas requirements which may limit their ability to substantially expand exports and, even when they choose to do so, they may, like those in the Middle East, prefer

the flexibility of LNG exports to the relative rigidity of destination which pipeline gas dictates.
- West African LNG supplies, specifically from Nigeria, are probably the best hope for a significant expansion beyond currently anticipated projects, but domestic politics may complicate a major expansion of exports.
- Increasing competition for LNG supplies with North American and Pacific importers may constrain European options regarding additional large-scale gas deliveries dedicated to Europe.

If the assumptions which underlie these propositions are correct, this paints a picture in which, after 2020, the source of the next supply increment of 50–100 bcm/year for European markets is not obvious. To repeat what was said in the introduction, this judgment is not related either to the existence of gas resources or to the commercial profitability of bringing these resources to Europe at current gas prices. There is an abundance of known reserves in countries with the potential to deliver gas profitably to Europe at prices well below those of 2008. This situation is entirely different from a past in which the main constraints on natural gas development appeared to be whether the industry could develop the technology to deliver challenging projects, and whether prices would be sufficiently high to allow such projects to be commercially viable.

The resource, supply/demand and geopolitical picture which has been painted here is not predetermined. Prior to 2020, the long-term time horizon of this chapter, there is still time for the outlook to change:

- new resources could be discovered in European countries, and the infrastructure built to deliver them to markets;
- political and geopolitical changes could create a more favorable environment for gas development and transportation to European markets, although some of the problems in the current political and geopolitical environment for gas supplies – particularly from Russia and the Middle East – appear relatively intractable; and
- gas demand (and therefore supply requirements) could be reduced by a combination of the adoption of non-gas-fired power generation, and reduction of demand in the non-power sector through efficiency measures driven by high prices.

To the extent that these developments do not happen, political constraints and increasing global competition for LNG may limit the prospects for European gas supplies particularly after 2020. This should not give rise to any immediate panic about security of European gas supplies.

Some European political and commercial reactions to perceived threats from exporters seem extreme. Despite the fact that there is no sign that an 'OPEC for gas' is on the horizon, any suggestion of collaboration between exporters, as in the 2006 discussions between Russia and Algeria, created an extreme reaction from some European governments including calls for EU intervention. It should not automatically be assumed that gas exporting companies and governments are intent on collective action to control volumes and prices to the detriment of EU importers.

Exporting countries have reason to believe that they have been subject to collective commercial decisions of importing countries. The introduction of EU gas liberalization and competition policies has not only increased commercial complexity for exporting countries, but requires them to conform to rules with which they may not agree. These measures have the aim of introducing gas-to-gas competition which, from the perspective of exporters, can only reduce their financial returns. In such circumstances, and given the recent high oil price environment, it is not surprising that exporters, who have a growing share of the European gas market, seek to retain oil-linked gas pricing (Stern, 2007).

In the mid-2000s, a combination of much higher export prices and growing internal demand is causing major gas suppliers to Europe to review their future plans. Much higher revenues than anticipated a few years ago have removed the pressure to increase export volumes, while countries with large populations are finding that the requirements of domestic energy markets are raising the issue of a limit on exports. Indonesia is the clearest example of a major gas exporter which, since the mid-2000s, has been unable to service existing long-term contracts to Asian customers due in part to increased domestic gas demand, and has made it clear that many contracts will not be renewed when they expire. Tension between rising domestic requirements and exports are becoming more common among suppliers to the European market and this will continue, particularly if price liberalization in exporting countries increases the commercial profitability of sales to domestic markets.

If not reversed, the combination of impending decline of indigenous production, political and institutional obstacles to gas export developments within gas supplying countries, and the worsening geopolitical environment between those countries and Europe, will place longer-term supply constraints on European gas consumption. Specifically these constraints threaten the expansion of natural gas as a fuel for power generation in Europe after 2020. From a broader European energy perspective, this may present no significant problems if other energy sources can be mobilized to fill any potential gap left by gas. However, from a carbon emissions perspective this has a serious consequence. If gas is unable to take a larger share of the power generation market then the gap is most likely to be filled by coal, unless a combination of demand reduction, new and renewable energies and nuclear power make much faster progress than currently anticipated. In those circumstances, the new security environment would mean not only that gas would fail to provide any part of a 'bridge' to a lower carbon electricity future, but also that after 2020, natural gas would become a 'sunset industry' in Europe.

NOTES

1. The technical and financial definition of 'reserves' is that they are commercially viable at current prices but this is not always how that term is used in general literature.
2. At around 120 bcm/year, Norwegian Petroleum Directorate (2005). In October 2007, the Norwegian Ministry of Petroleum and Energy refused to approve a proposal which would have accelerated production and exports of gas, http://www.regjer ingen.no/en/dep/oed/Press-Center/Press-releases/2007/Troll-Future-Development.html ?id=486412.
3. For an overview of European supply and demand see Honoré and Stern (2007).
4. A very brief overview of the past 25 years of this debate can be found in Stern (2005).
5. This figure, which is not compatible with data for previous years for statistical reporting reasons, does not include the three Baltic countries to which Gazprom exported 4.9 bcm in 2006 but which may also have received small additional quantities of Russian gas from others.
6. Gazprom exports to the UK in 2005 were 3.8 bcm (Gazprom, 2006a).
7. For details of this crisis and the subsequent reaction, see Stern (2006a, 2006b).
8. See, for example, the comments by the Moldovan president in July 2006 when the price of gas to Moldova increased to $160/mcm (million cubic meters) compared with a European border price of around $240/mcm in the same month (BBC Monitoring Service, 2006a).
9. In the Italian case, deliveries were still up to 15 percent below nominations at the beginning of March 2006.
10. The International Energy Agency (IEA) refers to '..the political cut-offs of gas supplies aimed at transit countries during negotiations over assets or tariff levels', despite the fact that these would seen to be economic and commercial issues (IEA, 2006, p. 35).
11. There are indications that confidential letters were sent from both the EU and the Energy Charter Secretariat to the Ukrainian government pointing out shortcomings in the latter's behavior; but, even if these existed, they stood in sharp contrast to the harsh and very public condemnation of Russia.
12. The details of how much gas was delivered and taken by which parties and on which days, in comparison to their rights and obligations, has never been agreed.
13. Those who hold this view of Russian foreign policy cite Section IV.3 of the Russian Energy Strategy 2003 where one of the stated strategic aims of gas industry development is to 'secure the political interests of Russia in Europe and surrounding states, and also in the Asia-Pacific region'. They also cite President Vladimir Putin's PhD dissertation (see Balzer, 2006).
14. For the history of the EU–Russia Dialogue and the Energy Charter Treaty in relation to Russian gas trade with the EU, see Stern (2005, pp. 134–9).
15. *Financial Times*, April 20, 2006.
16. See http://www.whitehouse.gov/news/releases/2006/05/20060504-1.html (accessed October 2006).
17. See http://www.iea.org/journalists/topstories.asp (accessed May 23, 2006).

18. The dates for the commissioning of the second line are unconfirmed. The first line has been complicated by environmental concerns and resistance from the littoral states and may not be completed until 2011.
19. Ukrainian nameplate (that is, design) transit capacity is 175 bcm but usable capacity is probably less than 130 bcm in 2006. Much of the nameplate capacity could be restored with a comparatively small investment – much less than that required for building a new export pipeline.
20. This pipeline may have a number of branches including a route via Serbia.
21. Russian Energy Strategy (2003, Chart 8, p. 51).
22. The Shtokmanovskoye field in the Barents Sea which will supply North America, and fields in Eastern Siberia and the Far East which will supply Asia. In relation to West Siberian gas supply to Asia see the section on China below.
23. 'China gas supplies to end Russia's European dependence – experts' (RIA/Novosti, March 21, 2006).
24. Gazprom (2006d). These data are not temperature-corrected and therefore it is difficult to see the underlying trend. The data are also not compatible with the same Gazprom publication for the previous year (Gazprom in Figures 2000–04, p. 27) where data for the same year are up to 19 bcm lower. There is considerable uncertainty about how much gas from independent producers is sold to customers and how much either directly or indirectly to Gazprom. With such a level of uncertainty about current data, projections are fraught with difficulty but are likely to be crucial to future export availability, particularly during the 2010s.
25. Price elasticity of Russian gas and electricity demand is a largely unknown and unaddressed issue.
26. This applied to contracts with Germany, France, Italy and Austria.
27. Shown by its desire to increase its share of the UK gas market to 10 percent by the early 2010s.
28. This is particularly the case for Middle East countries. In North Africa only Libyan reserves exceed 100 years, reserves of the other major producers range from 52 to 65 years (IEA, 2005).
29. Abu Dhabi started to export LNG in 1977 and was joined by Qatar in 1997; all other Middle East exports started more recently. Having taken a long time to get off the ground, Qatari exports will increase extremely rapidly during 2006–10.
30. Algerian LNG exports commenced in 1964 and pipeline exports in 1987; Libyan exports only became significant with the start of pipeline trade in 2004.
31. For example, by 2030, less than 40 bcm out of an anticipated total of 200 bcm of Algerian gas production will come from fields currently in production (IEA, 2005).
32. Calculated from the statement that the share of LNG in total MENA exports will not exceed 60 percent (p. 178) and the figures in Figure 5.6, p. 180 (IEA, 2005).
33. If the figures for Qatar include regional exports via the Dolphin pipeline system then these percentages will be somewhat lower.
34. The Saudi oil minister has been quoted as saying that the Kingdom will not consider gas exports until production reaches 120 bcm/year which may happen around 2020–25 (*Gas Matters*, 2006a).
35. The contract has a ceiling price based on Brent crude oil at $31/bbl (barrel) (LNG Focus, 2006).
36. IEA (2005, p. 365) shows that at a $28/bbl oil price – roughly equivalent to the price threshold in the Indian LNG contract – the value of gas is $75/mcm for the LNG project and nearly $350/mcm for reinjection.
37. In 2007, Iran signed gas contracts with Chinese and Malaysian companies which could lead to LNG exports.
38. The exports shown in Table 3.1 up to 2020 are only for regional consumption (Iraq has a contract for export to Kuwait). The volumes shown for 2030 would only be large enough for exports to Europe if they were combined with an additional source of exports.

39. EU (2006b) mentions a scenario in which 10–15 percent of EU gas supplies would come from the Caspian region by 2025 suggesting 2–3 Nabucco-sized pipelines by that date.
40. As illustrated by disruptions to the Iran–Turkey pipeline in 2006–07 (mainly) due to Kurdish terrorists.
41. See http://www.gecforum.org/ (accessed December 2007).
42. In particular, the declaration by DG COMP that joint sales, destination clauses and profit-sharing mechanisms in existing long-term gas contracts involving EU companies were violations of competition rules.
43. Neither has an organization of Eurasian (CIS) gas exporters, suggested by the Russian president and prime minister in 2002–03, made any visible progress.
44. The same argument was made by the CEOs of Suez and Gaz de France to support their merger, 'GdF highlights Gazprom threat' (*Financial Times*, August 29, 2006).
45. At least in Europe. Arguably the most serious gas security incident seen worldwide occurred in Australia in 1998 when an explosion at a gas processing plant deprived the entire state of Victoria of gas for nearly two weeks.
46. Little reliable public information is available about this incident. Some anecdotal accounts suggest that the flow was cut for 45 days and this was the trigger for the building of strategic storage in Italy.
47. 'Belgian king leads mourning for victims of Ghislenghien gas explosion' (*Gas Matters Today*, August 2, 2004).
48. Although this may depend on the exact definition of 'political instability'. Political instability has delayed or prevented a number of contracts from being concluded; but the only example of political instability – meaning the inability of a central government to maintain political control over a region – which this author can recall and which has caused any protracted disruption of supplies in an ongoing contract was Indonesian LNG deliveries from Aceh (Sumatra) to Japan and Korea in 2001.
49. Gazprom's European earnings fell from around 63 percent of total receivables in the early 2000s. Given the huge increase in European gas prices and volumes post-2004 this is significant and shows the importance of increased domestic and CIS gas prices during the same period.
50. For example, the possibility that in the current political climate, the US government might refuse to allow future imports of Iranian LNG.
51. If this accident had happened any earlier in the winter, there would have been even less supply to meet demand requirements; any later in the winter and the repairs could not have been made in time to pump gas back into the facility for the following winter heating season.
52. The UK market framework will provide adequate import capacity, probably backed by adequate supply, but 2–3 years later than was required by the market. Whether it is able to provide the storage which is needed – and whether this problem is more related to planning constraints than to market liberalization, is uncertain. Indeed whether liberalized and competitive markets provide more or less security than monopoly markets is a question requiring further investigation.
53. In 2005 North American (US, Canada and Mexico) gas demand was around 756–775 bcm compared with a 'Europe of 35' (the EU plus Central/Eastern Europe and Turkey, not including Ukraine and Belarus) demand of 536 bcm. Estimates from Cedigaz (2005) and BP (2006).
54. Only terminals on the east and Gulf coasts would compete directly with Europe for LNG. Terminals on the west coast would compete with the Pacific Basin. In addition there were 21 potential terminals of which 10 were on the east and Gulf coasts, four were in Canada and one in Mexico.
55. Nigeria, Oman, Qatar and also Malaysia.
56. Many were anyway of dubious commercial viability.
57. For an analysis which suggests a much lower price range than this, see Foss (2007).
58. This could be a different European location, but is most likely to be a location on the other side of the Atlantic.

59. For example, the impact of tropical cyclones Rita and Katrina on production in the Gulf of Mexico was still evident one year later (EIA, 2006a).
60. It is not certain whether the fall of US gas prices below residual fuel oil in the early months of 2006 was a 'blip' which finished in July, or whether it suggests some more complex price dynamics.
61. For details of the competitive position of LNG in China, see Miyamoto and Ishiguro (2006).
62. The date of 2009 is overoptimistic but one or two pipelines are likely to be operating by 2012 (BBC Monitoring Service, 2006b).
63. A protocol was signed in March 2006 between Gazprom and CNPC for deliveries of pipeline gas from Russia to China. However, subsequent progress on these projects has been slow and they have been overtaken by Central Asian pipelines.
64. In early 2006, the Indian Company GAIL called for expressions of interest from companies to develop ships to bring compressed natural gas (CNG) from Myanmar. Richa Mishra, 'GAIL to call for Eols to transport CNG from Myanmar' (*The Hindu*, February 15, 2006).

REFERENCES

Balzer, H. (2006): 'Vladimir Putin on Russian energy policy', *Energy Politics*, Issue IX, Spring, pp. 31–9.
BBC Monitoring Service (2006a): 'Moldovan president says ready to pay for independence', *BBC Monitoring Service*, July 29.
BBC Monitoring Service (2006b): Text of Turkmen–China gas pipeline deal, *BBC Monitoring Service*, April 4.
BP (2006): *Statistical Review 2006*, London.
Cedigaz (2005): *Gas Year in Review 2005*, Rueil-Malmaison, France.
Clingendael (2004):'Study of Energy Supply Security and Geopolitics', Report prepared for DG TREN, Clingendael International Energy Programme, The Hague.
Clingendael (2006): 'The paradigm change in international natural gas markets and the impact on regulation', Clingendael International Energy Programme, The Hague.
Correlje, A. and C. van der Linde (2006): 'Energy supply security and geopolitics: a European perspective', *Energy Policy*, Vol. 34, No. 5, pp. 532–43.
CPB (2006): 'Government involvement in liberalised gas markets', CPB document 110, M. Mulder and G. Zwart, The Hague.
DTI (2005): 'Secretary of State's first report to parliament on security of gas and electricity supply in Great Britain', Department of Trade and Industry, London, July.
EIA (2006a): 'The Impact of Tropical Cyclones on Gulf of Mexico Crude Oil and Natural Gas Production', Energy Information Administration, Washington, DC, June 2006, available at: http://www.eia.doe.gov/emeu/steo/pub/pdf/hurricanes.pdf.
EIA (2006b): 'International Energy Outlook 2006, Special Topics', Energy Information Administration, Washington, DC, available at: http://www.eia.doe.gov/oiaf/ieo/special_topics.html.
EU (2000): 'Towards a European Strategy for the Security of Energy Supply', Green Paper, COM (2000) 769 Final, November.

EU (2004): 'Council Directive 2004/67/EC of 26 April 2004 concerning measures to safeguard security of natural gas supply', *Official Journal of the European Union*, 29 April, L127/92.

EU (2006a): 'A European Strategy for Sustainable, Competitive and Secure Energy', Green Paper, COM (2006) 105 Final, 8 March.

EU (2006b): 'Commissioner Piebalgs welcomes agreement to accelerate Nabucco gas pipeline project', Press Release IP/06/842, Brussels, 26 June.

EU (2007): 'An Energy Policy for Europe', COM(2007)1 Final, Brussels, 10 January.

Foss, M.M. (2007), 'United States natural gas prices to 2015', Oxford Institute for Energy Studies Working Paper NG 18.

Fridley, D. (2008): 'Natural gas in China', in J. Stern (ed.), *Natural Gas in Asia: The Challenges of Growth in China, India, Japan and Korea*, Oxford: Oxford University Press, pp. 7–65.

Garriba, S. (2006): 'Dealing with gas supply disruption in Italy during winter 2005–06 and its aftermaths', Paper presented at the IEA Gas Security Workshop, Paris, June 12.

Gas Matters (2006a): 'Economic diversification boosts gas pressure in Middle Eastern Markets', *Gas Matters*, March, p. 13.

Gas Matters (2006b): 'The Chinese gas market – a growth forecast in search of affordable gas', *Gas Matters*, August, pp. 1–9.

Gazprom (2006a): *Annual Report 2005*, Moscow.

Gazprom (2006b): 'On results of Alexey Miller's meeting with ambassadors of the European Union countries', Gazprom Press Release, April 18.

Gazprom (2006c): 'Gazprom delegation visits Algeria', Gazprom Press Release, May 31.

Gazprom (2006d): 'Gazprom in figures 2001–05', p. 28, available at: www.gazprom.com (accessed July 2006).

Hallouche, H. (2006): 'The Gas Exporting Countries Forum: is it really a gas OPEC in the making?', Oxford Institute for Energy Studies Working Paper NG13.

Hallouche, H. (2007): 'Algeria's gas future: between a growing economy and a growing export market', Oxford Institute for Energy Studies, mimeo.

Honoré, A. and J. Stern (2007): 'A constrained future for gas in Europe?', in D. Helm (ed.), *The New Energy Paradigm*, Oxford: Oxford University Press, pp. 223–54.

IEA (2004): *World Energy Outlook 2004*, Paris: OECD.

IEA (2005): *World Energy Outlook 2005*, Paris: OECD.

IEA (2006): *Optimising Russian Natural Gas: Reform and Climate Policy*, Paris: OECD.

International Gas Report (2006): 'Italy appeals to Europe to ensure its winter gas', *International Gas Report*, August 11, p. 35.

Joshi, S. and N. Jung (2008): 'Natural gas in India', in J. Stern (ed.), *Natural Gas in Asia: The Challenges of Growth in China, India, Japan and Korea*, Oxford: Oxford University Press, pp. 66–115.

LNG Focus (2006): 'India warms to LNG', *LNG Focus*, June, pp. 23–6.

Martin, P. (2006): 'EIA's current view on LNG imports into the United States', Presented at EIA Energy Outlook and Modeling Conference, March 27, available at: http://www.eia.doe.gov/oiaf/aeo/conf/martin/martin.ppt (accessed September 2006).

Mitchell, J.V. (2006): 'A new era for oil prices', Chatham House, London, available at: http://www.chathamhouse.org.uk/pdf/research/sdp/Oilprices0806.pdf (accessed September 2006).

Miyamoto, A. and C. Ishiguro (2006): 'Pricing and demand for LNG in China', Oxford Institute for Energy Studies Working Paper NG9.

Norwegian Petroleum Directorate (2005): 'FACTS: The Norwegian Petroleum Sector 2005', Stavanger.

Russian Energy Strategy (2003): 'Energeticheskaya Strategiya Rossiya na period do 2020 goda', confirmed by the Russian government on August 28.

Shell International (2005): 'The Shell global scenarios to 2025', available at: http://www.shell.com/static/royal-en/downloads/scenarios/exsum_23052005.pdf (accessed June 2006).

Skinner, R. (2006): 'Strategies for greater energy security and resource security', Oxford Institute for Energy Studies, OIES Background Notes, June, available at: http://www.oxfordenergy.org/presentations/BANFF_June_06-1.pdf#search=%22energy%20security%22 (accessed August 2006).

Stern, J. (2002): 'Security of European Natural Gas Supplies: the impact of import dependence and liberalisation', RIIA Briefing Paper, Royal Institute of International Affairs, London, July, available at: http://www.chathamhouse.org.uk/viewdocument.php?documentid=4603.

Stern, J. (2004): 'UK gas security: time to get serious', *Energy Policy*, Vol. 32, No. 17, pp. 1967–79.

Stern, J. (2005): *The Future of Russian Gas and Gazprom*, Oxford: Oxford University Press.

Stern, J. (2006a): 'The Russian–Ukrainian Gas Crisis of January 2006', Oxford Institute for Energy Studies, available at: http://www.oxfordenergy.org/pdfs/comment_0106.pdf.

Stern, J. (2006b): 'Natural gas security problems in Europe: the Russian–Ukrainian gas crisis of 2006', *Asia-Pacific Review*, Vol. 13, No. 1, pp. 32–59.

Stern, J. (2007): 'Is there a rationale for the continuing link to oil product prices in continental European long term gas contracts?', Oxford Institute for Energy Studies Working Paper NG19.

Yafimava, K. and J. Stern (2007): 'The 2007 Russia-Belarus Gas Agreement', Oxford Institute for Energy Studies, available at: http://www.oxfordenergy.org/pdfs/comment_0107-3.pdf.

4. Natural gas and geopolitics
David G. Victor

1 INTRODUCTION

Natural gas is rapidly gaining in geopolitical importance. Gas has grown from a marginal fuel consumed in regionally disconnected markets to a fuel that is transported across great distances for consumption in many different economic sectors. Increasingly, natural gas is the fuel of choice for consumers seeking its relatively low environmental impact, especially for electric power generation. As a result, world gas consumption is projected to more than double over the next three decades, possibly each surpassing coal as the world's number two energy source and potentially overtaking oil's share in many large industrialized economies – although recent projections made in light of high gas prices have been less bullish (EIA, 2002, 2003, 2004; IEA, 2006).

Currently, most natural gas is transported by pipeline. Elaborate pipeline networks in North America and Europe connect consumers to production areas and provide an important source of energy. In Asia, liquefied natural gas (LNG) is the primary means of connecting end-users to supply, most of which originates in remote locations and must be compressed and refrigerated into liquid form, allowing easier transport by vessels across oceans. International trade in LNG, though limited in application, has been occurring for over 30 years and involves shipments from close to a dozen countries. Japan is by far the largest importer of LNG, consuming almost half of all LNG traded worldwide. South Korea is the second largest importer of LNG (EIA, 2006).

In the 1990s, roughly 5 percent of world natural gas consumption moved as LNG (BP, 2007). But this is expected to rise as mature producing basins in the industrialized West, particularly in North America, begin to decline.

About three-quarters of the world's proven gas reserves are located in the former Soviet Union and the Middle East – far from the areas where demand for gas is expected to rise most rapidly (USGS, 2000; BP, 2007). Indeed, construction of transportation infrastructure is currently the major barrier to increased world natural gas consumption. Cumulative investments in the global natural gas supply chain of $3.1 trillion, or $105 billion per year, will be needed to meet rising demand for gas between 2001 and 2030, according to the International Energy Agency (IEA, 2003). Exploration and development of gas fields will represent over half of this required investment, with more than two-thirds of new capacity needed to replace declines in existing fields. Investment in LNG facilities is expected to double after 2020.

The Energy Forum of the James A. Baker III Institute for Public Policy and Stanford University's Program on Energy and Sustainable Development sponsored a major study of the geopolitical impact of this transition to a gas-fed world (Victor et al., 2006).

The two-year Stanford University–Baker Institute study utilized seven historical case studies on the special challenges of investing in large-scale, long-distance gas production and transportation infrastructures. These studies concentrated on countries that do not have the long histories of cooperation and the stable legal and political environments that are often seen as essential to attracting private investors. The expansion of gas as a global fuel depends in large part on success in attracting investment within such political, institutional and economic environments. The studies examined the factors that explain why these projects were built and why alternative viable projects stalled. The case studies (later, book chapters) covered projects in Algeria, Russia, Turkmenistan, Indonesia, Trinidad and Tobago, the southern cone of Latin America and Qatar (for methods, see Hayes and Victor, 2004, 2006a).

Simultaneous to the analyses of historical case studies, a group of scholars at Rice University developed a dynamic spatial general equilibrium economic model to simulate the development of global gas markets between 2005 and 2030 based solely on commercial considerations of available supply and its development costs, transportation costs, the cost of capital, end-use demand, and interfuel competition. The model, the Baker Institute World Gas Trade Model (BIWGTM), found a schedule for the development of gas resources and transportation routes to satisfy consumer demands at least cost (Hartley and Medlock, 2006a). It allowed analysis of scenarios, such as possible effects on world markets of rising demand for gas in China. It simulated the exploitation of monopoly power by allowing key producers to earn monopoly rents by delaying the development of critical new sources of supply (Hartley and Medlock, 2006b),

The study findings include four broad conclusions that apply to the assumed shift to greater reliance on natural gas:

1. An integrated global gas market will emerge, in which events in any individual region or country will affect all regions.

2. The role of governments in natural gas market development will change dramatically in the coming decades.
3. The rising geopolitical importance of natural gas implies growing attention to supply security.
4. The rapid shift to a global gas market is not a certainty. It depends enormously on creating the context in which investors will have confidence to deploy vast sums of financial and intellectual capital; it requires finding solutions to the adverse social and political consequences of developing natural resources in countries where governance is weak; and it assumes a continued pull from the growing world electricity sector.

2 EMERGENCE OF AN INTEGRATED, GLOBAL GAS MARKET

A major conclusion of the joint study is that a shift is taking place today from a gas world of previously, regionally isolated markets to an international, interdependent, market of global gas. A series of developments – increasing demand, technological advances, cost reductions in producing and delivering LNG to markets, and market liberalization – is spurring this integration of natural gas markets. Such market interconnections will have large ramifications for both large gas consumers and producers.

Results from the study's economic modeling suggest that the shift to a global market will make each major consuming or producing region vulnerable to events in any region. Disruptions or discontinuities in supply or demand will ripple throughout the world market. Moreover, the timing of any major gas export project coming online will affect prices and project development in all regions. Policy makers now focus on the macroeconomic effects of variable oil prices; similar concerns will arise with the transition to gas.

Major consuming countries will have to learn to consider the interdependencies of a global gas market. While large gas importing countries have in the past been focused on key supply relationships (Victor et al., 2006, chs 3, 5 and 6), this point-to-point approach to project development is unlikely to prove as effective for the future where price and supply security in the gas market will become more like the commodity oil market of today (see also Jensen, 2004; Hayes, 2007a, b and c).

According to base runs of the BIWGTM, in a world of fully integrated natural gas markets, for instance, gas users in Japan will have a vested interest in the stability of South American gas from the southern cone reaching the US. West coast; those in the United States will have concerns about natural gas policy in Africa and Russia, and the EU will be compelled to monitor the political situation in gas producing countries as remote as the Russian Far East and Venezuela (Jaffe et al., 2006).

Russia will play a pivotal role in price formation in this new, more flexible and integrated global natural gas market, the model suggests. It was one of the first major gas exporters to the European market and could utilize the nascent European pipeline network taking shape alongside the rising Russian exports (see Victor and Victor, 2006). Russia benefits not only from its location and size of resources but also from its status as the key incumbent (Stern, 1993). Throughout the model period to 2030, Russia is expected to be a very large supplier to Europe via pipeline, exceeding 50 percent of total European demand post-2020. The model suggests that Eastern Siberian gas will flow to Northern China by the middle of the next decade. Strategically positioned to move large amounts of gas both east and west, the presence of low-cost Russian pipeline gas in both Asia and Europe will serve to link Asian and European gas prices. The model also suggests that Russia also will eventually enter the LNG trade via the Barents Sea, providing an additional link between gas prices in North America, Europe and Asia (Hartley and Medlock, 2006a).

Other resource-rich nations, such as Iran and Saudi Arabia, could also become major players. However, they will be disadvantaged because they must bear the fixed costs of market entry due to lack of existing infrastructure to carry their gas to the lucrative European and Asian markets. The model estimates that their entry is delayed until demand rises sufficiently to accommodate those incremental supplies. Neither Middle East resource powerhouse is expected to be a major gas player in the next two decades, according to study predictions. Prolific Turkmen gas may also be slow to come to market due to political and economic barriers in moving that gas across rival Russia (see Olcott, 2006).

The modeling work suggests that the US market will remain a premium region as North American production fails to keep pace with demand, and high prices pull gas supplies from around the world. Alaska is an important source of future supply, flowing to the lower US 48 states by 2015 and replacing dwindling supplies from western Canada. This new Alaskan source does not collapse North American prices that are, today, at all-time highs. Nor do Alaskan supplies eliminate the need for imported LNG, which in 2003 accounted for just 2 percent of US gas supplies and remains small. Policy makers have imagined that US regasification facilities could be utilized more fully, but the reality is that in a world of fungible markets regasification terminals tend to be used at historically low capacity factors because they are options to import not assured imports (Hayes, 2007b).

The international gas industry is already responding to this integration of supplies and major gas consuming regions. As liquidity in the market and the number of available supply alternatives have grown, the average distance between neighboring suppliers has declined, creating new opportunities for price arbitrage. In this new market context, there will be a reduced need for long-term bilateral contracts to hedge risks (see Hartley and Medlock, 2006a,b). Expectations about the future market evolution are influencing investment and trading decisions today, and this in turn is accelerating the change in market structure – a self-fulfilling prophecy.

Such a transformation is already taking place in the world gas market (Jensen, 2003). More international oil companies are investing in major natural gas infrastructure projects without the security of fully finalized sales for total output volumes. Instead, companies are counting on their own ability to identify end-use markets at some future time, closer in line to the investment pattern that characterizes development of multi-billion dollar oil fields. Expectations of a premium, liquid US market are a key factor encouraging this change as was liberalization of certain European markets which allowed gas sellers to bypass European state gas monopolies and sell directly to large gas customers and power generators (see Shepherd and Ball, 2006).

3 NEW MARKET STRUCTURES AND THE CHANGING ROLES FOR GOVERNMENTS

Throughout most of the historical development of the gas industry, government has played the central role in creating markets for gas as well as in directing gas supply projects. Government-owned enterprises have built and operated the infrastructures that were essential to distributing the large volumes of gas that have arrived with supply projects. Government-to-government agreements, usually backed with government controlled financing, have been essential cement for the producer–user relationships (Hayes and Victor, 2006b; Victor and Heller, 2007).

However, as market liberalization takes hold in many key gas consuming countries and global trading of natural gas expands, the role of government is changing – away from builder, operator and financier of gas projects and toward a greater role as regulator and creator of the context for private investment. Historical case studies have allowed examination of how this market-oriented structure – which itself is part of a broader trend in the organization of modern states and economies – will affect the incentives to create new, greenfield gas transportation networks that are essential if the world is to continue its rapid shift to gas.

In all the cases where gas has been supplied to a market that does not exist, study findings suggest that governments have played a central role in 'creating' demand for new import volumes of gas. Absent the state, very few, if any, of these projects would have been able to move ahead at the same speed or with the same volumes of deliveries (Hayes and Victor, 2006b).

Studies of the first-of-a-kind LNG export projects from Arun in Indonesia (1970s) and Qatar (late 1980s) to Japan show the importance of willing government to orchestrate the investment – in these cases, the government of Japan and a small coalition of Japanese buyers. The first of these projects – Arun – rested on the willingness of the Japanese government (through the Ministry of International Trade and Industry (MITI) and Japan's Export–Import Bank) to orchestrate the purchase of the gas and the timely construction of an infrastructure for utilizing the gas. The Japanese government provided crucial financial support as Japanese trading companies launched the Arun venture; the government's interest was rooted in its high priority on energy security and a desire to diversify energy supplies away from coal and oil. In the Japanese context, as an island nation, the government supported an infrastructure that was not a gas pipeline transmission grid (as seen in Europe) but, rather, a network of LNG receiving terminals, serving a cluster of relatively isolated local markets. Constraints on moving gas between those markets helped each local monopoly protect its position and thus invest with confidence in long-term returns. Lack of similar government backing for proposed sales of Arun gas to California meant that contracts to that market languished in the face of Japanese insistence that it be given the right of first refusal on any increased gas exports from Arun (see von der Mehden and Lewis, 2006).

Similarly, the role of the Japanese government and its buying coalition was important to Mobil Corporation's ability to get the Qatargas project off the ground in 1987. Although the strength of MITI and other crucial arms of the Japanese government had weakened considerably as part of a broader effort to expand the role for market forces in the Japanese economy, the role of a Japanese buying consortium along with access to existing import infrastructure was critical to Qatargas's success in gaining financing and sufficient sales contracts. The timing of the project coincided with a reduction in Japanese concerns about the political stability of supplies from the Persian Gulf with a rising US military presence in that region (see Hashimoto et al., 2006).

In the same vein, much of the variation in the outcomes of the two proposed projects to pipe gas across the Mediterranean in the late 1970s is also due to the starkly different roles that the Italian and Spanish governments took towards the prospects of starting to import large volumes of gas. Like

Japan, Italy was actively seeking gas imports and was willing to mobilize significant state resources to secure new energy supplies. Through its own export credit agencies, the government provided the bulk of financing for the Transmed pipeline project. And state-owned ENI was positioned at that time to orchestrate the Trans-Mediterranean ('Transmed') pipeline project as well as the development of Italy's domestic gas transmission grid. State backing allowed ENI to invest with confidence and provided cover for international lending. Spain, on the other hand, did not have supporting policies in place, and thus could not lead successful development of a major gas import project in the late 1970s and early 1980s (see Hayes, 2006).

Importantly, other chapters of the Victor et al. book show that the ready availability of large volumes of gas is not enough to create demand for gas in end-user markets. In markets where the state has avoided a central role in creating infrastructures, rapid gasification has not taken place. In 1990s Poland, for example, the large pipeline from Russia was constructed mainly to supply additional volumes of gas to the German market. Because it crossed Polish territory, large volumes were also available to Poland – yet the Polish market has used very little of that available gas – despite take-or-pay contracts for Polish offtake. The Polish gas market stalled in large part because no entity in Poland was prepared to build the infrastructure needed to distribute gas (Victor and Victor, 2006).

In looking at the role of the state in gas market development, it is also important to examine the role of government in market regulation. The book chapter on the southern cone (Mares, 2006) provides two contrasting examples. The GasBol pipeline, connecting Bolivia to Brazil, was a favorite of both governments and multinational development banks looking to support market reform, transparency and intra-regional trade in the aftermath of a bilateral peace treaty between Chile and Argentina. Under pressure from multinational organizations, market liberalizers and domestic trade groups, the Brazilian government forced state-owned Petrobras to contract for the bulk of gas purchases from the pipeline and also encouraged the company to provide financial support for the investments in field development in Bolivia to be sure that the project went forward. But the failure of demand for gas in Brazil to materialize – in part due to the failure of the Brazilian government to create a regulatory context that would allow gas-fired power plants to sell their electricity – meant that GasBol could not survive financially. Petrobras was left on the hook for volumes of gas it could not sell (see Mares, 2006; de Oliveira, 2007).

The GasAndes pipeline from Argentina to Chile indicates the type of project that seem likely to emerge in the absence of direct state support.

The GasAndes project, a small pipeline to connect gas fields in Argentina to a small number of power generators near Santiago, Chile, beat out its competitor, Transgas, because it was able to find private sector buyers and environmentally driven government support for a limited, strictly commercially viable project. The liberalizing electric power market in Chile along with the tighter air pollution regulations in badly polluted Santiago created favorable conditions for the project.

In contrast, the Transgas project sought to build a much more elaborate gas distribution network south of Santiago, seeking to supply gas to new distribution companies that would serve industrial and residential gas consumers, in addition to new gas-fired power generators. A rival project, GasAndes, sought to supply just large electricity plants in Santiago directly. The Transgas project was more costly; payback would have occurred over a longer period and with greater uncertainty. Transgas sought a concession from the government to allow it to recover investments in the gas distribution grid; as political efforts to get that concession foundered, the GasAndes project moved quickly ahead (see Mares, 2006).

On the supply side, the role of government has been equally important. Even where private firms have actually made the investments in developing gas fields and in building the transmission infrastructure, governments have been essential guarantors of long-term contracts that, historically, have underpinned most large-scale gas infrastructure investment. In the past, investor risk has been mitigated by 'take-or-pay' contracts. But new, more flexible contracting is being pressed upon the industry as gas markets become more global and akin to a commodity. Gas-on-gas competition, new gas resale contract clauses and joint investor/host country spot marketing strategies are creating new uncertainties that are creating a new market structure for gas.

While the role of the state weakens, the key anchoring role for gas projects is shifting to the private sector. In the old world, the governments had deep pockets and a strategic vision that was organized around serving national markets and developing national resources. The development and implementation of this vision was often inseparable from the state-owned and supported enterprises whose charge it was to supply energy to the national market. In that world, projects were national ventures (see Victor et al., 2006, chs 3, 8 and 10).

In the new world, a handful of large energy companies with deep pockets and a similar strategic vision are taking over the role as creator and guarantor of the implementation process. These players are largely private, but they also include national energy companies that are now playing a larger role in the *international* marketplace – ENI, PetroChina, Petrobras and others. This shift to large energy companies, however, is likely to mean that infrastructure development will increasingly be driven by commercial interests rather than national energy security objectives (see Shepherd and Ball, 2006).

The advent of new, more commercially oriented players dominating the gas scene will also change the nature of how contracts are negotiated and enforced. In the regulated, state-controlled environment, it was relatively easy for governments and their bidders to tailor the terms of gas trade agreements for political ends. But as gas markets liberalize – especially in Europe, where countries are small and borders are plenty – directed gas trade is harder to sustain, especially as provisions such as destination clauses are undone. In the emerging commercially driven environment, the role of courts as enforcers has grown – made possible, in part, by legal reforms that have accompanied the shift to markets and given courts and quasi-judicial bodies, such as regulators, greater authority. Although the industry press is just now focusing on the implications of this shift, investigation on this issue suggests that this shift has been under way for more than a decade (see Hayes, 2006; Shepherd and Ball, 2006).

Ironically, the importance of existing contracts may lie less in their enforceability and more in their ability to coordinate the 'sinking' of investment. By facilitating the creation of sunk costs, existing relationships act as a deterrent to others and a binding agent for the project investors. Once Italy had partnered with Algeria and had begun to lay pipe, the deal was sunk and there were huge incentives to continue cooperation (see Hayes, 2006). Russia's contract with Poland partly deterred alternative (more costly) suppliers to that market, but the most effective deterrent existed only once the contract had focused investment on Russia's pipeline. The ultimate deterrent to Norwegian supplies to Poland was the fact 'on the ground' of Russia's pipeline (see Victor and Victor, 2006).

With the exception of Russia, various cases show that private commercial players have been better placed than state gas concerns to position themselves as first movers. Owners of Trinidad LNG were able to push Algeria's Sonatrach from lucrative US East Coast markets by creating lower costs (see Shepherd and Ball, 2006). Nimble GasAndes beat out slow-paced Transgas, which had hoped to tap government support to create a market (see Mares, 2006). A topic that remains to be explored is whether government-owned entities will be able to act as strategic players in the more competitive gas world or whether private commercial players will be able to organize competitive supplies to get to market more effectively, thereby leaving state monopolies to wait for long-term market growth to make space for them to enter without the pressure of innovation. Gazprom's troubles in getting the Shtokman field into operation on its own suggest that Western market-oriented companies continue to play important roles as suppliers of technology, project management, and market savvy.

4 GLOBAL GAS AND SECURITY OF SUPPLY

The shift from the highly structured gas world of government-backed bilateral, fixed-price contracts to a new world of private, market-related contracts raises questions about national security of supply. Private sector participants have different interests from countries; they cannot be expected to consider automatically the energy security concerns of client nations as they are driven mainly by commercial considerations.

One area of attention is the potential formation of a gas cartel similar to OPEC. Concern for maintaining a secure supply of reasonably priced natural gas, which until now has taken a back seat to its oil sister, will increasingly be viewed as a vital national interest. In the past, gas users have feared interruption in vital gas supplies for a variety of reasons such as contract disputes between Algeria and its customers (see Hayes, 2006), to political unrest in Indonesia (see von der Mehden and Lewis, 2006) to transit country risk such as Ukraine and Belarus for Russian exports (see Victor and Victor, 2006). In addition to supply interruption fears, major gas-consuming countries or regions worry that a key exporter such as Russia (to Europe) or group of exporters could exercise monopoly power to extract inflated rents for their product.

In May 2001 the Gas Exporting Countries Forum (GECF) held its first ministerial meeting in Tehran with the aim to enhance coordination among gas producers. Although the GECF ministers announced that they did not intend to manage production or set quotas, certain individual members of the group have debated the merits of exercising some form of market influence or control. Such ideas have nonetheless gained momentum.

The GECF has already tried, unsuccessfully, to exercise some collective influence in the European market. GECF helped to catalyze formation of a working group headed by Russia and Algeria who sought to resist EU attempts to outlaw destination clauses that prevent buyers from reselling gas. (The option to resell gas is a pivotal mechanism for market arbitrage and efficiency as it helps to prevent segregation of markets that allows gas sellers to exert monopoly power.) In another example, Egypt has sought a change in gas pricing systems that would end the link to crude oil prices with the aim of easing the penetration of gas into European markets. Both of these efforts, so far, have generated little practical change; a gas exporters' cartel remains at a theoretical stage (see Jaffe and Soligo, 2006).

The GECF has too many members with diverging interests to exert effective constraints on capacity expansion projects in the near term. It is likely to be a decade or more before they can assert sustained monopoly power in world gas markets, leaving consumer countries ample time and opportunity to adopt countermeasures. It will take many years to work off a plethora of supplies from within major consuming regions and small competitive fringe producers.

Gas suppliers might be able to extract short-term rents in particular markets by manipulating supplies into markets where alternative supplies are not available. Algeria used this position to force higher prices on the Italian and French markets in the 1970s, but Algeria quickly suffered when circumstances changed. Over the long term, Algeria has paid a high cost due to the reputation it gained as an unreliable supplier (see Hayes, 2006). The same Algerian effort to lift prices also contributed to its loss of share in the US market, which created an opening that new export projects from Trinidad eventually filled (see Shepherd and Ball, 2006).

Over the long term, gas exports may eventually concentrate in the hands of just a few major producers, which could make it more feasible for a group of gas producers to restrain capacity expansion to gain higher rents. The overall distribution of world natural gas reserves is more concentrated than the distribution of oil reserves. The two countries with the largest gas reserves, Russia and Iran, have roughly 45 percent of world natural gas reserves, while the two countries with the largest oil reserves, Saudi Arabia and Iraq, have just 36 percent of world oil reserves. Indeed, the base case of the model estimates that Russia will become a very large supplier to Europe via pipeline, exceeding 50 percent of total European demand after 2020. This dominance could leave Russia in a position to curtail capacity additions and boost rents for its gas.

Policy responses to the risk of cartelization are numerous. Among them is the privatization of gas reserves and the gas transport networks in producer countries. All else equal, it is probably easier for national, state-owned, producers to participate in a cartel than for privately owned firms that might have different objectives from the state. If numerous private Russian gas producers emerge, for example, it will be more difficult to reconcile their conflicting corporate ambitions with those of a cartel – especially if pipeline operators are constrained through effective regulation for using their network for market manipulation (Hayes and Victor, 2006b; Victor and Victor, 2006).

As *Natural Gas and Geopolitics* (Victor et al., 2006) shows, diversity of supply is an important protection from rent-seeking behavior both of both gas exporters and transit countries. When Ukraine first interrupted Russian gas exports in 1995, European buyers who redoubled their efforts to diversify found many alternative suppliers, confirming the importance of market reforms that encourage multiple supply sources and gas on gas price competition. Moreover, the declining costs of LNG and pipeline trade mean that markets will be contested by ever-distant arbitrage potentials.

Gas suppliers who dream of extending their powers forget that it is harder to corner gas markets when users have a choice. Algeria learnt that lesson in 1981 when it left a key pipeline empty in a pricing dispute with Italy – extracting a better price at the time but losing billions of dollars for the future by destroying its reputation as a reliable supplier. Gas infrastructure is costly to build and buyers can afford to be choosy (Victor, 2006).

5 RISKS TO THE GREATER GAS VISION

For many analysts, the assumption that the world will shift to gas is rooted in current trend lines and economic modeling that, understandably, do not fully reflect the myriad of political and institutional factors that often play a large role in determining where gas investments occur. Thus, the bright gas future is by no means assured.

First, the vision for gas depends enormously on investor confidence and the supply of vast sums of financial and intellectual capital. A plethora of studies has confirmed that world gas resources are abundant, but many of those resources are not in countries that have traditionally been attractive for private investors. The capital-intensive nature of gas and the long payback periods typical of gas projects – 15 to 20 years or longer for some of the most complex projects – makes investors especially wary. 'Useful reserves' are those where large amounts of gas combine with a political and governance system that is conducive to such projects. Where those conditions do not exist – such as in Iran and possibly in Russia – large amounts of gas will be left in the ground.

Second, developers of gas resources may run afoul of concerns about mismanagement of gas revenues, intra-state disputes over rents, harm to indigenous communities and other afflictions that often get the label: 'resource curse'. While the Arun case concludes, for example, that non-governmental organizations and social discontent had less impact on Arun development in the 1970s because critics had yet to organize themselves sufficiently politically to provide significant impediments to the Arun operation. By 1998, agitation in Aceh where Arun was located became so severe that operations were temporarily suspended and led finally to full-scale central government military action against local armed groups (see von der Mehden and Lewis, 2006).

The case of Arun may be a telling sign of an era coming to an end – an era where developers of these resources faced much less external scrutiny on their operations and where states, themselves, directed many resource development projects. It is plausible to argue that neither of those two conditions will hold in the future. With the advent of revenue management schemes on the Chad–Cameroon pipeline, in Azerbaijan, and other such arrangements emerging for oil resources, it is plausible to expect that gas projects could some day face similar intervention.

In addition to more challenging local politics, visions for gasification may also run afoul of difficulties in siting major gas infrastructures, especially amid emerging worries about terrorism. LNG is the key to the shifting structure of the world gas market – toward a global market – and the US market is a keystone to that development. Yet today the developers of LNG projects are facing a string of failures and political difficulties in siting LNG regasification facilities in nearly every part of the US market except the Gulf coast.

Finally, *Natural Gas and Geopolitics* also underscores that since around 1990 much of the dash to gas has depended on expectations about electric power markets (see Newbery, 1995; Victor and Heller, 2007). The conventional wisdom that gas is favored for electricity has been shaped by the experiences in England and Wales, the United States, and several other markets. In many, gas has gained due to tighter environmental rules. It has also gained because liberalization has created additional pressure to select the least-cost options. But close attention must be given to markets where gas-fired generation is not the current low marginal cost supplier or where electricity demand might be constrained by other factors.

In Poland, the dominance of incumbent coal-fired power plants, the vast oversupply of electric generating capacity and the lack of strong government incentives for gas have made it difficult for Russian gas to enter the market (see Victor and Victor, 2006). In Brazil, a darling for potential investors in the 1990s, the recent collapse of economic growth, combined with the dominance of incumbent hydropower and an unfavorable regulatory setting, has impeded the entry of gas (see Mares, 2006).

It is not yet clear whether gasification in other emerging markets – such as China and India – will follow the examples set in the United States and England (where electrification and liberalization favored gas for electricity) or Poland and Brazil where governments failed to institute the incentives for a push to gas. We end, thus, with a note of caution, especially when projections such as the IEA's *World Energy Outlook* (IEA, 2006) envision that more than half of the incremental demand for gas will come from electric power.

REFERENCES

BP (2007): 'Statistical review of world energy', London.
de Oliveira, A. (2007): 'Political economy of Brazilian power industry reform', in Victor and Heller (eds), pp. 31–75.
EIA (2002): *International Energy Outlook*, Washington, DC: US Energy Information Administration.
EIA (2003): *International Energy Outlook*, Washington, DC: US Energy Information Administration.
EIA (2004): *International Energy Outlook*, Washington, DC: US Energy Information Administration.
EIA (2006): *World LNG Imports by Origin, 2005*, Washington, DC: US Energy Information Administration.
Hartley, P. and K.B. Medlock (2006a): 'The Baker Institute World Gas Trade Model', in Victor et al. (eds), pp. 357–406.
Hartley, P. and K.B. Medlock (2006b): 'Political and economics influences on the future world market for natural gas', in Victor et al. (eds), pp. 407–38.
Hashimoto, K., J. Elass and S.L. Eller (2006): 'Liquefied natural gas from Qatar', in Victor et al. (eds), pp. 234–68.
Hayes, M. (2006): 'The Transmed and Maghreb projects: gas to Europe from North Africa', in Victor et al. (eds), pp. 49–90.
Hayes, M. (2007a): 'Monthly gas trade in the Atlantic Basin circa 2015', Program on Energy and Sustainable Development, Stanford University.
Hayes, M. (2007b): 'Flexible LNG supply and gas market integration: a simulation approach for valuing the market arbitrage option', Program on Energy and Sustainable Development, Stanford University.
Hayes, M. (2007c): 'Institutions and gas market security', Program on Energy and Sustainable Development, Stanford University.
Hayes, M. and D.G. Victor (2004): 'Factors that explain investment in cross-border natural gas transport infrastructures: a research protocol for historical case studies', Program on Energy and Sustainable Development (PESD) Working Paper No. 8.
Hayes, M. and D.G. Victor (2006a): 'Introduction to the historical case studies: research questions, methods and case selection', in Victor et al. (eds), pp. 27–48.
Hayes, M. and D.G. Victor (2006b): 'Politics, markets, and the shift to gas', in Victor et al. (eds), pp. 319–53.
IEA (2003): *World Energy Investment Outlook 2003*, Paris: OECD.
IEA (2006): *World Energy Outlook*, Paris: OECD.
Jaffe, A., M. Hayes and D.G. Victor (2006): 'Conclusions', in Victor et al. (eds), pp. 467–83.
Jaffe, A.M. and R. Soligo (2006): 'Market structure in the new gas economy: is cartelization possible?', in Victor et al. (eds), pp. 439–64.
Jensen, J. (2003): 'The LNG revolution', *The Energy Journal*, Vol. 24, No. 2, pp. 1–45.
Jensen, J. (2004): 'A global LNG market: is it likely and if so, when?', Oxford Institute for Energy Studies Working Paper NG 5.
Mares, D.R. (2006): 'Natural gas pipelines in the southern cone', in Victor et al. (eds), pp. 169–201.

Newbery, D. (1995): 'Regulatory policies and reform in the electricity supply industry', Cambridge Working Papers in Economics 9421, Department of Applied Economics, Cambridge University.

Olcott, M.B. (2006): 'International gas trade in Central Asia', in Victor et al. (eds), pp. 202–33.

Shepherd, R. and J. Ball (2006): 'Liquefied natural gas from Trinidad & Tobago', in Victor et al. (eds), pp. 268–318.

Stern, J. (1993): *Oil and Gas in the Former Soviet Union: The Changing Foreign Investment Agenda*, London: Royal Institute of International Affairs.

USGS (2000): *World Petroleum Assessment*, Washington, DC: United States Geological Survey.

Victor, D.G. (2006): 'Gas and oil do not mix in the chaotic world of energy policy', *Financial Times*, London, May 9.

Victor, D.G. and T.C. Heller (eds) (2007): *The Political Economy of Power Sector Reform: Experiences in Five Major Developing Countries*, Cambridge: Cambridge University Press.

Victor, D.G., A.M. Jaffe and M.H. Hayes (eds) (2006): *Natural Gas and Geopolitics: From 1970 to 2040*, Cambridge: Cambridge University Press.

Victor, N.M. and D.G. Victor (2006): 'Bypassing Ukraine', in Victor et al. (eds), pp. 268–318.

von der Mehden, F. and S.W. Lewis (2006): 'Liquefied natural gas from Indonesia: the Arun project', in Victor et al. (eds), pp. 91–121.

PART II

Nuclear power

5. European electricity supply security and nuclear power: an overview

William J. Nuttall and David M. Newbery

This chapter is based upon the Nuclear Energy Policy Brief prepared for the European Sixth Framework Programme Project 'CESSA – Coordinating Energy Security in Supply Activities'. Here we present a range of issues relating to nuclear energy and in particular new power plant construction in the European Union (EU). In doing so we are conscious of the words of Professor Gordon MacKerron, contributor to the CESSA conference held in Berlin in 2007. He reminds us that 'nuclear power is special'. As will perhaps become clear in the pages that follow, while sound economics is an essential prerequisite for a European nuclear renaissance it is not, and will not be, sufficient to ensure the success of such an endeavor.[1]

1 THE EUROPEAN NUCLEAR RENAISSANCE

There are two major drivers for renewed interest in nuclear power in many countries of the EU: the need for secure electricity supplies and the need to reduce greenhouse gas emissions. During the course of the CESSA project the scope of ambition for new nuclear power plant construction in Europe has broadened. In 2007 the most ambitious nuclear new build plans for Europe would not have done more than replace exiting nuclear energy capacity. However, in early 2008 several countries (such as the UK, Italy and Romania) were considering measures that raise the possibility of a long-term net growth in European nuclear energy capacity, now made more attractive by the large increase in fossil fuel prices.

Concerns for global climate change have led to the European Emissions Trading Scheme and other policy measures that act in favor of nuclear new build by internalizing a key externality and rendering nuclear power more cost competitive. However, several measures at both an international and a national level either continue to exclude nuclear power (for example, Kyoto Protocol Clean Development Mechanism) or are reserved for renewables only (for example, UK Renewables Obligation Certificates). Such policies may be defended as supporting emerging low-carbon technologies, but uncertainties about their likely penetration in electricity may have adverse impacts on the economics of nuclear power. For instance, high levels of intermittent wind power capacity could force down power prices on windy days with low demand and cause very high prices on still evenings with high power demand, reducing the profitability of inflexible capital-intensive plant (such as nuclear power) relative to cheaper flexible plant (such as gas turbines). If renewables play only a minor role, prices are likely to be much more stable. Uncertainty on such matters makes the economics of future nuclear power more difficult to assess. It would be unwise to assume that the future supply demand balance will be similar to that operating today.

2 A DIVERSITY OF EU MEMBER STATE OPINIONS ON NUCLEAR POWER

Several EU countries (including Ireland and Austria) remain resolutely opposed to nuclear power. The growth of the EU from 12 states to 27 has reduced the proportion of member countries with an anti-nuclear stance.

The salience of electricity security differs greatly across the member states of the EU. In western member states, history has provided a robust and flexible electricity system and market liberalization is generally well advanced. These countries enjoy a diverse range of energy sources and much investment is underway to expand this range of supply options. Investment in nuclear energy represents one such option. For many countries in Central and Eastern Europe there are extremely high levels of dependency upon natural gas imported from Russia. These countries fear both long-term high prices (arising from contracts linked to oil prices) and risks of supply interruptions with a wide range of possible causes (ranging from the technical to the geopolitical and upstream market power). Some of these countries also face poor electrical interconnections, combined with problematic gas supply pipeline routes and strong policy pressure to decommission legacy nuclear plant. It is unsurprising therefore that several of these EU states are among those most interested in new nuclear power programs.

3 THE NUCLEAR OPTION IN CONTEXT

For the world electricity industry as a whole the fuel of the future is coal – this presents enormous challenges to climate change policy. Generally worldwide nuclear new build is in competition with new pulverized coal plant. In the future there could be a competition with clean coal (for example, integrated gasification combined cycle technology perhaps with carbon capture and storage). In Europe today the main generation choice is still between new nuclear build and new investment in combined cycle gas turbine plant, although there are signs that coal is returning to favor as gas prices rise.

Across the European Union much consideration there is being given to the question of how, and whether, to replace aging nuclear power plants with new nuclear power plants. However it is also important to recognize the need to replace aging conventional power plants (particularly coal-fired power plants). If carbon capture and storage can be developed and deployed at scale, then coal may have a significant place in the future European electricity mix. If not, there may be increasing pressure to replace coal-fired power plants with nuclear energy, depending on the success or otherwise of renewables generation and the ability to retain enough flexible plant to manage intermittency.

In considering a technology choice between gas and nuclear one should note that, while it is true that modern nuclear power plants can operate flexibly, generally nuclear power is much more likely to play a base-load role. This is primarily an economic rather than a technical matter owing to the more capital-intensive nature of the cost structure of nuclear generation and the very low variable costs. There is some evidence that nuclear power can supply short-term balancing services to compensate for sudden increases in wind output.

4 TWO REGULATORY PRESSURES

Nuclear power policy is shaped by two regulatory pressures: the regulation of electricity markets and the safety regulation of a hazardous and politically contentious technology. While the benefits of a single European electricity market are widely recognized, progress on the question of pan-European safety regulation is much less developed. International project collaboration is emerging, particularly in Eastern EU member states. CESSA would support moves towards the regionalization and the eventual Europeanization of safety regulation.

Economics is central to the future of nuclear power. We stress that nuclear power plants can be developed in a liberalized electricity market with no direct subsidy. This possibility is favored by stable long-term carbon prices; sustained high oil and gas prices and regulatory approval for grid reinforcement by monopoly transmission companies similar to that put in place to assist new renewables projects. During the CESSA project the relative economic attractiveness of nuclear energy investment has improved significantly, such that economic risks now appear less daunting, although important issues of economic risk do remain, notably arising from the recent rapid escalation in construction costs and remaining uncertainties about the time before commissioning.

5 NUCLEAR NEW BUILD AND THE SUPPLY CHAIN

We observe significant interest from private companies in new nuclear power plant construction. In Finland the approach has been based on a consortium of companies, with none having majority control. In Romania, despite substantial private sector enthusiasm, the current approach is for public sector leadership and control. In the UK it seems likely that leadership will come from large international electricity companies. The UK has arguably the most liberalized electricity sector in Europe. In such circumstances, at least for first-of-a-kind (FOAK) plants, we expect large diversified electricity companies to hold the economic risks themselves via 'corporate finance' rather than by creating new businesses specifically for the new build project, that is, 'project finance'. Much thought is being given to novel financial structures for new nuclear power plants which are better suited to the economic project risks. These risks change with time, for example, as an FOAK plant is completed, risks decrease and the new financial structures become attractive. Furthermore, when a sequence of plants is planned, later plants can be collatoralized against an operational FOAK plant, reducing financial risks and costs.

One possible threat to nuclear energy security in Europe relates to the global supply chain of key components for the construction of new nuclear power plants (for example, nuclear pressure vessels and non-nuclear turbo-generator components). For the first few new plants in Europe these concerns are not pressing, as components are already reserved. Also in the very long term new engineering firms can be expected to enter the market and existing firms will increase their production capacity, removing the problematic constraints. However, there is a real risk of a problem for new European nuclear power plant projects with construction start dates planned in the medium term, approximately 2015–20.

6 FUEL SECURITY AND THE NUCLEAR FUEL CYCLE

Nuclear energy was the first large-scale energy system to attempt to fully manage its wastes. European countries adopt different approaches to the fuel cycle. For instance, France strongly advocates nuclear fuel reprocessing whereas Sweden takes the view that spent fuel should be regarded as a waste for disposal. The economic benefits of direct disposal of spent fuel can improve if associated with a medium-term option to retrieve. Such an option might be exercised if nuclear fuel reprocessing became overwhelmingly attractive in the future.

While uranium yellowcake prices have risen steeply in recent years there is no immediate or medium-term prospect of fuel resource scarcity, nor do rising prices have much impact on the economics of nuclear power. CESSA notes, however, that a global shift from *replace nuclear with nuclear* towards policies of *nuclear expansion* could put uranium supply under pressure and hence greatly improve the economics of reprocessing.

Reprocessing-based fuel cycles raise a set of special issues such as policy for separated plutonium. CESSA has links to the EC Red-Impact project. That project has considered options for plutonium management.

7 FUTURE NUCLEAR ENERGY SYSTEMS

CESSA noted the relative lack of European Commission (EC) research support for next generation nuclear energy systems. CESSA notes that Generation III nuclear power systems have only modest research needs and existing EC efforts appear sufficient. It is important, however, for Europe to consider its role in the longer-term future of nuclear energy generation. If Europe is to ensure the option of Generation IV nuclear systems then EC-sponsored research would be appropriate. Such a policy would help provide a Generation IV option for member states, but would in no sense represent an obligation on them. It is important to stress that such research could extend the benefits of nuclear energy beyond electricity generation, for instance to include hydrogen production. This aspect of our work benefited greatly from the structure of the CESSA project with its thematic emphasis on hydrogen.

Noting the ITER (International Thermonuclear Experimental Reactor) machine under construction in Cadarache, France, CESSA has considered Europe's leading position in fusion energy research. CESSA researchers have expressed concern for the very long timescales of fusion research and the very high sunk costs involved. This presents Europe with a policy choice: either wholeheartedly to assume global leadership in this field or cut losses and divert fusion research funds to other opportunities with possibly better prospects. CESSA was not best placed to answer this question, but we believe it to be important and hence we recommend a high-level European review of fusion research policy.

8   NUCLEAR WEAPONS PROLIFERATION

The proliferation of nuclear weapons rightly remains a major global concern. While there may be geopolitical merit in European countries leading by example (for instance, by unilaterally removing highly enriched uranium and separated plutonium from their civil nuclear activities), it is important to note that all EU countries are robust in their measures to prevent proliferation. Furthermore, any European member state decision to expand nuclear power would not raise the risk of nuclear weapons proliferation in any direct way. CESSA expresses no opinion on proliferation, safety and nuclear fuel cycles issues outside the EU.

9   A EUROPEAN ROADMAP FOR NUCLEAR ENERGY?

It is unlikely that Europe will be able to 'speak with one voice' on matters of energy policy. This is especially true of electricity generation mix and nuclear power. Formally all member states accept that the generation mix is a sovereign matter for each state consistent with the subsidiarity principle. This is only tempered by a need formally to show sensitivity to the concerns of neighboring states. In considering long-term European electricity security it can appear perverse that a technology enthusiastically endorsed by some EU member states (for example, France) is not legally permissible in others (for example, Ireland).[2]

The relationship between national choices for the electricity generation mix and European electricity security are clearly described by the 2006 Green Paper from the Commission, which states:

> Tackling security and competitiveness of energy supply: towards a more sustainable, efficient and diverse energy mix.
>
> Each member State and energy company chooses its own energy mix. However, choices made by one Member State inevitably have an impact on the energy security of its neighbours and of the Community as a whole, as well as on competitiveness and the environment. (EC Green Paper, 2006)

The Green Paper refers specifically to the regional effects arising if one member state chooses to permit the construction of a new nuclear power plant:

> Decisions by Member States relating to nuclear energy can also have very significant consequences on other Member States in terms of the EU's dependence on imported fossil fuels and CO_2 emissions.

Noting that both climate change and EU energy import dependency are both issues that face the EU as a whole, CESSA has repeatedly returned to the question as to whether issues of the energy mix would be better handled at a European rather than at a member state level. While there is a widespread sense that such a shift could have real merit, CESSA recognizes that the political realities militate against such an approach. We note with interest the following observation in the 2006 EC Green Paper:

> Furthermore it might be appropriate to agree an overall strategic objective, balancing the goals of sustainable energy use, competitiveness and security of supply. This would need to be developed on the basis of a thorough impact assessment and provide a benchmark on the basis of which the EU's developing energy mix could be judged and would help the EU to stem the increasing dependence on imports. For example, an objective might be to aim for a minimum level of the overall EU energy mix originating from secure and low-carbon energy sources. Such a benchmark would reflect the potential risks of import dependency, identify an overall aspiration for the long term development of low carbon energy sources and permit the identification of the essentially internal measures necessary to achieve these goals. It would combine the freedom of Member States to choose between different energy sources and the need for the EU as a whole to have an energy mix that, overall, meets its core energy objectives. The Strategic EU Energy Review could serve as the tool for the proposal and subsequent monitoring of any such objective agreed by the Council and Parliament.

CESSA accepts that the energy mix will not be determined at a European level anytime soon. Despite this, we note that the European Union has developed useful policy roadmaps for other energy technology and policy issues (for example, renewables). We suggest therefore that, as a minimum first step, the Commission should initiate a process for a European Nuclear Energy Roadmap.

The 2006 Green Paper was an encouragement in this direction: 'The [proposed] Review should also allow a transparent and objective debate on the future role of nuclear energy in the EU'. As we noted in connection with R&D policy for future systems, such a policy (for a nuclear energy roadmap) would help provide a policy option for member states, but would in no sense represent an obligation on them. As such the recommendations expressed here do not seek to overturn the accepted position concerning subsidiarity and the generation mix, but we do propose that a full range of options should be made available to all EU member states. Stronger efforts from the Commission to support pre-competitive research and to provide a technology roadmap would be important steps in developing the nuclear option as a realistic alternative for member states struggling to meet the combined challenges of secure, affordable and low CO_2 electricity production.

NOTES

1. The authors are most grateful to the numerous colleagues and associates who kindly provided inputs to the CESSA conference series. In addition, the authors are most grateful to CESSA management colleagues François Lévêque, Jean-Michel Glachant, Pippo Ranci, Christian von Hirschhausen, Franziska Holz, Julian Barquín and Ignacio Pérez-Arriaga. The opinions expressed in this chapter are not necessarily shared by those who have provided assistance and all responsibility for errors and omissions rests with the authors. The authors are most grateful to the European Commission Sixth Framework Programme Project CESSA – Co-ordinating Energy Security in Supply Activities – EC DG Research contract number 044383 for financial support. The authors also acknowledge the assistance of the ESRC Electricity Policy Research Group.
2. In Ireland in 2008 the relevant minister is required to approve all new power stations under the Electricity Regulation Act of 1999, but he or she is barred by statute from granting such permission to a nuclear fission-based power plant (Ireland, 1999).

REFERENCES

EC (2006): 'A European Strategy for Sustainable, Competitive and Secure Energy', Green Paper, Commission of the European Communities, Brussels, (SEC(2006) 217), Section 2.3, March 8.

Ireland, Irish Statute Book, Electricity Regulation Act 1999, section 18 item 6.

6. Contractual and financing arrangements for new nuclear investment in liberalized markets: which efficient combination?

Dominique Finon and Fabien Roques[1]

1 INTRODUCTION

All the nuclear power plants operating today have been developed by vertically integrated regulated utilities. Many developed countries, and an increasing number of emerging countries, are in the process of moving away from an electric industry structure built upon vertically integrated regulated monopolies to an industry that relies primarily on competitive generation power plant investors. Under traditional industry and regulatory arrangements, many of the risks associated with construction costs, operating performance, fuel price changes, and other factors were borne by consumers rather than suppliers. The insulation of investors from many of these risks had significant effects on the cost of capital used to evaluate alternative generation options. While vertically integrated monopoly utilities could pass on costs to consumers, and had no problem in financing capital-intensive investments, utilities in liberalized markets have to bear the construction and operating risks associated with new investments in power generation.

The current context for new nuclear build is significantly different from that in the days of the vertically integrated monopolies. The electricity industry structure has been transformed by gradual liberalization in developed and developing countries over the past 20 years. In a competitive market, investors bear the risk of uncertainties associated with obtaining construction and operating permits, construction costs and operating performance. Part of the electricity price risk can be shifted to electricity marketers and consumers through long-term contracts in vertical integration. Depending on the proportion of the construction and operating risks which are borne by the power plant investors, they will ask for a different return on investment. This in turn will affect the optimal financing arrangement of the project and the relative competitiveness of nuclear compared to other technologies.

There have been few nuclear plant orders in liberalized markets over the past decade –the Finnish and French plants under construction being the exceptions – but rising fossil fuel and CO_2 prices are reviving interest in nuclear power. A potential nuclear power renaissance in liberalized electricity markets will face a number of hurdles associated with the specificities of the technology and the legacy of past experiences. Indeed, nuclear power suffers from some specific risks: (i) the regulatory risk associated with the instability of safety regulations and design licensing; (ii) the policy risk where electoral cycles could undermine the commitment to nuclear power and the development of nuclear waste disposal facilities; and (iii) the construction and operation risks associated with the necessary relearning of the technology. Furthermore, the large size of a nuclear project and the capital intensity of the technology make it relatively more sensitive to some critical market risks such as electricity price and volume risks.

The key factor in the success of nuclear power in liberalized markets will be the ability of the power industry to engage with regulatory and safety authorities, plant vendors and consumers to allocate risks to parties that are best able to manage them. By shifting part of the pre-construction, construction, operating, and market risks to other parties (regulators, plant vendors, creditworthy consumers, and so on), electricity producers are in a better position to attract potential investors (lenders, and so on).

The allocation of the different construction, operating and market risks in turn influences the selection of the financial arrangements among different options. While in the past regulated utilities financed their investments using corporate financing with recourse debt and bonds, a wide range of options ranging from project finance with non-recourse debt and with high gearing to corporate and hybrid financing approaches are now available to investors in power markets (Esty, 2004). Project finance and hybrid financing approaches have been widely used to finance large and capital-intensive infrastructure projects in the past decade. In theory, modern project finance fits perfectly well with the business model of the pure power producer, but interest in the so-called 'pure merchant plant' model without long-term contracts has collapsed, with the bankruptcy of many independent producers that financed gas plants in the US and the UK in the late 1990s in this way. On the other hand, the standard corporate financing approach is often seen as an ineffective way for lenders to control risks associated with a company's project, given that managers are more likely to subsidize the new investment from other corporate assets rather than choose project finance. This approach risks bankruptcy by the special entity via default on debt payments for the new investment (ibid.). Given the risks specific to nuclear power and the alternative contractual risk allocations, it is important to identify the possible combinations of financing arrangements.

The objective of this chapter is to study how the risks specific to a nuclear power investment in liberalized markets can be mitigated, how they can be allocated to the different stakeholders, and what financial arrangements are consistent with the alternative allocations of the construction, and operating and market risks.

The chapter is organized as follows. Section 2 details how the risks specific to nuclear power can be mitigated or transferred away from the plant investor to other parties. Section 3 contrasts the different possible financing arrangements and how these are intrinsically linked to the contractual risk allocation between the different parties. Section 4 illustrates through four different case studies how different combinations of contractual and financing arrangements among the electricity producer, the plant vendor, the consumers, the public authorities and the lenders are viable, depending on the local institutional and regulatory environment and the industry structure. Section 5 concludes.

2 HOW CAN THE RISKS SPECIFIC TO NEW NUCLEAR BUILD BE MITIGATED OR SHIFTED AWAY FROM THE INVESTOR?

We consider independently the different risks specific to nuclear build and different ways to *ex ante* mitigate them or to shift them away from the producer–investor onto another party when they are relevant in terms of economic and social efficiency. Indeed, if the government assumes a major part of, or all, investment risks, it tends to transform private sector projects into public projects, with the same drawbacks that have been observed with such projects in the past (an immature blueprint, bad planning, soft budgetary constraints, large delays, and so on).

We consider successively regulatory and political risks, construction and operating risks, and finally market risks (volume and price risks). We consider them independently in order to identify relevant arrangements in each case. Thus risks associated with regulatory action or political choice which amplify construction risks and which are to be considered are those that are exogenous and not inherent in the management of plant planning and realization. For instance, a company that does not respect the specifications of the planning license, submits sloppy planning material to the safety authority, and has poor communication with the public or the administration, is at risk of not receiving a planning or operating license in sufficient time to prevent project interruptions.

2.1 Regulatory and Political Risks

While all power generation technologies are subject to the risk of changing regulations on environmental protection, nuclear projects face specific regulatory and political risks. Sponsors face such risks in gaining necessary public support before the project is undertaken. The uncertain outcome and likely complexity and length of the public inquiry add to the licensing phase uncertainties. Furthermore, political and regulatory requirements may change during the design and construction phase, adding to the above risks (for example, following a change in government). There are also regulatory and political risks during the operating phase (such as retroactive regulations, political phase-out decisions and so on). In the past, disputes about licensing, local opposition, cooling water source, redesign requirements, quality of control, and so on have delayed construction and completion of nuclear plants in a number of countries, in particular in the USA and Germany (Bupp and Derian, 1978). The licensing process is specific to each project, and safety regulation changes can be imposed on projects during their construction.

More generally, political and judicial risks are related to the 'politicization' of nuclear energy and the difficulty of winning widespread social acceptance. Levy and Spiller (1994) highlight for network industries how the credibility and effectiveness of a regulatory framework – and hence its ability to facilitate private investment – vary with a country's political and social institutions. In this perspective, countries that want to reopen the nuclear option need strong political leadership in order to reduce regulatory and licensing risks at different levels (Delmas and Heiman, 2001). These risks include:

- safety regulations, both for the certification of reactor technology and for the stabilization of safety regulations;
- the definition of a legitimate solution to the nuclear waste disposal issue;
- the stability of the legal framework on limited liabilities and insurance provision in case of nuclear accidents; and
- the political process for building acceptability on plant siting and nuclear waste management.

In this perspective, governments and regulatory and safety agencies have a critical role to play in setting out clear and consistent procedures for licensing design and authorization procedures for siting. The mitigation of the key risks in the regulatory and licensing process requires close cooperation of regulators, utilities, and nuclear plant vendors, in ensuring, respectively, an appropriate plant siting and licensing process, a clear design certification procedure and the stability of the safety rules.

In countries such as the USA, where in the past safety regulation has generated large construction cost risk and long lead-times, new streamlined licensing procedures should help reduce regulatory risk. In addition, governments might want to provide investors with additional guarantees that they will shoulder any unforeseen costs due to regulatory changes or delays. In the USA, a complementary guarantee against regulatory risk has been introduced in the 2005 Energy Policy Act for the first new nuclear projects.[2] Similarly, concerning protection against political risks and electoral cycles, the British financial community has suggested that given the long lead-time of nuclear projects it would be economically efficient for the government to guarantee the state's commitment in favor of the nuclear option.

2.2 Construction Risks

All large-scale complex projects are characterized by above-proportional levels of completion and financial risks (Esty, 2002). In a review of 60 large $1 billion engineering projects, Miller and Lessard (2000) show that the critical factors of poor performance are a high proportion of public ownership due to soft budgetary control; extra-large-scale size (complexity and management problems); and if they are first of a kind (FOAK) or one of a kind (lack of experience, design risks, and so on), these last two factors will be at play in nuclear projects.

Compared to other power generation technologies, new nuclear build is characterized by long lead-times (three years for project preparation, five to six years for construction), and high front-end cash outflows (€3–4 billion for a FOAK plant of 1500 MW, €2 billion for a standard plant, to compare to an investment cost of €200 million for a large combined-cycle gas turbine (CCGT) of 600 MW). It is also likely to have high cost estimation and schedule risk around the forecast baseline lead-time, based on past experience construction cost overruns. Nuclear plant construction risks are amplified by the capital intensity inherent in such large and complex projects. An International Energy Agency study (IEA, 2006) shows, for instance, that a construction delay of 24 months will increase the levelized cost of nuclear kWh by 9.6 percent instead of 2.6 percent for a gas CCGT and 6.6 percent for a coal generation plant. Furthermore, industrial relearning associated with advanced reactor designs increases not only the construction cost, but also the construction risk for the first units. Investors will need to gain confidence in the maturing 'Generation 3' evolved nuclear technologies (the Advanced Boiling Water Reactor: ABWR; the European Pressurized Water Reactor: EPR; the Westinghouse AP1000; the Advanced CANDU Reactor: ACR; and so on) proposed by nuclear plant vendors.

One critical aspect of assessing project construction risk is the quality of project management – more precisely, the interaction between the plant vendor, the utility, and the engineering and construction (E&C) company. Past experience shows a large difference of efficiency in project management between countries (Thomas, 1988): large utilities benefiting from their own engineering and procurement capacity are in a better position to: (i) limit the overall engineering costs of each project; (ii) develop industrial programming and standardization on series; and (iii) maintain a bargaining power with the reactor vendor. Électricité de France (EDF) has been able to leverage such advantages by maintaining a large engineering department, while German utilities have been relying on the engineering services of the reactor vendors, and US utilities have historically been dependent on architect engineers such as Bechtel, Ebasco, and so on; exceptions include Duke Power and TVA.

Former electricity monopolies traditionally assumed nuclear plant construction risks, which represented less of a burden in a regulated industry. But for nuclear build in liberalized markets the difficulty is twofold: nuclear technology is at the stage of industrial relearning in a number of countries with new designs to be tested, and the producers have to support construction and market risks on a very large investment.

Different solutions are possible to mitigate construction risk by spreading the risk across different parties, or transferring part or all of the project risk to the plant vendor. One solution is to create a consortium comprising the reactor supplier, the E&C company and the investor. The consortium can collectively commit to a firm construction price contract, as presented in the Texas University study on the South Texas nuclear plant project (TIACT, 2005). Such a fixed-price contract would prompt vendors and the E&C company to better control the manufacturing and engineering costs. A more direct solution is a 'turnkey' contract which shifts a substantial part of the construction risk onto the vendor. From the perspective of initiating a renaissance of the nuclear market, plant vendors might be more inclined to bear part of the construction risk than in the past, in order to demonstrate their evolved new designs and build confidence. This is the strategy adopted by AREVA which carries the major part of the construction risk for the first unit of its EPR design to be built, the Finnish Olkiluoto3 reactor with a total project fixed price of €3.2 billion. The total construction costs will ultimately be higher than the agreed fixed price.

It is, therefore, unlikely that nuclear plant vendors will agree to bear all of the construction cost risk in turnkey contracts in the future after the FOAK. Some states might want to subsidize the first new nuclear units by shouldering part of the construction cost risks in order to help the relearning process of nuclear power technologies. The relearning cost for the first units could indeed deter investment and some argue that government support is necessary to help demonstrate the technology. Public subsidy for early units could be justified by the future social benefits that can be expected from cumulative learning. A key benefit would be avoided CO_2 emissions achieved at reasonable cost by nuclear electricity generation. In the USA, for instance, the 2005 Energy Policy Act (EPACT) creates a federal support scheme which includes a provision of loan guarantee for the first 6 GWe of nuclear plants ordered before a deadline, as well as a production tax credit, of $18/MWh for eight years which is a response to these learning costs and risks for the first 6 GWe of nuclear plants ordered and commissioned before precise deadlines. It also includes a loan guarantee up to 80 percent of the investment cost if the investment is financed by credit The Department of Energy is allowed to issue this guarantee to several projects for a total budgetary envelope of $18.8 billion.

2.3 Operating and Performance Risks

From the perspective of a financial investor, operating, performance, design and construction risks can be regarded as layers of the same category of risks, as they represent the same underlying uncertainty about successful operation of a given technology and design, in particular when a technology has been dramatically improved.

The extent of technological uncertainty relating to the FOAK depends on whether established designs have been used, or whether relatively new designs have been put forward. At the operating stage, this may also affect technical reliability. In the case of a nuclear plant, considerable complexity and highly specific engineering both add to the problem of limited understanding of those risks by external investors. In theory, financial investors should demand a very high premium for informational asymmetry arising from limited understanding of these risks; in practice, investors may be unwilling to assume these risks at all – at least until confidence has been established in the performance of the technology. Experience of exchanges of nuclear assets on the US electricity industry between 1998 and 2001 shows that creditors did not want to assume any portion of nuclear performance risk even when there is an established track record (Esty, 2002; Scully Capital, 2002).

In response to this phenomenon, contractual arrangements have been developed in different industries to mitigate and transfer these risks away from uninformed parties. Performance risk could be allocated to the equipment vendor, for example, through a guaranteed lifetime load factor. In the case of CCGT projects, the large vendors (General Electric, Alstom, Siemens, and so on) agree to bear the performance risk during the lifetime of the plants. In the case of a nuclear project, the Finnish contract contains provisions for the vendor, AREVA, to assume part of the operating risk: the contract is based on a nominal load factor of 91 percent on the lifetime of the equipment.[3] Based on empirical data from existing reactors, this appears to be a risky bet for a FOAK reactor, and it is unlikely that the other nuclear vendors will be ready to assume such a risk in their future FOAK projects.

2.4 Market Risks

Market risks are sell-side risks arising from highly fluctuating power prices. While these risks are specific to the power sector, investors have generally been willing to accept them in the case of financing new nuclear build, as well as purchases of existing generators. Market risk is not specific to nuclear projects, but the large size of a nuclear plant exposes the investor to greater risks than other smaller-size modular generation technologies. Indeed, despite low capital intensity and the benefit of relatively stable net cash flows through highly correlated gas and power prices in many markets (Roques, 2008), market risk also exists for CCGT plant. A large number of pure CCGT producers failed in the US in 2002–03 when the gas price increased threefold, because they were displaced from base- to mid-load. A nuclear plant is not exposed to the same 'dispatchability risk' as a CCGT plant, because its low variable costs ensures that it is dispatched as a base-load generator, provided that that is available. On the other hand, with a cost structure symmetrical to that of the CCGT (60 percent of capital investment in total cost against 25 percent for the CCGT), the capital intensity of a nuclear plant makes it vulnerable to the risk of low electricity price which could lead to a loss and prevent the nuclear generator from reimbursing the debt cost for a long period.

One major aspect of this price risk is the risk associated with the CO_2 price. Indeed the attractiveness of a nuclear plant as a power producer will increase as a result of the additional cost placed on fossil fuel generation technologies by climate policies, which is reflected in the marginal price on hourly electricity markets. But CO_2 policy based on a quantity instrument such as 'cap and trade', rather than a price instrument (CO_2 tax), introduced a fundamental uncertainty to CO_2 price. The risk is also largely political and results from uncertainty regarding the stringency and long-term commitment of climate policy.

Different options are possible for investors and producers to securitize any generation investment in electricity markets by transferring part of the market risk to other parties, such as vertical integration, long-term contracts, or the combination of horizontal integration and vertical arrangement in a consortium. Such arrangements can help to shift the market risks onto players other than the producers, in particular retailers and consumers:

- *Long-term contracting between new nuclear generators and large credible buyers* The interests of generators and large wholesale buyers converge as they seek to manage their market risks (Chao et al., 2008). Indeed, producers and buyers have a natural incentive to insure each other against volatile spot prices over a long period. But their interests diverge in two respects: first, the producers' need for an off-take guarantee contrasts with the suppliers' need for flexibility because of the variation of their loads and market shares; second, the risk of opportunistic behavior by the buyers who are less committed to the transaction than to a new generator, and hence could be tempted to break the contract in the event of a market downturn. In fact, suppliers do not wish to be bound by power purchase agreements (PPAs) with fixed prices (or any clause of price indexation on fuel price) at least for a long time. But to commit to fixed-price contracts, wholesale buyers (distributors, industrial consumers) must be quite sure that power prices will not drop to a low level (Neuhoff and de Vries, 2004). Recent literature studying the conditions of generation investment and vertical arrangements has shown that the required contractual credibility could be reached if there are guarantees that limit the opportunistic behavior of the other party (Joskow, 2006a; Michaels, 2006; Chao et al., 2008; Finon, 2008). In the case of suppliers, the guarantee could result from the possibility of partly shifting their risks onto their customers, either because they retain a large core consumer base or because they benefit from a supply franchise on households. In the case of industrial consumers, the guarantee could be common ownership of the new generation equipment in partnership with a producer in a consortium. Similarly, it has been suggested that CO_2 price risk could be transferred to the government via long-term option contracts which would be auctioned in order to guarantee minimum revenue for new non-carbon capital-intensive equipment, rather than to marginal fossil fuel generation units (Newbery, 2003; Ismer and Neuhoff, 2006).

- *Model of cooperative generation* In some industries, such as the world oil and gas industries, producers are accustomed to jointly develop some large projects in order to share costs and risks. Joint interests of different stakeholders could lead to the creation of a consortium to develop a new nuclear project in order to share costs and allocate (market and construction) risks by mixing horizontal and vertical arrangements. Different types of consortium structure can be envisaged: first, a consortium of end users and suppliers; second, a consortium of end users, large suppliers and power producers; or, third, a consortium comprising nuclear business (reactors vendors, E&C companies), end users and power producers.[4] Arrangements would need to be organized by means of PPAs between the consortium and its member end users and suppliers to securitize repayments of debt. In particular, as end users are unlikely to be as risk averse as the non-regulated suppliers and seek a high return on investment, the joint company could sell them nuclear output at cost, plus a reasonable margin, as with the Finnish EPR project. These contributions could help consolidate the transfer of the different risks organized in the different contracts and be perceived as a source of efficiency that could make the consortium structure an attractive organization model.[5]
- *Combination of vertical and horizontal arrangements* Partial or complete vertical integration is another option to secure investment in generation by guaranteeing off-take quantities and sales prices of the project power production and by passing the fuel risk on to the internal wholesale buyer. When vertical integration is associated with a diversified portfolio of generation equipment, the latter gives a complementary advantage in terms of investment project hedging within a vertically integrated company in comparison to a merchant plant project – even backed by a long-term contract with a credible party – as pointed out by Chao et al. (2008). Since such large companies benefit from a large and diversified asset base, they are able to obtain loans under corporate financing arrangements. Typically, owing to a 50/50 debt–equity ratio and good ratings, such firms can save on capital costs and any risk premiums. But a number of successive nuclear projects in an ambitious strategy might alter the credit rating of the company and its average capital cost for the large volume of capital of the company.

Companies benefiting from a diversified portfolio of generating stations can, during low price periods, rely on 'portfolio bidding', that is, occasionally bid at prices below the generation cost (investment and fuel) of their capital-intensive equipment. For instance, if one company adds one nuclear power station to its portfolio, it would be able to protect its investment if prices decrease below the completion costs, that is, when net cash flow does not cover annual amortization. Finally, integrated companies can generally leverage a large and diverse set of customer relations. This combination of advantages is likely to be critical when considering potential candidates for a new nuclear plant project with specific market and construction risk mitigation arrangements.

3 THE COMPATIBILITY OF CONTRACTUAL, ORGANIZATIONAL AND FINANCING ARRANGEMENTS

The different contractual and organizational arrangements detailed in the previous section in order to mitigate or transfer some risks specific to nuclear plant to other parties, have in turn an impact on the attractiveness of alternative financing structures for a nuclear plant. This section explores the compatibility of the different contractual arrangements with different financing arrangement alternatives, ranging from project financing to corporate financing and hybrid financing. In project finance, which is based on non-recourse debt, each new project is separated from the developers' other assets. New separate entities are created to share the risks of the project at the construction and operations stages, without any interactions with the parent company's balance sheet, the main sponsor for the project.[6] A 'special purpose entity' (SPE) is the borrower and the asset is on its balance sheet. Lenders look only to the specific assets in order to generate the cash flows (net of operating expenses) which provide the sole source of debt payments. These payments are secured by cash flow prospects and the assets of the SPE. Conversely, a corporate finance project is financed by recourse debt and equity. Intermediate options are known as 'hybrid finance'. In such cases a newly created power-generating company, which is the borrower, will use the backing of its parent company or long-term contracts with consortium members to improve its creditworthiness.

3.1 Financing Arrangements for New Nuclear Build

In theory, there are a large variety of financing structures that might be considered for a nuclear plant project. A precise answer as to which exact structure would be optimal is likely to involve a detailed investigation of all the possible pros and cons of different designs. Since the financing structure will have important implications not only for the costs of financing and risk allocation, but also for rules of operation and contingent control over assets, careful consideration of the suitability of different financial arrangements should be made in parallel with the choice of technology design and reactor vendor.

The two basic types of financing are equity and debt. Equity capital acts as a buffer for absorbing variability in cash flows and is necessarily influenced by the risk profile. Considerable uncertainties associated with successful implementation of the construction phase are likely to make it difficult to raise high levels of debt for the initial part of the project in the absence of government support and if the nuclear industry is in the phase of relearning, and if the future owner–operators are not backed by a parent company with a strong balance sheet. The government involvement in this situation is likely to raise familiar questions about the degree of public authority support for private investments, potential bailouts for private creditors, or accounting issues such as classifications for the public sector borrowing requirement.

Overall, the exact level of project gearing will need to be optimized according to various considerations, including the need for new capacity to follow consumption growth and equipment closures, anticipation of price spikes in relation to the competitors' technology mix on the market, anticipation of the trend of fuel price and CO_2 allowance price, predicted financial characteristics of revenues, and allocation of risk between different parties.[7] Nevertheless, financing choices are not constrained to a simple dichotomy between equity and debt. Typical business financing models are now diversified; they not only rely on the canonical model of project finance, but they are also adjusted to fit the particular purposes and needs of the project. Although in general rather complex, project finance solutions can be value creating and particularly applicable in situations where certain business characteristics of the project are unique and can be exploited for the mutual benefit of operators and capital providers alike.

The issue of equity investment is common to both project and corporate finance. In any financing structure there could be a single sponsor or a consortium of sponsors. Since the participation of more than one sponsor usually involves creating a separate company with split ownership, such arrangements are more typical of project finance, although they can also be adopted in hybrid structures with both corporate and project finance characteristics. While it is common for an electricity utility to be the sole sponsor of a new plant development, minority participants might co-sponsor the project. Engineering and construction companies often participate in new, large-scale investment as sponsors. This participation might take the form of a direct equity contribution. Such arrangements are usual for large-scale projects outside the electricity sector (infrastructures, oil and gas projects), and are gaining popularity in the power generation business. Given the fact that the amount of equity required for the project can be considerable, creating a broad consortium of equity holders can be critical to the success of the project (OXERA, 2004).

Candidates for sponsors include specific nuclear technology providers and others with particular interests in the nuclear sector. It could be an incentive for a reactor vendor to reduce lead-time and construction cost, in particular at the end of the construction process. Given the unique nature of this development and its potential importance for nuclear technology providers, the latter could become substantial equity holders in projects for the first one or two reactors that they would sell in order to benefit from future industrial recommendations (ibid.). An alternative route to involvement in a nuclear build is the turnkey contract which allocates a major part of the construction risk to the nuclear plant vendor, an arrangement that AREVA chose for the first EPR plant that it sold, the Finnish EPR. It certainly helps the sponsors to obtain cheaper loans. But note that no constructor has an interest in bearing the construction risk for later reactors. Even beyond the first two reactors, one risks encountering a design mistake correlated across a series of new stations, and if such a design problem implies long repairs, then this can easily bankrupt the vendor that has provided guarantees.

3.2 Corporate Finance versus Project Finance

Given the risks and investment characteristics outlined above, financing construction of a new power plant poses unique challenges. As described earlier, the two main approaches to financing the development of such a project can be referred to as corporate finance and project finance. Between the polar extremes of corporate and project finance lie a multitude of hybrid options.

The crucial feature of corporate financing is the importance of the project developer and its direct involvement in taking the risk of the project onto its own books. Under such an arrangement the new asset (the power plant) remains an integral part of the sponsor's entity, and hence of the sponsor's balance sheet. Therefore, from the financial perspective, the critical aspect of corporate finance is that neither the new asset nor the liabilities to the creditors financing the new asset are legally separated from the remainder of the sponsor company's assets and liabilities. Implicitly, new creditors purchase an option on cash flows from the company's other assets because managers are more likely to subsidize the new investment from other corporate assets than to risk bankruptcy of the company as a whole by defaulting on financing the new investment.

However, such purely theoretical financial considerations in practice need to be nuanced for two reasons. First, producers can hedge their investments in a technology by diversifying their risks between different technologies in the same market (Roques et al., 2008). A portfolio theory approach helps identify the best risk-return portfolio of power plant assets for a de-integrated producer, with the optimal share of nuclear power depending on the degree of risk aversion of the producer (Bazilian and Roques, 2008).[8] Portfolio valuation approaches make nuclear power relatively more attractive to investors than when they are valued on a standalone basis, as nuclear power (as well as renewable technologies) can mitigate some of the fossil fuel and CO_2 price risk exposure of the producer mix of plants (Roques et al., 2008).[9] Second, the optimization of the dispatch of a producer portfolio of plants can yield significant benefits. Companies benefiting from a diversified portfolio of plants during a low price period can rely on 'portfolio bidding', for example, by occasionally bidding at prices under the generation cost (investment and fuel) of their capital-intensive items. Such a strategy represents another way of hedging a capital-intensive investment by leveraging the company's other power generation assets.

The critical point in the modern finance perspective is that corporate financing is not asset specific but represents the sponsor company's general borrowing. It is therefore driven by the sponsor's general financial situation as its terms are based on the sponsor's credit rating and leverage in addition to pure investment factors. In the case of a new nuclear power plant, the sponsor's financial circumstances might therefore uniquely determine the terms and conditions of the new borrowing, which is viewed as negative from the modern finance perspective.

The key feature of project finance is the legal separation from sponsors' other assets of what is most typically a single large asset constituting a new, self-contained, well-specified investment by the sponsor(s). The legal separation ensures that the project entity's creditors – the lenders to the independent power producer (IPP) – have no recourse to the parent. The 'project' in project finance is not simply a group of assets based on a self-contained and highly focused investment, but is also a set of contracts governing the use of that investment. These contractual arrangements can significantly alter allocations of risk among different entities involved in the project. Specifically, selected risks can be transferred away from the project finance vehicle and onto sponsors. For the construction of a new power plant, these contracts typically include: (i) a construction and equipment contract with an E&C company, and several different contractors and technology providers (reactor and turbo-alternator vendors) which could include some turnkey principles; (ii) a long-term fuel supply contract; (iii) one or several long-term power purchase agreements with electric suppliers or large consumers at a fixed price; and (iv) an operations and maintenance contract. During the stage of institutional and industrial relearning of nuclear technologies, government could also assume some risks, in particular by way of a loan guarantee, as the US government is doing for the first 6 GWe of plants to be ordered. In other words, project finance could be conceivable for new nuclear builds if built around a special vehicle entity. A situation of a PPA at a fixed price, turnkey contracts and government loan guarantees greatly assists with the securing of the investment. In practice two items from this list would be sufficient as in the Finnish example.

While project finance deals are characterized by a significant degree of complexity and thus high transaction costs, they have been popular for new projects in the power industry in liberalized markets in the 1990s and the early 2000s with low capital-intensive CCGT projects. However, following the bankruptcy of many IPPs' CCGT merchant plants in the US after the 2002 gas price upheaval, lenders have become much more cautious and financing for new merchant plants has dried up. Most of the merchant plants installed in the US liberalized markets did not have long-term off-take power purchase contracts and were exposed to significant volumetric risk. As the gas price dramatically increased, the IPP generators' CCGT plants were displaced from base-load demand supply to semi-base-load supply on the spot market, which generated much less revenue and net cash flow (Joskow, 2006a; 2006b; Michaels, 2006).

New hybrid finance arrangements have emerged in which project finance is combined with long-term fixed-price/indexed-price contracts. Project financing for merchant nuclear could be envisaged, but only under two possible schemes close to corporate financing. The first scheme is project finance with one or several long-term fixed-price contracts with creditworthy buyers, and a low degree of leverage of 50 percent; but it could eventually be increased to a level of 75-80 percent, with the addition of a government loan guarantee as we see in some US projects, or else with turnkey contracts and performance guarantees as is the case for every CCGT project. The second possible scheme is project financing structured as corporate financing, that is, in which the power-generating company is the borrower with the backing of the parent company, a corporate structure that combines a power generation company and an electric distribution company.

In recent experience, non-recourse project financing for power generation assets has become a much less attractive financing option for power generation financing. Scully Capital (2002) identifies two sources of this trend: (i) the widening of the project–corporate spread (the cost differential between project and corporate financing) due to declines in credit quality among deregulated generation companies in the USA and a simultaneous market decrease in overall corporate spreads; and (ii) rating agencies' incorporation of project-financing debt into the balance sheet of the corporate parent. Hudson (2002) also observes a preference among nuclear plant vendors for corporate financing for new nuclear developments in the form of 'conventional owner financing using the balance sheet of a strong and integrated generation and utility company'.

The key advantage of the corporate finance methodology is its simplicity. No special legal, financial or administrative structures are required. This is likely to diminish the transaction costs substantially in comparison with project financing. Also, since financing in the latter case is done on the basis of the existing corporate balance sheet, it can build on previous financing arranged for this entity, enjoying the same market name recognition, reputation, investor familiarity with the risks involved, and past performance record. This explains the recent trend in the power industry to come back to corporate finance, with investment risks managed through a diversified plant portfolio and vertical integration inside a large firm – or else through such a portfolio combined with a set of long-term PPAs with creditworthy buyers in the case of pure producers as in some US cases (Exelon, Constellation, NRG Energy, and so on). A combination of corporate finance and vertical integration would be an appropriate solution for financing future nuclear plants because this combination is quite well aligned with risk management requirements for a new nuclear build. Most importantly, corporate finance could be the only available option if the project is seen by the investors as too risky to be financed on a standalone basis.

3.3 The Impact on Cost of Capital

Limited leverage is an important drawback of any corporate finance funding arrangement (OXERA, 2004). Leverage in corporate financing is likely to be as low as 50 percent, with a substantial amount of equity required for the project. However, some possible benefits of leverage, including low capital commitment and high debt tax shields, would no longer be available relative to a comparable project-finance transaction for CCGT. But this drawback needs to be nuanced in the case of a nuclear plant given that gearing in a project finance arrangement for a nuclear plant is likely to reach 50 percent because of the high ratio of investment cost in the cost price.

Given the probable remaining concern for nuclear stations once the first few have been built, lenders will prefer to focus not only on the risk exposure of projects, but also on the financing profile of companies (size and structure of its balance sheet). Because of this concern, when equity investment in a nuclear project goes beyond 15 percent of market capitalization, they will be worried about the effect on the shareholder value. The financial equation of nuclear investment for companies of any size below a €20 billion market value is more difficult to balance. A nuclear power investment of €2–3 billion is likely to put stress on such a company's credit rating and its stock share value. This value can be altered by two factors:

- diluting capital: if 50 percent of equity capital is required, the company's net cash flow and owners will need to provide about $1 billion and, given the long lead-time it will not yield revenues for a number of years. As a consequence, the company's stock valuation will be affected. Raising the equity capital required would dilute existing shareholders' equity and earnings per share. This would lead to lower stock prices, further reducing the company's attractiveness to the financial community; and
- placing substantial capital investment at risk for an extended period of time. Under current law in the USA, nuclear generating units are treated as 15-year property. This depreciation period (which is appropriate for a regulated, low-risk, cost-of-service regulation business environment) is not suitable for a competitive, high-risk commodity business environment (NEI, 2006).

That does not mean that the weighted average cost of capital (WACC) of large companies will not be altered if they want to build a number of nuclear plants. Each project, if it is perceived as risky, will add a small risk premium to the WACC of the companies by decreasing the credit rating, but ultimately, applied to a large volume of capital, it could have an important effect, as already mentioned above. Moreover, the total equity investment in several nuclear builds could eventually reach the same precautionary threshold of 15 percent of market capitalization (for instance €6 billion in equity for four plants for a market cap of €40 billion) as one nuclear build for a small company.

But one important parameter which could enhance the credit rating of a company in the future is likely to be the ownership of several existing nuclear assets inside a diversified portfolio of generating assets if the confidence of the financing community in nuclear technology is restored. Indeed, on electricity markets where the market price is driven by the marginal cost of fossil fuel generation plants, ownership of already amortized nuclear plant is a guarantee of stable cashflows.[10] The existing nuclear plant cashflows could pay for a new nuclear build with a good prospect on future return on equity, in particular in a probable scenario of a high CO_2 price.

Table 6.1 Comparison of WACC (nominal after tax) and beta coefficient in some European companies in 2000

	Endesa	E.ON*	RWE	Iberdrola	Electrabel
Nuclear share in their capacity on home market (%)	12	25	14	15	40
WACC (%)	6.3	6.2	6.4	6.1	6.6
Beta coefficient in CAPM**	0.81	0.61	0.58	0.62	0.54

Note: * Average of VEBA's and VIAG's WACC and in 2000 before merger. ** Capital asset pricing model.

Sources: Lautier (2003) for WACC, and Vernimmen (2000) for beta coefficient.

However, to date, the relation between nuclear plant ownership and the companies' credit rating in the rating agencies is not obvious, as nuclear plants have long been perceived as a source of risk for a company rather than as a potential hedge for new investment. In theory, companies with a large number of nuclear plants in their asset portfolio should benefit from a lower correlation of their share value with the market value compared to rival electricity companies with no nuclear plants ('beta coefficient'). There is as yet no systematic research published on the effect of ownership of nuclear assets on the 'beta' of electricity companies in Europe (Table 6.1). In the US, a 2005 study by Bloomberg Financial Markets on the largest energy companies operating nuclear plants (Exelon with 17 reactors in 2005, Entergy with 10 reactors, Dominion Resources and the FPL group) shows that they have far outperformed the overall stock market performances in 2004 and 2005 (Gray, 2005).[11]

Lenders are likely to prefer to lend money to companies with a large balance sheet with a large and diversified asset base and vertical integration. Corporate financing by large European companies (the so-called 'seven sisters' with more than €35 billion of market capitalization) is likely to be the dominant form of financing for new nuclear plant in Europe. But smaller-size companies are also candidates for new nuclear build in liberalized markets, in particular in the US where the industry is more fragmented than in Europe. In addition to some vertically integrated medium-sized companies, some independent producers with a balance sheet of less than €7 billion ($10 billion) such as Constellation or NRG Energy and a less-diversified asset portfolio have recently announced plans to build new nuclear plants in the US (Table 6.2).

Table 6.2 Market capitalization of some large- and medium-sized electricity companies in Europe and the USA in March 2008

Market valuation	Companies	Comments
€50–100 bn	EDF (€108 bn); E.ON (€80 bn); Suez-Electrabel (€52 bn)	
€30–50 bn	Iberdrola (€50 bn); RWE (€40 bn); ENEL (€40 bn); Endesa (€35 bn); Exelon (€35 bn)	In this range, Exelon is the sole US company
€10–30 bn	TVA (€20 bn); FPL Group (€18.5 bn); Duke Energy (€15.2 bn); Entergy (€13 bn); Texas Utilities TXU (€15 bn); Vattenfall-Europe (€11.9 bn); Detroit Edison (€9 bn)	As TVA is a government company, its capitalization comprises the balance sheet total TXU was bought by hedge funds for $45 billion but had debts of $25 billion
Less than €10 bn (non-integrated companies)	NRG Energy (€6 bn); Constellation (€3 bn); AES (€6.5 bn); Calpine (€4.8 bn); Mirant (€3.5 bn)	AES, Calpine and Mirant are not candidates to invest in nuclear plants

Note: Exchange rate US$1.5 to €1.

Source: Stock market quotation in mid-March 2008.

The large size and strong balance sheet of the dominant European companies ensures that these companies are likely to have access to better borrowing conditions than smaller-scale companies. This is highly important for a project as capital intensive and of such a scale as a nuclear plant. As a consequence, it is likely that any small- or medium-sized company would have to support a higher cost of capital than a large company, which might make nuclear uncompetitive as compared to other generation technologies.

4 CASE STUDIES: OTHER COMBINATIONS OF CONTRACTUAL AND FINANCING ARRANGEMENTS FOR NEW NUCLEAR BUILD

When comparing different nuclear investment case studies, it is important to emphasize how the local industrial organization and institutional environment will make some organizational and contractual arrangements more suitable in one country than another (Delmas and Heiman, 2001; Bredimas and Nuttall, 2008). Countries in which the operators remain vertically integrated between generation and supply and where nuclear option meets sufficient political legitimacy, could benefit from the better position of such companies to manage the risks specific to a nuclear power investment. Furthermore, in some countries such as France, the slow pace of electricity market reform has been partly motivated by the objective to preserve the capacity of the incumbent company to invest in large-scale and capital-intensive projects such as nuclear plant (Finon and Staropoli, 2001).

In the case of new nuclear build, the local institutional environment and the industrial organization of the power and equipment companies will therefore play a determinant role in enabling different types of risk transfers from the utility to other parties. In this section, we illustrate through four case studies how the local environment leads electricity companies to favor one set of organizational and contractual arrangements. Similarly, we review how the financing arrangements (corporate finance, hybrid finance, project finance) are aligned with these contractual arrangements, the specific institutional environment and the industrial structure of the electricity industry (including the size of the investor company and its vertical and horizontal integration to benefit from scale and scope economies in the management of its risks).

This section successively reviews four cases of nuclear new build in different market structures: first, the US nuclear merchant South Texas Project of the NRG Energy group; second, the consumers' consortium project of Okililuoto in Finland; third, the large vertical company nuclear project (EDF's Flamanville EPR); and fourth, some hypothetical projects from oligopolisitic medium-sized vertically integrated companies such as are likely to occur in the UK.

4.1 The Conditions of Viability of a Nuclear Merchant Project in Liberalized US Markets

In addition to the nuclear plant orders in US states which are still regulated, some announcements of nuclear projects by non-vertically integrated producers in liberalized US markets in 2006–07 seemed to signal the renaissance of nuclear orders in such markets in a merchant framework. The rationale to invest in nuclear build in liberalized markets lies in the opportunity to earn potentially greater revenues than under the cost of service regulation (Lacy, 2006).[12] As well as the South Texas Project (STP) which is studied next, four other projects have been announced in these liberalized markets.[13]

The case of the STP

The South Texas Project of two ABWRs, each of 1,200 MW, is promoted by an independent producer: the NRG company. NRG has been the first unregulated company to apply for a joint construction and operation license (COL in September 2007. The project financing arrangement of the STP is made possible by different federal guarantees, which aim to alleviate the construction and regulatory risks, and also by a series of PPAs with creditworthy parties. Indeed the project will be backed by long-term fixed-price contracts with municipalities and historic suppliers.

A consortium of producer and suppliers to share costs and risks The best way to achieve this condition is to associate historic suppliers or monopolist distributors to the project with a consortium structure. This is possible in Texas in which there are monopoly 'islands' comprising large municipalities (Austin, CPS Energy of San Antonio, and so on) which cover 20 percent of the retail markets (Adib and Zarnikau, 2006).[14]

The promoter of the project, NRG Energy, is an IPP company with a diversified portfolio of 23 GW in different technologies (CCGT, open cycle gas turbine: OCGT, coal plants and a nuclear plant) and operating on different markets (Texas, South-Central, North-East and outside the USA in Australia and Brazil).[15] It creates a consortium with two monopolist municipalities with the following shares: NRG 44 percent, Austin Energy 16 percent, and CPS Energy of San Antonio 40 percent. The last will contractually off-take 56 percent of the electricity on a cost-price basis. So risks are shared between NRG and the two municipalities, while NRG also benefits from its asset portfolio for risk management.

Mitigation of risks on construction costs and performances Importantly the investor reduces the risks on siting and construction costs by building them on, or adjacent to, an existing nuclear site, in part because local communities already accept the plants. It also chose the General Electric ABWR technology already developed and tested in Japan and Taiwan by GE's licensees Hitachi and Toshiba. This approach reduces both construction and operational risks. The experience with the previous construction of ABWRs implies that the constructor can rely on existing manufacturing lines, and thereby reduce the FOAK engineering cost, thanks to the construction partnership between GE and Hitachi. This cautious technological choice combined with the federal standby insurance against the regulatory risks can explain why a turnkey contract is not judged necessary.

Learning costs and risk bearing by the federal government As mentioned above, US federal support includes a production tax credit (PTC) of

$18/MWh for the first 8 years of operation allocated to the first 6 GWe of nuclear plant. It is meant to compensate the learning costs which affect FOAK projects and to enhance the financial attractiveness of such a project. However, it does not address financing challenges before and during construction. If lenders require it, securitization can turn these guaranteed revenue streams from government into lumps of capital in the special purpose vehicle of the project financing. Another support defined in the 2005 EPACT allows a US federal loan guarantee of up to 80 percent investment of projects of unregulated companies.[16] The bankers get guarantees to receive their payments in case the electric company defaults on a new nuclear plant developed with project finance after the commissioning of the equipment. So, a large part of the learning cost and construction risks is borne by the federal government.

Regulatory risk bearing by the US federal government All the new nuclear projects benefit from the mitigation of regulatory risks by the safety certification of Standard Plant Design, the new early site permitting process, and the new streamlined procedure of license established at the end of the 1990s in order to limit the cost and delays associated with licensing new commercial plants. The major part of remaining regulatory risks would be borne by the federal government if the project applies for a COL before the end of 2008, a deadline that is defined in the 2005 EPACT in order to be eligible for federal support. A complementary element of federal support is the limitation of regulatory risks by the federal government by the standby insurance for regulatory delays for the four first projects ($500 million for the first two and $250 million for the next two).

Market risks: securing the investment by long-term contracts The consortium is likely to benefit from favorable financial terms because of the presence of municipal utilities which will be a proof of predictability of the customer base and stability (TIACT, 2005). They could transfer risks to local consumers via their tariffs. Moreover 75 percent of the NRG's energy share (44 percent) will be sold by long-term contracts with historic suppliers in Texas. Only the production of the remaining 400 MW will be sold in the market in order to seize opportunities to retain benefits from future carbon policies in the medium term.[17]

Project financing Project financing relies in theory on a set of long-term contracts. But the biggest help in this case comes from a loan guarantee for up to 80 percent of the project. This guarantee is an important subsidy as it has a double effect on the financing structure of a nuclear plant project. First it allows access to guaranteed debt, which has therefore a lower interest rate (for instance, 5 percent instead of 8 percent in real terms). Second, since loan guarantees cover up to 80 percent of a project, it is possible to increase the leverage of the project, using up to 80 percent of debt as compared to 50 percent in the case in which debt is not guaranteed. These two factors combined have a large impact on the WACC. It has been calculated that the effect of the federal loan guarantee on the cost is a decrease of the total cost of $70/MWh by $11/MWh (Deutch and Moniz, 2003).

Reproducibility

The STP shows that NRG is likely to be interested in investing in nuclear because it can shift most of the risks onto other parties through contracting arrangements and through federal guarantees, making 'merchant financing' possible in this specific institutional context. Four other companies are candidates to develop some so-called 'merchant' nuclear generators in liberalized US markets.[18] Constellation Energy considers that in its Maryland project, 'some of the output may be sold under long term contract, but [its] project could in fact be built with all the output being sold into the wholesale market'.[19] But it is doubtful whether it will succeed without PPAs for the major part of the off-take. Some of these merchant projects could succeed under the same conditions as the STP, in particular the loan guarantee and the production tax credit in the terms of the 2005 EPACT, and if they could trigger interest from historic suppliers and municipalities to a long-term contract because they are credible counterparties, which is a key condition to long-term contracting.

Could this merchant model be reproduced after the ending of the federal support? Banks are likely to agree to lend only in a hybrid finance type arrangement, that is, with the backing of PPAs. There seems to be little demand for such long-term PPAs from 'pure' suppliers, such that the most likely arrangement will involve corporate financing with vertical suppliers and with large IPPs able to contract with historic suppliers that have a stable customer base.

4.2 Model of a Consumers' Cooperative in a Decentralized Market (with reference to Finland)

New nuclear build can be promoted by a cooperative of large consumers and suppliers which look to manage their risks and control their cost of sourcing by installing equipment with a production cost not exposed to risks which usually determine the electricity price volatility on a market, that is, fuel price risk, CO_2 price risk or hydraulic inflow risk on a hydro-dominated market. If consumers or suppliers anticipate high fossil fuel and CO_2 prices in the coming decades, one way to hedge such risks is to build and operate nuclear power plants. In this context we analyse the case of the Finnish nuclear project ordered by a cooperative of large consumers before drawing some general lessons on the opportunity to invest in nuclear plants in this institutional environment.

The Finnish case
The Finnish Okiluoto III project developed by an existing cooperative of consumers has three main characteristics: it is developed in a political environment of consensus; it is the benchmark of a consumers' consortium project in which consumers share equally project costs and risks; and the reactor vendor assumes the construction risk via a turnkey contract. It relies typically on two contractual structures for electricity price risk and construction cost risk – a set of PPAs with the consortium members and a turnkey contract with the vendor. This type of arrangement makes possible a corporate financing approach, in which the cooperation between the borrower (with the backing of the shareholding companies) and the set of PPAs allows for an unusual high gearing ratio of 75/25.

A consortium of consumers and suppliers The promoter, TVO, is an electricity generation cooperative of pulp and paper companies and some electricity supply companies. Indeed the cooperative is 60 percent controlled by PVO, which is itself a cooperative owned by different forestry companies (42 percent UPM, 16 percent Stora, 42 percent other), and already owns and operates two nuclear reactors and a thermal plant. The other shareholders are the main Finnish production and supply company Fortum (25 percent), a distribution company EPVO (6.6 percent), and Helsinki city (8.1 percent).

Mitigation of political and regulatory risks The Finnish policy and institutional environment guarantees stability of the government commitment to nuclear power and limits the political risks. The nuclear plant order signed in 2005 was preceded by a long democratic process to determine the national energy policy, the siting of the plant and the development of a nuclear waste storage facility. The vendor has implicitly accepted to bear the exogenous regulatory risks, by signing a turnkey contract without provision of revision of the price in the case of unanticipated regulatory difficulties.

Reallocation of construction and performance risks to the vendor: a turnkey contract The turnkey contract with AREVA allocates the construction risk to the reactor vendor above a cost level which includes unforeseen learning costs (€3.2 billion, that is, €2,000/kW). As a consequence of the ongoing difficulties and delays in the construction of the plant, AREVA has set aside provisions of around €800 million in 2006 and 2007 for construction delays and E&C cost increase due to safety controls. A number of reasons have been brought forward to explain the problems and delays, including the inexperience of AREVA in E&C and the specificity of the Finnish style of control of the safety criteria. Another peculiarity of the contract is that the operational risk, which is important for a FOAK project, is also completely shifted onto AREVA, with a penalty to be paid when the performance is lower than an average 91 percent load factor in 40 years.[20]

Market risk: a set of power purchase agreements at cost price Long-term PPAs with a fixed price have been signed *ex ante* with the members of the consortium for a period of 60 years, that is, the lifetime of the reactor. What makes the PPAs an obvious solution is the fact that the purchasers are the owners of TVO. Contracts at cost price without reference to the market price will link the cooperative with its members – the company will sell nuclear output at cost to its shareholders in proportion to the number shares.[21] The fixed price transfers the market risk to the purchasers in the sense that they will support an opportunity cost if the Nordic market price decreases below the fixed price. But this risk appears limited given the need for new power generation capacity and the likely increase of the average Nordic power price as a result of the rising CO_2 price in the European market. The associates in the cooperative are likely to avoid the effects of CO_2 price volatility and benefit from the CO_2 rent. They are also freed from the price effects of long-term market power exercised by incumbent generators in terms of capacity development restrictions.[22]

A particular example of corporate finance The project relies on corporate financing with a very high leverage of 75–25, thanks to the double hedging of the PPAs and the turnkey contracts. This allowed the owners to finance the project with only 25 percent of equity and to get 75 percent of the financing by loans at preferential rates: 2.6 percent nominal for the €1.85 billion loan from the Bayerische Landesbank during the construction before refinancing; and a credit of €0.6 billion from the French 'export' credit bank COFACE. Refinancing will be at the rate of 4.6 percent after commissioning.[23]

With the combination of a low debt cost, a high leverage, and a low return on equity, the project has a low WACC of about 5 percent. According to some studies this leads to a cost price as low as €24/MWh for the owner–operator (Tarjanne and Luostarinen, 2003).[24] This low level of cost assumes a load factor of 91 percent, a performance which is guaranteed by the turnkey contract with AREVA.

Reproducibility

As the turnkey contract is an essential pillar of the Finnish nuclear project, it represents a major unknown for the reproducibility of this model of generation cooperative with projects based on new technologies. Long-term contracting with large consumers at cost price appears to be the other main condition of success of the Finnish model and the cheap financial arrangement. There are a number of new projects which attempt to reproduce some of the key characters of the Finnish project. In June 2007 a consortium of Finnish industrial and energy companies, Fennovoima, launched a new nuclear plant project of 1,000–1,800MW for commissioning in 2016–18.[25] In April 2008, British Energy envisaged a project to be developed by a consortium with several large consumers. One issue is the time period of the contractual agreement with the nuclear producer. It is unlikely to replicate this cooperative scheme in globalized industries where prompt relocation could occur, if they are not locked-in by a determining advantage to operate in the country such as some natural resource endowment, like forestry is in Finland for pulp and paper companies. Large industrial consumers (aluminum smelters, steelworks and so on) are unlikely to be willing to commit to a long-term PPA for as long a period as 40–65 years because they face potential relocation and market risk. Moreover, in the case of a consortium which regroups suppliers to buy electricity on a long-term basis, the stability of such a consortium assumes that the regulator accepts the *entente* of these retail competitors for part of their procurement, which will not be possible above a certain total market share. The same restriction could occur if successive industrial projects develop in the same market, reducing the competitive market share on a short-term basis.

4.3 Nuclear Development by Large Vertically Integrated Firms (with reference to France)

Electricity reforms in a number of countries have not been so radical as to dramatically alter vertical and horizontal industrial structures. In many European countries, incumbent companies were allowed to retain their vertical integration between generation and supply, and were not obliged to divest some of their production units. Furthermore, over the years such companies have expanded abroad and have thereby generally increased their horizontal integration by mergers and acquisitions in other markets.

Large vertical firms which benefit from a large base of 'sticky' consumers on their home markets are in a good position to invest in large and capital-intensive equipment such as nuclear plants, because their size and vertical integration makes it possible to limit market risks and lower capital cost. Large vertical companies are therefore generally likely to benefit from better financing conditions than medium-sized vertical firms for their generation projects and *a fortiori* large independent producers.

The Flamanville 3 reactor project
The EDF's EPR project, the Flamanville 3 reactor project, was started in 2006 after a long-lasting political debate to address industrial relearning of APWR technology in view of the progressive replacement of the French nuclear fleet of 59 PWRs built in successive series in the 1970s and 1980s. The 1,650 MW EPR reactor was sold by AREVA without a turnkey contract, and EDF is its own E&C service provider, and bears the risks associated with the construction cost.

Mitigation of regulatory and political risks in the French environment   The French political and judicial environment allows strong governance which is reflected in the stability of safety regulation. The recent democratic modernization of the decision process for the siting of large industrial equipment such as nuclear plants adds some political legitimacy (Bredimas and Nuttall, 2008). That is a key advantage for limiting regulatory risks during the long lead-time of the construction of a set of reactors and for guaranteeing stability for decommissioning and waste management requirements. Moreover, the combination of EDF's large engineering capacity and the French regulatory style in nuclear safety limit regulatory risks. Indeed, being its own architect–engineer, EDF avoids the potentially costly effect of an E&C company's intermediation between the electricity company and the safety authority during the plant construction which is observed in other countries (for example, Germany and the USA). Indeed, given that the E&C payment is made on a cost-plus basis, there are few incentives to balance the requirements of the safety regulator. At the end of the process, any remaining residual regulatory risks could be borne without major problem by EDF.

Market price and CO_2 price risks   EDF's position in the French power market as the historically dominant supplier with a large segment of 'sticky' consumers, a large set of written-off nuclear assets, a diversified portfolio of activities (generation assets, large supply business, national markets diversification, and so on) allows it to manage market risks on capital-intensive nuclear investments in several advantageous ways, for example, allocation of risk to consumers and portfolio management. In particular, ownership of a large fleet of written-off nuclear plants gives EDF a stable source of cash flow. The integration of the French market with the electricity continental markets ensures that fossil fuel marginal plants set the electricity price at levels which reduce the market price risk for new nuclear plants. Ultimately in a low probability context of low gas and CO_2 prices, EDF could internally subsidize the EPR investment cost recovery through the cash flows of existing nuclear plant.

Control of construction cost and risks The size of EDF and its vertical integration allow the company to bear the construction risks and not to seek the protection of a turnkey contract with AREVA for the nuclear reactor. Beyond its large size, EDF benefits from the capability of its important engineering department which is the architect–engineer for the project and which gives it a strong bargaining power with the reactor vendor and the safety authority. EDF is also likely to benefit from AREVA's experience with the Finnish reactor. Moreover, exceptional risks (such as relinquishing the project after serious problems of misconception or a political U-turn after a nuclear accident in the world) could be borne by EDF as historical precedents (such as the cost of closure of the large EDF fast breeder reactor (FBR) demo plant SuperPhenix) tend to indicate.

Corporate financing EDF has reduced the cost of the project through an association with ENEL which finances 12.5 percent of the investment cost.[26] EDF finances all its investment needs in corporate financing and it does so for its Flamanville 3 reactor as it would for any other project. It benefits from a good credit rating, which allows it to borrow at around 5 percent. Financing large investment is, however, more costly for EDF than before the market reform and its partial privatization (15 percent of stocks were private in 2008). A government report places the new standard for return to equity at 13.7 percent nominal (IGF-CGM, 2004). With a 50/50 percent financing split, this results in a weighted average cost of capital of 9.3 percent when the cost of debt is at around 5 percent. Because of this relatively high cost of capital, the levelized cost calculated for the Flamanville EPR (€46–48/MWh) is much higher than the price charged to the members of the Finnish TVO cooperative (€25/MWh) calculated with a WACC of 5 percent, a higher level of performance, and a longer lifetime (65 years instead of 40 years).

4.4 Nuclear Investment in an Oligopoly of Medium-sized Vertical Companies (with reference to the UK and Eastern Europe)

Let us now consider other candidates for new nuclear plant investment: other large companies Suez-Electrabel, EON, RWE and ENEL on the one hand, and some medium-sized or small companies in Eastern Europe and in US liberalized markets.

The nuclear ambitions of other large companies under national restrictions

Other European companies are likely to benefit from their size and diversified portfolio when investing in nuclear power. Suez-Electrabel (merged with GDF) is attempting to install an EPR in France to be commissioned soon after 2012 and to be a partner in some projects in Eastern European countries (Romania). E.ON and RWE also plan to invest in nuclear plants in Great Britain (perhaps with the Westinghouse technology AP1000) and in different Eastern European countries (Bulgaria, Romania) in association with local producers: ENEL possibly in Italy, but also in Slovakia (with Slovenske Elektrarne, the historic company acquired in 2007, which aims to build two Russian VVER nuclear pressurized water reactors in Mochovce at a cost of €2 billion) and in Romania.[27] Given their size and the portfolio of existing assets, all of these companies have the financial capabilities to develop such projects and to benefit from vertical integration to control market risks.

However, these possible nuclear investments differ from the French EDF model in three respects. First, the legal restriction to develop nuclear plants in their home market (respectively, Belgium, Germany and Italy) constrains these companies to proceed in other markets with fewer restrictions: the necessity of acting in an institutional environment less familiar to them induces some political and regulatory risks. Second, their weak engineering capability compared to EDF might prevent them from being their own architect–engineer and thereby reducing investment costs. And third, these companies can shoulder fewer regulatory and political risks than EDF, given their less important position in their respective home markets. This explains why they mostly concentrate on partnerships with local firms.

In the perspective of the lenders, financing a large nuclear investment by medium-sized companies is most likely to be via corporate finance, provided that they have a portfolio of various assets to diversify risks, and a high degree of vertical integration to control market risks. In a more fragmented industrial structure like those in the US or the UK markets, companies are less likely to meet the risk capital requirement that would accompany the authorization of a new nuclear plant. Moreover, a large upfront cost investment financed with a large debt ratio would alter their stock value (Lacy, 2004). If 50 percent of equity capital is required in a corporate finance project, owners with a stock value of less than 10 to 20 billion dollars will need to provide about $1 billion for a 1,000 MW reactor.

As widely reported in the British debate (Nuclear White Paper, 2008) and the rising American one (as evoked by Joskow, 2006b), a clear policy is needed to stabilize the carbon value of a nuclear investment in one way or another. It is noteworthy that such a claim has not been expressed in Finland or France when the respective nuclear plants have been authorized because of a belief in the competitiveness of a new nuclear plant, even if the CO_2 price is quite low and does not give a supplementary advantage to the nuclear project.

Cost and risk sharing in a consortium of producers

Even in a competitive environment, some medium-sized companies willing to invest in nuclear technology will seek to share the construction costs and risks by establishing a consortium to order, own and operate a reactor – all the more so if they have to absorb the learning costs of a new kind of reactor. Such a consortium might be allowed by its competition authority provided that the market share of the associates is not too great. A producers' pool is envisaged in the development of some new nuclear plants in the integrated Baltic and Polish markets in association with historic suppliers. Consortiums are likely to be created between large Western European companies and local public companies in Bulgaria and Romania.

In the British market the nuclear projects announced by medium-sized companies are not exactly of this type because the companies are subsidiaries of the main European companies, for example, EDF and E.ON, with large balance sheets, good credit ratings and experience of nuclear ownership and operation. British Energy, which owns and operates the eight more recent reactors as a pure producer on the British market cannot develop projects on its own because of the financial aftermath of its quasi-bankruptcy of 2003, resulting from its nuclear specialization within a volatile market (Taylor, 2007), unless it could organize a consortium with some large industrial consumers. Several consortiums have been envisaged to share the costs and risks of nuclear projects: a consortium of E.ON-UK and RWE Npower to share the learning cost of the Westinghouse–Toshiba technology AP1000, two consortiums with British Energy which owns the available sites and, respectively, EDF-Energy and E.ON-UK to solve the problem of planning and siting. But the acquisition of a 35 percent share of British Energy by one of these European majors is a way for them to benefit from the rents of existing nuclear assets and to acquire some of the available sites which constitute very valuable assets.

4.5 The Choice between Institutional and Financing Arrangements

To conclude from these four case studies, it appears that requirements for managing specific risks associated with new nuclear build in liberalized markets with new advanced light water reactor (LWR) technologies are so important that there is a natural selection of industrial organization and institutional arrangements by which investment in nuclear plants is made possible. In fact, poorly liberalized markets without major changes in industrial structures and with the preservation of large vertical incumbents appear to be the best configuration for the development of new nuclear plants, provided that there are no political restrictions and regulatory risks as is the case in France and Eastern European countries. Reintegrated oligopolies such as the UK market follow on the ladder of industrial structures favorable to nuclear investment. Nuclear projects are likely to be promoted by energy companies first in their home market or in other markets where they have vertical subsidiaries or where they could form a consortium with local historic companies. In such cases, corporate finance would appear as the most appropriate arrangement to benefit from the strong balance sheets of medium-sized and large companies, and from their diversified portfolio of plants.

In other types of industrial organization prevailing in markets that have been comprehensively reformed, there are possibilities to develop nuclear projects in the contractual framework of a consumers' cooperative (on the model of the Finnish EPR project) or with the backing of long-term contracts at a fixed price with credible parties (historic suppliers, municipalities in particular) as in the South Texas Project case. Banks seem to be reluctant to commit to project or hybrid finance without strong complementary guarantees at this stage of industrial relearning: long-term and turnkey contracts in Finland, loan guarantees, standby insurers against regulatory risk and PPAs with credible parties in Texas.

Moreover, whereas in corporate finance companies will avoid the addition of a risk premium in the cost of debt, in a project financing, the higher construction and operating risks of nuclear would lead lenders to demand a greater return on investment of 2–3 percent, both on the debt cost and the return on equity (ROE) as a risk premium, as both the MIT and the Chicago University studies argue (Deutch and Moniz, 2003; Tolley and Jones, 2004). The resulting 'risk premium' on the weighted average cost of capital would in turn penalize nuclear against existing technologies by increasing levelized cost (by around €17 /MWh ($26/MWh) in the MIT study.[28]

Finally, despite the difference in institutional arrangements and financing structure, the cost of capital is not so very different in the different cases unless substantial support is given for new nuclear build through risk transfers to the government or regulator, given that gearing is likely to be limited to about 50/50 in project finance projects for nuclear plants (Table 6.3). In the specific cases where significant risk is transferred to governments, there is a substantial advantage to project or hybrid financing schemes which enable higher leverage and lower the global project

Table 6.3 *Different combinations of risk allocation and financing arrangements on nuclear projects in liberalized and regulated markets*

Type of reform	Decentralized market industries with IPP companies	Decentralized market industries	Liberalized industries with large vertical companies	Liberalized industries with medium-sized vertical companies
Reference case	South Texas Project	Finnish plant Olkiluoto III	French EPR Flamanville 3	UK projects US project Eastern European projects
Allocation of construction risks	On government: Standby insurance governmental loan guarantee on 80%	On vendor: Turnkey contracts	On producer	On producer consortium
Allocation of market risks on consumers	PPA with municipalities / historic suppliers	PPA with large industrial users / historic suppliers	Large base of sticky consumers	Large base of sticky consumers
Structure of financing	Project finance	Hybrid finance	Corporate finance	Corporate finance
Capital structure ratio debt/equity	70/30	75/25	50/50	50/50
WACC In nominal	9.2%*	5%	9.3%	NA

Note: *Hypothesis: Normal financing conditions equivalent to those on coal and gas generation projects with 12 percent ROE and 8 percent interest rate on debt in nominal and after tax.

cost of capital. This would be the case in the South Texas Project for which the government loans guarantee would allow a high gearing of 70/30 and a WACC of 9.2 percent, given that financial investors will not require a risk premium. Nevertheless a consumers' consortium with creditworthy participants is clearly the most favorable arrangement. It combines the possibility of borrowing at low rates, to obtain high gearing and to allow sponsors to buy the offtake directly (avoiding the need for conventional investment returns). Such a consortium would not participate in the wholesale market, as exemplified by the Finnish project. That project's capital cost (5 percent) is much lower than the capital cost of large balance sheet companies such as EDF (9.3 percent). This alters the cost price of nuclear generation, as shown in the difference of the levelized cost calculated for the Flamanville EPR (€46–48/MWh) and Finnish TVO (€25/MWh) projects.

5 CONCLUSION

The risks specific to nuclear power investment in liberalized markets – regulatory, construction, operation and market risks – can be mitigated or transferred away from the plant owner–operator through different institutional, contractual and organizational arrangements. We argue that in liberalized markets significant risk transfers from plant investors to consumers, plant vendor and government are needed to make nuclear power projects attractive to investors, and bankable for lenders. Based on four case studies, we show that there are a range of alternative consistent combinations of contractual and financial arrangements for new nuclear build. The suitability of the different alternatives depends largely on factors specific to the industrial organization of the electricity market and the institutional environment which shapes the nuclear policy in a country.

In the first phase of nuclear relearning, the likely range of viable contractual and financing arrangements appears quite limited. The most likely financing structure will be based on corporate financing or some form of hybrid arrangement backed by the balance sheet of one company or a consortium of large vertically integrated companies. In the perspective of project financing of new nuclear plants in liberalized markets, the minimal conditions are loan guarantees by government, and PPAs at a fixed price for almost all the offtake. Turnkey contracts for the FOAK reactors could also provide a guarantee during the construction phase followed by refinancing for the plant operation phase. During the first phase of nuclear relearning, banks and lenders are therefore likely to favor corporate financing by firms with a strong balance sheet, which can shoulder a greater share of the risk through a diversified asset portfolio and vertical integration.

This implies that countries where electricity reform has been partial and which have preserved industrial champions could be the most favorable for new nuclear investment. This does not exclude nuclear development in countries with a more fragmented industry, but more original models for risk pooling and/or risk transfer are likely to emerge in such countries, such as a consortium of consumers and suppliers with original arrangements to lower the cost of capital and increase leverage, as in Finland.

The four case studies presented have highlighted that there remain many critical factors specific to a country's industrial and regulatory environment, such as the reproducibility of some current innovative approaches, for example, the consortium of industrial users in Finland or the 'merchant' project in Texas backed by US federal loan guarantees. For liberalized markets there is not one optimal 'once-and-for-all' contractual and financing arrangement for investing in capital-intensive equipment with risks as specific as those facing new nuclear plant projects. The optimal combination of contractual and financing arrangements is likely to be determined on a case-by-case basis depending on the specific local industrial organization, the market position of the investing company and the institutional environment prevailing in the country.

NOTES

1. This chapter is based upon work presented at the 2nd CESSA conference held at the Judge Business School, University of Cambridge, December 13–15, 2007. The authors would like to thank William Nuttall and participants of the conference for their useful comments.
2. Under this scheme a standby insurance for regulatory delays is provided for the four first projects: $500 million for the first two and $250 million for the next two.
3. Personal communication with an AREVA manager.
4. The consortium of owners created to manage the new investment could take one of several forms – the simplest being a corporation, which is a distinct company created solely with the purpose of managing the project. Another possible form of a legal entity for sponsors is a general partnership, which operates as a distinct legal entity for contractual and financing purposes.
5. Such a consortium would also reduce the market risks. But they present different performances in terms of organizational issues to control costs and performances, financial issues and required rate of return on investment. Three consortium options for a nuclear power plant project in Texas have been compared in this sense by the University of Texas (TIACT, 2005).
6. In the terminology of modern finance, we distinguish the following categories: the developers who promote the project, the operator, the lenders, the project sponsors, that is, the parent company in simple projects, but also eventual associates in a consortium project as equity sponsors; and other interested parties as fuel vendors.
7. White (2006) shows that the gearing ratio of debt to equity for a nuclear plant is mechanically much lower than for a CCGT which has a much lower ratio of fixed to fuel costs. While the gearing ratio might easily reach 80 percent of debt for a CCGT whose fixed cost is only 20–25 percent of the total cost, we can calculate that the nuclear plant gearing ratio is no more than 50 percent, given that the investment cost is 65 percent of the total cost.

Contractual and financing arrangements for new nuclear investment 151

8. If there are two or more possible projects among which the investor can choose, the investor will get a better rate of return for a given risk, or a lower risk for a given rate of return if it holds a combination of these projects than if it holds any one on its own.
9. It is noteworthy, however, that such hedging value of nuclear plant decreases very rapidly with high degrees of correlation between power, gas, and CO_2 prices, as observed in most liberalized electricity markets, as shown in Roques et al. (2008).
10. This is one reason why a number of written-off nuclear plants in the US markets have been sold by regulated utilities to some merchant companies specialized in nuclear plant operation. These latter companies can indeed extract greater benefit from the large margins of these plants than regulated companies with cost-of-service tariffs (Lacy, 2006).
11. Bloomberg study 2005, available at: http://www.bloomberg.com/markets/rates/.
12. 'As a merchant we have to be careful, but also as a merchant the reward is at a much higher level of return compared with regulated utilities' (M. Shattuck, Constellation chairman at the Merril Lynch Power and Gas Leaders Conference in New York, September 2007).
13. The four are: the Texas power company (TXU) project of two advanced pressurized-water reator (APWR) builds, two projects of the specialized nuclear producer Constellation-Unistar with EPR projects in Calvert Cliffs in Maryland, and Exelon's project in Clinton, Illinois.
14. In the US states in which electricity markets have been liberalized, there are a variety of electricity firms: historical suppliers which have retained part of their generation assets, independent producers which sell their electricity on the power exchange and bilaterally, municipal utilities which retain their legal monopolies, and new suppliers which compete with historical suppliers in specific market segments.
15. NRG (2007), 'NRG Strategy: platform established for multiple growth opportunities', Website of NRG Energy.
16 On December 17, 2007, the US Congress voted the budget for loan guarantees to non-carbon technologies, including $18.4 billion for nuclear reactors. This means that up to 12 $3 billion nuclear projects with a debt ratio of 50 percent could benefit from this loan guarantee.
17. 'NRG CEO: Nuclear projects may fit merchant model best', September 2007.
18. In the USA, five companies have developed specialization in nuclear generation with existing assets which have been sold by utilities or acquired with the help of mergers and acquisitions in the liberalized and regulated regional markets: Florida Power and Light FPL (which owns four reactors), Constellation (Unistar) (four reactors in Maryland and New York), Dominion (six reactors), which have all experienced restructuring in their home states; and Exelon (14 reactors) and Entergy (nine reactors), which both purchased a relatively large number of plants.
19. 'NRG CEO: Nuclear projects may fit merchant model best', September 2007.
20. Personal communication with AREVA managers. In fact, since the AREVA contract with TVO is confidential, the information on this clause has never been precisely disclosed.
21. Shareholders are committed to pay TVO's fixed cost regardless of whether they take their portion of electricity produced by TVO. The variable costs are paid by the owners in accordance with the amount of electricity they have taken from TVO.
22. Interview with TVO managers in *Enjeux Les Echos*, November 2005.
23. Such low rates on loans, which have been challenged by opponents as a state aid before the European Commission, are in fact explained by the bankers' confidence in the collaterals and in the guarantee offered by the PPAs.
24. A very optimistic cost price of €16/kWh was calculated in 2003 when the decision was made. It had been calculated with a discount rate of 3 percent and a load factor of 91 percent. On present estimates, the price is €25/MWh, with a discount rate of 5 percent. In the two cases, the calculation assumes a very optimistic load factor of 91 percent guaranteed by the vendor AREVA.

25. Fennovoima comprises various industrial companies – Outokumpu, Boliden, Katterna, Rauman Energia and E.On, which is not present in Finland.
26. ENEL will benefit from equivalent drawing rights on the reactor production by paying only variable costs thereafter.
27. It is noteworthy that in 2008 Suez, RWE and ENEL competed to be partners of the state company Nuclearelectrica in the new Cernavoda projects envisaged in Romania, and RWE and EON competed to be partners of the Bulgarian public company NEK in the two VVER projects of the Belene plant.
28. The Chicago University study (Tolley and Jones, 2004) estimated that the risk premium required by lenders and equity holders (before tax) for financing new nuclear plants could be 3 percent higher than for other technologies (respectively, an ROE of 12 percent instead of 9 percent after tax, and an interest rate of 7 percent instead of 4 percent for coal and gas equipment). Similarly, the MIT study (Deutch and Moniz, 2003) assumes that merchant financing of nuclear power would require a 15 percent nominal ROE (as compared to a 12 percent nominal ROE for gas and coal), but a similar interest rate. In the two cases the WACC before tax for a nuclear project is about 3 percent higher than for a CCGT plant.

REFERENCES

Adib, P. and J. Zarnikau (2006): 'Texas: the most robust competitive market in North America', in F.P. Shioshansi and W. Pfaffenberger (eds), *Electricity Market Reform: An International Perspective*, London: Elsevier, pp. 383–418.

Bazilian, M. and F. Roques (eds) (2008): *Analytic Methods for Energy Diversity and Security: Applications of Mean Variance Portfolio Theory. A Tribute to Shimon Awerbuch*, London: Elsevier.

Bredimas, T. and W.J. Nuttall (2008): 'A comparison of international regulatory organizations and licensing procedures for new nuclear power plants', *Energy Policy*, Vol. 36, No. 4, pp. 1344–54.

Bupp, I. and J.C. Derian (1978): *Light Water: How the Nuclear Dream Dissolved?*, New York: Basic Books.

Chao, H.-P., S. Oren and R. Wilson (2008): 'Reevaluation of vertical integration and unbundling in restructured electricity markets', in F.P. Sioshansi (ed.), *Competitive Electricity Markets: Design, Implementation, and Performance*, London: Elsevier, pp. 27–65.

Delmas, M. and H. Heiman (2001): 'Government credible commitment to the French and American nuclear power industries', *Journal of Policy Analysis and Management*, Vol. 20, No. 3, pp. 433–56.

Deutch, J.M. and C. Moniz (eds) (2003): *The Future of Nuclear Power: An Interdisciplinary MIT Study*, Cambridge, MA: MIT Press.

Esty, B. (2002): 'An overview of project finance – 2002 Update', Cambridge, MA: Harvard Business School Publishing, 9-202-105.

Esty, B. (2004): 'Why study large projects?', Cambridge, MA: Harvard Business School Publishing: 9-203-031.

Finon, D. (2008): 'Investment risk allocation in decentralised electricity markets. The need of long-term contracts and vertical integration', *OPEC Energy Review*, Vol. 32, No. 2, pp. 150–83.

Finon, D. and C. Staropoli (2001): 'Institutional and technological coevolution within the French electronuclear industry', *Industry and Innovation*, Special

Issue 'Renewal of the French Model: Industry, Innovation and Institutions', Vol. 8, No. 2, pp. 179–99.

Gray, T. (2005): 'Nuclear power is poised to shake off its stigma', *The New York Times*, December 10.

Hudson, G. (2002): 'Opportunities for new plant construction: evaluating the viability of future nuclear developments', America's Nuclear Energy Symposium, Miami, FL, October 16–18.

IEA (2006): *World Energy Outlook 2006*, Paris: OECD, ch 12.

IGF–CGM (2004): 'Étude sur la formation des prix de l'électricité dans un marché concurrentiel, Inspection Générale des Finances et Conseil Général des Mines', Report to the French Ministry for the Economy, Finance and Industry, Paris.

Ismer, R. and K. Neuhoff (2006): 'Commitments through financial options: a way to facilitate compliance with climate change obligations', Cambridge University, EPRG 0625.

Joskow, P.L. (2006a): 'The future of nuclear power in the United States: economic and regulatory challenges', AEI Brookings Joint Center for Regulatory Studies Working Paper 06-25, Washington, DC.

Joskow P.L. (2006b): 'Competetive electricity markets and investment in new generating capacity', in D. Helm (ed.), *The New Energy Paradigm*, Oxford: Oxford University Press, pp. 76–122.

Lacy, B. (2004): 'Nuclear power plant and corporate financial performance in a liberalized electric energy environment', World Nuclear Association Annual Symposium, London, September 9–10.

Lacy, B. (2006): 'Nuclear investment: performance and opportunity', World Nuclear Association Annual Symposium, London, September 7–9.

Lautier, D. (2003): 'Les performances des enterprises électriques européennes', *Économies et Sociétés*, Série EN (Économie de l'Énergie), No. 9, pp. 403–32.

Levy, B. and P. Spiller (1994): 'The institutional foundations of regulatory commitment: a comparative analysis of telecommunications regulation', *Journal of Law, Economics and Organization*, Vol. 10, pp. 201–46.

Michaels, R.J. (2006): 'Vertical integration and the restructuring of the U.S. electricity industry', *Policy Analysis*, No. 572, pp. 1–31.

Miller, R. and D. Lessard (2000): *The Strategic Management of Large Engineering Projects*, Cambridge, MA: MIT Press.

NEI (Nuclear Energy Institute) (2006): 'Investment stimulus for new nuclear power plant construction', Washington, DC.

Neuhoff, K. and L. de Vries (2004): 'Insufficient incentives for investment in electricity generation', *Utilities Policy*, Vol. 4, pp. 253–68.

Newbery, D. (2003): 'Contracts for supply of nuclear power', Nuclear Issues Group, OXERA, July 10.

Nuclear White Paper (2008): 'Meeting the Energy Challenge. A White Paper on Nuclear Power', Department of Business, Enterprise and Regulatory Reform (DBERR), London, January.

OXERA (2004): Private communication on 'Project Finance and Corporate Finance Solutions to Financing Construction of New Nuclear Electricity Generation Capacity', April.

Roques, F. (2008): 'Technology choices for new entrants in liberalized markets: the value of operating flexibility and contractual arrangements', *Utilities Policy*, Vol. 16, No. 4, pp. 245–53.

Roques, F., D. Newbery and W.J. Nuttall (2008): 'Fuel mix diversification incentives in liberalized electricity markets: a mean–variance portfolio theory approach', *Energy Economics*, Vol. 30, No. 4, pp. 1831–49.

Scully Capital (2002): 'Business Case for Early Orders of New Nuclear Reactors', Report for the Department of Energy, Washington, DC, July.

Tarjanne, R. and K. Luostarin (2003): 'Competetiveness Comparison of Electricity Production Alternatives', Research Report EN N-56, Lappeenranta University of Technology, Finland.

Taylor, S. (2007): *Privatization and Financial Collapse in the Nuclear Industry: The Origins and Causes of the British Energy Crisis of 2002*, London: Routledge.

Thomas, S. (1988): *The Realities of Nuclear Power: International Economic and Regulatory Experience*, Cambridge, MA: Cambridge University Press.

TIACT (2005): 'Nuclear Power 2010 Texas Gulf Coast Nuclear Feasibility Study', Texas Institute for the Advancement of Chemical Technology.

Tolley, G. and D. Jones (2004): 'The Economic Future of Nuclear Power', A Study Conducted at the University of Chicago, available at: www.ne.doe.gov/reports/NuclIndustryStudy.pdf (accessed May 2008).

Vernimmen, P. (2000): *Finance d'entreprise*, Paris: Dalloz.

White, A. (2006): 'Financing new nuclear generation', available at: www.climatechangecapital.com (accessed May 2008).

7. Nuclear power and deregulated electricity markets: lessons from British Energy
Simon Taylor[1]

1 INTRODUCTION

The British government privatized the more modern UK nuclear power stations in the form of the company British Energy plc in 1996. The company was unusual in being a wholly nuclear merchant power generator in a deregulated power market. It was also unusual in having full financial responsibility for its back-end nuclear liabilities. The company initially raised output and profits and saw its shares rise strongly. But by 2002 it had run out of cash and had to get emergency financing from the government to avoid going into administration. The subsequent financial restructuring saw shareholders lose most of their investment.

This episode, and the contrast between the company's initial success and subsequent financial collapse, offer an interesting case study in the viability of nuclear power in a deregulated market. But the facts do not support a simple conclusion that nuclear power cannot survive in such markets. A restructured British Energy Group plc was re-listed on the London Stock Exchange in 2005 and continues to trade, albeit with a lot of volatility owing to unreliable power station availability.

A detailed examination of the British Energy story suggests that the roots of the crisis were complex and historically deep (Taylor, 2007). The management had to contend with a unique type of technology and with fixed-price contracts for fuel reprocessing arising from government decisions taken decades before. The company distributed cash to shareholders which, with hindsight, was unwise and reflected a general misunderstanding of the riskiness of the company. The company's corporate strategy – to vertically integrate as a hedge against falling power prices – was sensible but badly executed. And the company's overall management of risk seems inadequate.

But in this author's opinion, none of this amounts to an indictment of nuclear power's ability to survive in liberalized markets. The rest of this chapter argues that the events at British Energy were historically unusual and to a large extent specific. It goes further in suggesting that the various risks associated with running a privately owned nuclear power generator in a liberalized market are not unique to nuclear power and that similar risks are routinely handled in other industries and markets without state intervention.

The structure of the chapter is as follows. Section 2 describes the events leading up to the financial crisis in 2002 in more detail. Section 3 then examines the proximate cause of the collapse, the fall in wholesale power prices from 1999 to 2002. Section 4 examines what is distinctive about nuclear power generation compared with fossil generation. Section 5 analyses what liberalization means for power markets. Section 6 brings these points together to suggest what a nuclear power company should logically do in a liberalized market. In Section 7, we compare the a priori analysis with British Energy's actual decisions, to show where and why the company became vulnerable to the power price fall. Section 8 looks further at the underlying risks in a privatized nuclear generator and argues that all are routinely handled in other privately owned industries. Section 9 then concludes.

2 NARRATIVE OF EVENTS

After an initial failed attempt to privatize nuclear power with the rest of the British electricity industry in 1990, the government put the nuclear stations into two state-owned companies, Nuclear Electric for the English and Welsh stations, and Scottish Nuclear for the Scottish stations. In 1995 the more modern advanced gas cooled reactor (AGR) stations plus the new pressurized water reactor (PWR) at Sizewell were privatized in the form of a new company, British Energy plc. The older Magnox reactors were retained in a company called Magnox Electric.

Figure 7.1 shows the share price of the company from its initial listing in June 1996 at a price of £2.03, to a peak of £7.33 in early 1999 and then a decline to less than £1 after the company sought government financial help in September 2002.

After a controversial sale and initially poor share price performance, the company became highly regarded on the back of strong cashflow generation and profit growth. By 1999 the company was able to pay back £432 million to shareholders, about 10 percent of its market capitalization. But then the company's profitability declined on the back of lower power prices and increasingly unreliable station operating performance. After the management failed to get a sufficient cut in reprocessing costs from British

Figure 7.1 British Energy share price, 1996–2003 (£, current prices)

Nuclear Fuels Limited (BNFL) the board concluded on September 5, 2002 that the company needed emergency financial support to keep operating. The government then provided a loan of £450 million and became the senior creditor in a financial restructuring of the company, leading to a debt for equity swap. The new company was listed on the London Stock Exchange in January 2005.

3 THE PROXIMATE CAUSE: POWER PRICES

The 'obvious' cause of British Energy's financial crisis was the fall in wholesale electricity prices which began in 2000 and continued to mid-2003 (Figure 7.2). Prices fell from around £22/MWh to about £17/MWh, or about a quarter. The British Energy goal was to break even at a price of £16. By the autumn of 2002 the company was making accounting and cash losses and facing an imminent loss of investment grade credit rating. The immediate need for government funding was to allow the company to post collateral in the electricity trading market, without which it could not sell its power.

Other electricity companies suffered badly from the power price collapse. The US electricity companies AES and Edison International both

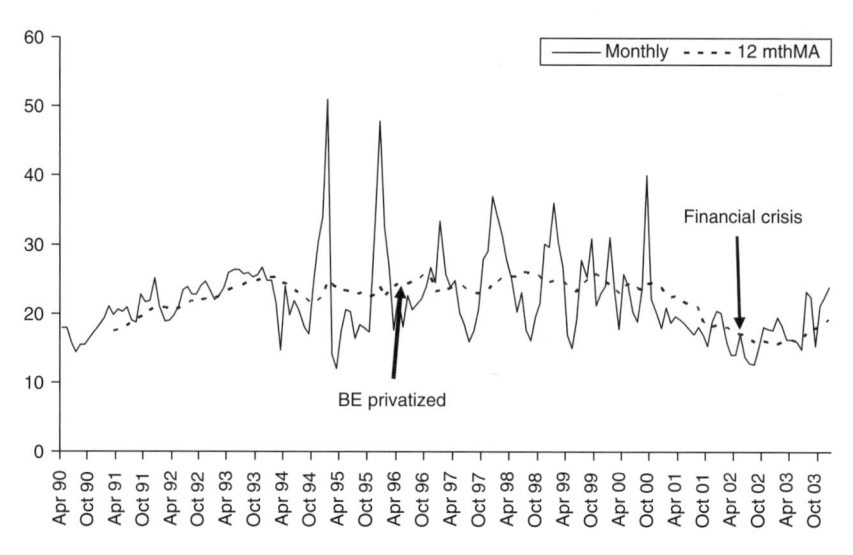

Figure 7.2 England and Wales: spot power prices, 1990–2003 (£/MWh, current prices)

Sources: Pool; Datastream; UKPX.

lost substantial amounts on coal power station investments and the company TXU Europe (a subsidiary of Texas Utilities) went into liquidation in 2002.

4 WHAT'S DISTINCTIVE ABOUT NUCLEAR?

The key economic points about nuclear power generation compared with fossil generation are shown in Table 7.1.

Compared with conventional thermal generation, nuclear plants typically have much higher fixed costs and lower marginal costs. A nuclear plant requires around 10 times as much capital investment as a combined cycle gas turbine plant (Roques et al., 2005). This means that they have an economic incentive to run at maximum load, that is, baseload (Pouret and Nuttall, 2007). It also means that small changes in the selling price lead to magnified changes in profits, known as 'high operating leverage'.

The other distinctive physical feature of nuclear plants is that they produce waste products with very long lives and requiring costs lasting decades or more for treatment, storage and disposal. In the UK this physical feature has important economic consequences, because nuclear

Table 7.1 Characteristics of nuclear generation

Characteristics	Implications for management	Relevance to British Energy	Relevance to other nuclear operators
High fixed, low marginal costs	High operating leverage* Run at baseload	Highly relevant	Highly relevant
Large deferred liability costs	High financial leverage Financial complexity	Highly relevant	Less relevant in US because government has responsibility for waste fuel

Note: * Extent to which a change in sales causes a change in operating profits.

companies are required to account for and pay for these waste treatment processes. In the US the federal government takes physical and economic responsibility for these costs in exchange for a fixed 0.1 c/kWh levy on output, which is a normal operating cost. By contrast, British Energy must provide for future waste storage and disposal costs, which lie in the future, representing a form of non-interest-bearing debt. In both countries the nuclear generator is responsible for the costs of decommissioning the power stations. The combined effect of spent fuel and decommissioning liabilities is that a nuclear company like British Energy has significant financial leverage, even if it has no interest-bearing debt.

High operating leverage and high financial leverage combine to make a company's net cashflows to investors more risky than average (implying a high beta in a capital asset pricing model framework). British Energy should therefore have been regarded from the start as an intrinsically high-risk company, unlike the monopoly utility companies that also traded on the London Stock Exchange.

5 WHAT DO LIBERALIZED POWER MARKETS MEAN?

The UK wholesale electricity market was liberalized when the industry was privatized in 1990–91. By abolishing barriers to entry in generation and by allowing first large customers (above 1 MW demand, from 1990) and then medium customers (above 100 kW demand, from 1994) freedom to choose supplier, the government allowed the electricity market to function more or less like other commodity markets. Generation ownership

Table 7.2 Characteristics of selected commodities

Commodity	Quality variation	Cyclicality	Seasonality	Derivative markets exist?	Distinctive features
Oil	By sulfur	Moderate	Yes (US driving season)	Extensive	Slow but cheap to move
Corn	Standard categories	Low	High	Futures	Slow but cheap to move
PVC	No	High	Low	Limited	Expensive to move
Coal	Sulfur, energy content	Moderate	Moderate	Limited	Regional rather than global market
Natural gas	No	Low	High	Extensive	Regional rather than global market
Electricity	No	Low	Very high	Limited but growing	Non-storable, limited international trading

remained highly concentrated until the two main incumbents, National Power and PowerGen, sold much of their coal plant at the end of the 1990s so competition was initially muted. But by the time British Energy was privatized in 1996, electricity was a substantially liberalized market.

A 'commodity' is something of a homogeneous, well-defined quality that is demanded by customers, normally for transformation into something of higher value. The traditional commodities are agricultural (soy beans, corn, orange juice) or industrial (coal, copper, zinc, oil). Gas and electricity are also commodities but this was less clear because they were typically not traded in competitive markets until relatively recently. Commodities and commodity markets have the characteristics shown in Table 7.2.

Electricity shares the key commodity features that it is a homogeneous, undifferentiated product of well-defined quality. Demand is less cyclical (that is, related to GDP fluctuations) than for industrial commodities such as metals and petrochemicals. But electricity demand is highly seasonal with very inelastic demand. When combined with the impossibility of large-scale storage, this makes electricity prone to very volatile short-term prices in a competitive market.

Commodity prices are typically volatile both intra-year and over several years, reflecting shifting demand and supply curves. For industrial commodities there is a traditional 'cycle' of interaction between GDP-driven demand fluctuations and lags in supply which leads to pronounced boom–bust pricing variations. This is especially true in industries with a high minimum efficient scale (MES) of capacity such as oil refining and petrochemicals, where new plant may add materially to industry supply, leading to a big fall in prices.

Electricity has a relatively low MES of capacity, especially since the advent of combined cycle gas turbines which are viable at levels of 250 MW (for example, compared with total UK installed capacity of around 78,000 MW (National Grid, 2007)). Annual demand variation is also much lower than for industrial commodities since a large part of demand is relatively insensitive to the state of the economy (heating, lighting, domestic use).

But the electricity market was very new in the mid-1990s and it is not at all clear that policy makers or the key market participants had adjusted to thinking of electricity as a commodity.

Commodity markets bring pricing risk for buyers and sellers. Well-established commodity markets have evolved futures markets and sometimes options markets too, in response to the demand for risk management. Sellers of corn or orange juice can hedge their positions efficiently using futures contracts. Similar markets have evolved for oil and some petrochemicals markets and now for natural gas. In electricity these markets have been slower to evolve, partly because of the limited physical integration of networks which has kept markets relatively small. In the UK, the concentration of ownership of generation undermined the scope for derivatives markets through the 1990s, so that the nascent electricity forward agreement (EFA) market only achieved low volumes (Herguera, 2000).

6 IMPLICATIONS FOR NUCLEAR POWER: INDICATED STRATEGIES

Nuclear power is commercially more exposed to commodity price risk because it has high fixed costs. If nuclear liabilities are regarded (as they should be) as de facto debt, then British Energy (BE) also had high financial leverage. This made the company's profitability highly sensitive to the price of power.

Short-term price volatility can be dealt with easily through contracts.

Most large buyers and sellers of power in the UK in the 1990s bought on contracts of one-year duration. This left the exposure to longer cycle variations in price. Demand for power is relatively non-cyclical, which leaves supply (capacity) variation as the main cause of long-term price variation.

Given the inherent commercial risk of nuclear power in a liberalized market there are a number of logical strategies for managing that risk. Risk cannot be reduced to zero and if it is costly to manage it then the optimal amount of hedging from a shareholder's point of view is not necessarily high. The main arguments for some hedging are the costs of financial distress and convexities in the tax system. Table 7.3 shows the options for a nuclear generator in a liberalized market.

7 BRITISH ENERGY'S APPROACH TO THESE STRATEGIES

Table 7.3 also shows BE's actions in relation to the range of options available for a nuclear generator managing risk in a liberalized market. The overall verdict must be that the company failed to execute a vertical integration strategy, tried but failed to implement commercial risk management (owing to the lack of demand) and pursued the wrong financial strategy. The upshot was that the company was very badly positioned to cope with the fall in power prices from 1999 and therefore ran into financial crisis in September 2002.

The crisis was made more likely by the existence of the long-term fuel reprocessing contracts with BNFL, which added to the company's fixed costs. But the company's corporate strategy made things worse too by adding to the company's exposure to the electricity price by: (i) buying a coal power station in 1999; and (ii) buying a portfolio of power offtake contracts with the acquisition of the Swalec supply business in the same year (Taylor, 2007, p. 110).

The fact that BE failed financially in 2002 reflects its financial and corporate decisions, not the inherent risks of nuclear power in a liberalized market.

8 NUCLEAR RISKS EXAMINED

The risks of a privately owned nuclear power generator are decomposed into categories in Table 7.4, which gives examples of other industries and markets that manage very similar risks.

Table 7.3 Risk management strategies for nuclear generation in a liberalized market

Activities	Comment	BE's actions
1. Corporate strategy		
Vertical integration	Questionable: downstream assets have intrinsic value; risk of overpaying	Failed
Diversified generation	Questionable: investors can diversify the risk themselves	Costly acquisition of coal station (Eggborough)
2. Commercial strategy		
Sell power on long-term contracts	Depends on demand existing	Limited success: lack of demand
Sell options to raise value of commodity power	Depends on demand and/or markets existing	Limited success: lack of demand
Maximize reliability of stations, back-up power sources or contractual equivalents		Underinvestment; reliability fell
3. Financial strategy		
Maintain strong balance sheet		Paid out too much cash in 1999
Have variable dividend policy or share buybacks (like steel)		Wrong dividend policy
Choose long-term debt to avoid liquidity crunches		Failed/bad luck: attempt to refinance bond in early 2002 hit by Enron fallout

Source: Taylor (2007).

Table 7.4 Component risks in nuclear power generation

Risk type	Other industries experiencing similar risk
Commodity price volatility	Steel, petrochemicals, oil, banks
Operations risk	Manufacturing, process industries
Very long-term liabilities	Extractive industries
Third-party accident risk	Chemicals
Political, litigation & regulatory risk	Oil, banks, tobacco
Catastrophe risk	None

The only type of risk that is unique to nuclear power is the risk of a catastrophe such as the Chernobyl disaster of 1996. The potential third-party liability of such events is so high that such risks are uninsurable in normal markets. The US introduced government insurance of nuclear plants with the Price–Anderson Act of 1957 (Rothwell, 2002). In the UK, nuclear operators' liability is capped under the provision of the Nuclear Installations Act 1965 and the Energy Act 1983, which implement the international convention on third-party liability signed in Paris in 1960 and Brussels in 1966 (OECD, 2003).

This means that nuclear generation, even in a liberalized market, does not present any new form of risk management beyond those already used in other industries, in the private sector.

9 CONCLUSIONS

The financial collapse of the nuclear generator British Energy plc in 2002 does not 'prove' that nuclear power is unworkable in a liberalized power market. The combination is certainly risky, mainly nuclear generation combines high operational leverage with (in the UK context at least) high financial leverage arising from the long-term liabilities. Liberalized power markets behave much like other commodity markets and the price volatility is a big challenge for risk management.

But none of the risks in nuclear power is unique, except for the catastrophe risk which is automatically borne by governments under international treaty. BE mismanaged its risks, resulting in costly financial restructuring, but this should not be taken as evidence against nuclear power more generally. The 'new' British Energy company, floated on the London Stock Exchange in 2006, has a much more appropriate financial strategy (chiefly low leverage and a variable payout policy) and is paying due attention to the operational risks of the aging British reactors. Investors understand the company better and the shares, while volatile, trade successfully like any other power company.

NOTE

1. I am grateful for comments from an anonymous referee. All remaining errors are mine.

REFERENCES

Herguera, I. (2000): 'Bilateral contracts and the spot market for electricity: some observations on the British and the NordPool experiences', *Energy Policy*, Vol. 9, No. 2, pp. 73–80.

National Grid Group (2007): *Seven Year Statement 2007*, London.

Organisation for Economic Co-operation and Development (OECD) (2003): 'Nuclear Legislation in OECD Countries – Regulatory and Institutional Framework for Nuclear Activities – UK', available at: http://www.nea.fr/html/law/legislation/uk.html (accessed April 2007).

Pouret, L. and W.J. Nuttall (2007): 'Can nuclear power be flexible?', EPRG Working Paper 07/10, Electricity Policy Research Group/University of Cambridge.

Roques, F., W.J. Nuttall, D. Newbery and R. Neufville (2005): 'Nuclear power: a hedge against uncertain gas and carbon prices?', EPRG Working Paper 05/09, Electricity Policy Research Group/University of Cambridge.

Rothwell, G. (2002): 'Does the US subsidize nuclear power insurance?', Policy Brief, Stanford Institute for Economic Policy Research.

Taylor, S. (2007): *Privatisation and Financial Collapse in the Nuclear Industry: The Origins and Causes of the British Energy Crisis of 2002*, Abingdon: Routledge.

8. Nuclear energy in the enlarged European Union
William J. Nuttall[1]

1 INTRODUCTION: 50 YEARS OF NUCLEAR POWER IN THE EU

Nuclear energy has a special place in the history of the European Union. Concerns for European collaboration on nuclear energy matters was one of the founding motivations of the European project. Specifically, in April 1956, following the 1954 failure of the European Defence Community, an international committee, under the Presidency of P.H. Spaak, the Belgian Minister for Foreign Affairs proposed:

- the creation of a general common market; and
- the creation of an atomic energy community.

These in turn became the two 'Treaties of Rome' signed in March 1957.

The first treaty established the European Economic Community (EEC) and the second the European Atomic Energy Community, better known as 'Euratom'. These two treaties entered into force on January 1, 1958. The EEC Treaty has been modified numerous times, most recently with the Lisbon Treaty ratified by the 27 member states of what is today known as the European Union (EU).

The absence of amendments to the Euratom Treaty, in contrast to the decades of haggling and deal-making surrounding the EEC amending treaties, should not be taken as an indication that all EU member states have a common opinion on nuclear energy matters. While the EEC treaty, and its amending treaties, have moved incrementally towards the aim of 'ever closer union', the Euratom framework has moved forward much more slowly. The individual member states, rather than agreeing on all things nuclear, have taken a broad range of occasionally almost irreconcilable positions on what has become a most politically contentious energy technology. One area of progress, however, has been in the area of nuclear installation safety and radioactive waste management. Fernando de Esteban has explained:

When the authors of the European Atomic Energy Community drafted the EURATOM Treaty, thoughts of nuclear installation safety and radioactive waste were not uppermost in their minds. For several years there was no Community activity directly dealing with nuclear installation safety. It was not until 1975 that the Community woke up to the seriousness of the issue. By then, nuclear power programmes in its then Member States had progressed and diverged along very different routes. Moreover, not only were many of the installations very different, but the national systems regulating them were also very different.

Furthermore:

> [A]s a result of co-operation between the main actors in the EU since the 1970s, there is a 'non-binding acquis' that is built on fundamental common principles. These form the basis of all the EU national nuclear safety regulations. (de Esteban, 2002)

This chapter explores issues of nuclear energy policy in the particular context of EU enlargement.

Climate change is a global threat. The bulk of its impacts occur outside the EU and the EU is only partly responsible for the anthropogenic harm caused by greenhouse gas emissions. Also, the whole European Union faces growing fossil fuel import dependency and near total uranium import dependency. Both these major drivers of energy policy affect the EU as a whole and involve important factors external to the EU. Given the external nature of the issues, it might seem sensible for the EU to seek to shape the fuel mix at a European level. Such a policy would improve economies of scale in research and deployment of new technologies, reduce the need for duplicative and wasteful policy development at the member state level and ease the development of a single European market in energy products and services. There is no prospect, however, that this will happen and the reasons are political. Notwithstanding notions of liberal European electricity market, the fuel mix remains a sovereign matter reserved for each member state to develop as it sees fit, subject only to the constraint that it should be respectful of the concerns of neighboring states. Arguably recent European binding commitments on renewable energy and biofuels erode the notion that the fuel mix is a national concern, but it is for the issue of nuclear energy where the desire to protect national discretion is most strongly expressed. Interestingly, and perhaps even somewhat paradoxically, those states that are usually most strongly Euro-Federalist on other aspects of policy (for example, Germany) are among the first to defend notions of 'subsidiarity' on matters relating to nuclear power and the fuel mix for electricity (European Energy Forum, 2006).

Table 8.1 summarizes the current situation for those EU member states that have ever operated a commercial nuclear power station.[2] Only one

Table 8.1 Nuclear stations in EU-27

Country	Power reactors operating May 2008[2]	Power reactors building & planned May 2008[2]	Closed by end 2007[3]	First kWh[1]	GWh 2007[2]	% of electricity generation 2007[2]
Belgium	7	0	1	1962	46.0	54.0
Bulgaria	2	2	2	1980[3]	13.7	32.0
Czech Republic	6	0	0	1985	24.6	30.3
Finland	4	1	0	1977	22.5	29.0
France	59	1	11	1959	420.1	77.0
Germany	17	0	17	1961	133.2	26.0
Hungary	4	0	0	1982	13.9	37.0
Italy	0	0	4	1963	0.0	0.0
Lithuania	1	0	1	1983	9.1	64.4
Netherlands	1	0	1	1968	4.0	4.1
Romania	2	2	0	1996[3]	7.1	13.0
Slovakia	5	2	1	1972	14.2	54.0
Slovenia	1	0	0	1981	5.4	42.0
Spain	8	0	1	1968	52.7	17.4
Sweden	10	0	3	1964	64.3	46.0
UK	19	0	25	1957	57.5	15.0

Sources:
1. Anthony Froggatt, 'Nuclear power the European dimension', in *Nuclear or Not?*, edited by D. Elliot, Palgrave (2006) except Bulgaria and Romania.
2. World Nuclear Association (WNA) http://www.world-nuclear.org/info/reactors.html and 3 except Italy.
3. Relevant WNA country briefings: http://www.world-nuclear.org/info/info.html#countries.

country (Italy) has actually eliminated nuclear energy from its electricity system, although several have at various times put forward policies for a nuclear power moratorium or phase-out (for example, Belgium, Germany, the Netherlands, Spain, and Sweden) (de Esteban, 2002). Recently with the return to power of Silvio Berlusconi, Italy has renewed its interest in nuclear energy.

2 NUCLEAR POWER IN THE EU-15

It is not the purpose of this chapter to seek to review the entire history of nuclear energy in the European Union, nor is it appropriate to attempt to address all of the drivers of past and present European energy policy. Rather it is perhaps sufficient to mention:

- The UK and France were the first countries to develop civil nuclear energy in Europe building upon their separate experiences with gas cooled reactors devoted to military plutonium production. In the 1960s France altered its technology policy to favor pressurized water reactors while the UK did not make an equivalent policy choice until 1979 with policy implementation spanning the 1980s. France and the UK are the only EU-15 countries ever to have been nuclear weapons states and both states continue to maintain nuclear weapons capacity.
- In the 1970s issues of nuclear waste became prominent and in some EU-15 states (notably the UK and Germany) policy progress on the expansion of nuclear energy became linked to a perceived need to resolve the waste question.[3] Waste then assumed a special significance in the wider policy debate surrounding nuclear energy.
- Following the severe accident at the Three Mile Island plant, Harrisburg PA, USA in 1979 and the disaster in Chernobyl, Ukraine in 1986, some European countries including Germany, Sweden and Italy established policies for nuclear phase-out although only in the Italian case was this policy taken to completion. Sweden is uprating nuclear power plants at Ringhals and Oskarshamn and this will offset the loss of capacity caused by the closure and decommissioning of Barseback units 1 and 2 (WNA-Sweden). There is much political discussion in Germany concerning life extensions of existing nuclear power plants.
- Some countries, such as Ireland, having had initial ambitions for nuclear power, have since moved to exclude the option formally from policy consideration. In Ireland, in 2008, the relevant minister is required to approve all new power stations under the Electricity Regulation Act of 1999, but he or she is barred by statute from granting such permission to a nuclear fission-based power plant (Ireland, 1999).
- Finland is unusual among the EU-15 in having spent the Cold War looking both west and east seeking to maintain balanced relations with both sides. Austria arguably adopted a similar approach although its western leanings were more obvious. While Austria resolutely avoided nuclear energy, Finland adopted nuclear power using technologies drawn from the west (from Sweden, and deployed at Olkiluoto) and the east (from the USSR, and deployed at Loviisa).
- In the 1970s the UK and the Netherlands developed indigenous offshore natural gas resources whereas in contrast France lacks

Table 8.2 *EU-15 member state opinion concerning nuclear energy in 2006 – author's assessment*

Strongly positive	Weakly positive	Neutral	Weakly negative	Strongly negative
Finland	UK	Luxembourg	Italy	Ireland
France	Netherlands	Denmark	Germany	Austria
	Spain		Sweden	
	Portugal		Belgium	
			Greece	

significant fossil fuel assets. Partly as a consequence of the oil shocks of the early 1970s, France moved heavily into nuclear energy such that today roughly three-quarters of France's electricity is supplied from nuclear energy, with the balance mostly being supplied from hydroelectricity sourced in mountainous regions.

These differing national experiences across the EU-15 states resulted in a remarkably balanced range of national opinions on nuclear energy ranging from the enthusiastic (for example, France) to the clearly hostile (for example, Austria). This balance of opinion is summarized in Table 8.2 as assessed by the author. Noting significant movement towards nuclear energy in the last two years, the table is perhaps best regarded as presenting the situation pertaining in the year 2006. Key criteria used to establish a given member state's position in the table include formal current government policy, the extent to which policy is a consensus across major political parties, the level of acquiescence and public acceptance of policy and the scale of operating infrastructures, such as power plants and/or research reactors.

Table 8.2 confirms the impression that in 2006 the EU-15 states were almost exactly balanced in their opinion of nuclear energy. One must concede that since 2006 some countries have become more pro-nuclear, for example, Italy and the UK, but generally opinion is still finely balanced in 2008 with roughly half the EU-15 states uncomfortable with the prospect of nuclear new build.

3 NUCLEAR ENERGY IN THE NEW MEMBER STATES

Elsewhere in this volume we seek to explore the range of energy security issues facing EU member states in the early twenty-first century. While

Table 8.3   EU-12 newest member states' opinions concerning nuclear energy – author's own analysis

Strongly positive	Weakly positive	Neutral	Weakly negative	Strongly negative
Lithuania	Poland	Malta	–	–
Romania	Latvia	Cyprus		
Bulgaria	Estonia			
Czech Republic	Slovenia			
Hungary				
Slovakia				

there are numerous points of comfort concerning EU energy security as a whole, there are notable differences between on the one hand, Western Europe, and the other, Central and Eastern Europe. Perhaps simplistically, one might argue that energy security is best assured by those energy systems that make use of a wide diversity of fuel types, drawn from a wide diversity of sources, via diverse transit routes and open to a plurality of trading opportunities. While fuels and electricity in Western European countries, such as the UK, tend to measure up well, most Central and Eastern European countries have weaker grounds for comfort. In time the energy security jeopardy faced by these countries can be alleviated via the improvement of transmission infrastructures. However, at present the logistics of the energy supply chain is far from ideal and there is much reliance, and even more perceived reliance, on natural gas effectively controlled by Russian state-controlled corporations. The end of the Cold War in 1991, less than 20 years ago, motivates a high level of distrust of Russia in several of the new EU member states, and the 2006 gas crisis in Ukraine and the 2008 Georgia crisis have done nothing to improve trust of the Russian Federation. These factors coupled with high global fossil fuel prices, increasing concern for global climate change (with its associated and tradable EU greenhouse gas emission reduction targets) and strong electricity demand growth have in several cases motivated significant interest in an expansion of nuclear energy.

Table 8.3 presents an impression of policy opinion in the 12 newest members of the European Union as assessed by the author using the same criteria as developed for Table 8.2. Table 8.3 shows that over time, public opinion has become more positive about nuclear energy. Comparing Tables 8.3 and 8.2 it becomes clear that the balance of opinion within the EU towards nuclear has now shifted dramatically. Furthermore, in order to understand the position of nuclear energy within the EU it is essential to consider the experience and opinions of member states in Central and Eastern Europe at least as much as those in Western Europe and perhaps even, in some respects, more so.

Rather than seek to consider nuclear energy policy in each of the 12 most recent members of the EU we shall devote the rest of this chapter to considering two specific examples, Romania and Lithuania, which between them illustrate some of the most resonant and provocative policy insights.

These case studies draw upon insights gained from CESSA-funded research visits to Romania (June 2008) and the Krynica Economic Forum in Poland (September 2008). This chapter represents merely the first in a planned series of research publications relating to nuclear energy in this most interesting of regions. In time it is hoped that it might be possible to complement the Romanian research visit with a similar visit to Lithuania. In the absence of such experience, the Lithuanian case has been informed by numerous helpful interactions with colleagues from the Lithuanian nuclear industry.

4 EXTENDED CASE STUDIES

In this section we consider in greater detail two topical case studies concerning nuclear power in recent member states of the European Union. The examples are chosen in part because they represent extremes concerning the relationship with Russia during the Cold War. The first example considered is Romania, which as we shall see, pursued a policy of national independence including significant distance from the policies of its ally – the Soviet Union. The other example will be the Baltic states, with particular emphasis on Lithuania. These states are noteworthy because they are the sole examples of former territories of the Soviet Union now in membership of the EU. These differing histories concerning the relationship with Russia yield very different nuclear infrastructures and also continue to shape both current and future energy policy choices.

4.1 Romania

Romania occupies a special place in twentieth-century European history. The early twentieth century was characterized by shifts of geopolitical allegiance and territorial gains and losses. At the close of the Second World War Romania fell under the influence of the Soviet Union. However by 1958 the departure of Soviet troops had been agreed and the country had a new leader, Nicolae Ceaușescu. Preserving communism, he ushered in an extended period of national independence verging on autarky. The Ceaușescu regime relied on a Stalinist authoritarianism for power, although it must be acknowledged that some of the worst human rights abuses occurred in the immediate post-war years before Ceaușescu's rule. Through the 1970s and 1980s authoritarianism became blended with a growing cult of personality reminiscent of that in North Korea for Kim Il Sung. Ceaușescu knew and admired North Korea, having visited in June 1971. The influence of the North Korean conception of socialism on Ceaușescu and his policies for Romania has been summarized and explored by Adam Tolnay (Tolnay, 2002). Through the 1980s the economy deteriorated, partly as a consequence of Ceaușescu's isolationist drive to repay international debt (Turnock, 2007, p. 33). In 1989, communist regimes fell across Central and Eastern Europe. At the very end of the year, political tensions boiled over in Romania and a violent revolution occurred culminating in the execution of Ceaușescu and his prominent, and widely disliked, wife Elena. While characterized as a popular revolution, it is noteworthy that in the years since the revolution of 1989 Romanian politics has repeatedly featured the figure of Ion Iliescu, once a close colleague of Ceaușescu and one of the small group that travelled to North Korea in the summer of 1971 (Tolnay, 2002). Key to Iliescu's position was the use of miners to break up anti-government protests, particularly in 1990. These aggressive interventions, known as 'mineriads', remain controversial to this day. Nevertheless in the twenty-first century Romania has emerged as a functioning democracy. Iliescu was defeated in genuine elections in 1996, returning to power in the elections of 2000. In 2004, Romania became a full member of both the European Union and NATO. Since December 2004[4] Romania has been led by the anti-communist former mayor of Bucharest, Traian Băsescu.

Ceaușescu's grandiose and foolhardy ambitions left Romania with an unbalanced legacy of industrial infrastructure. David Turnock notes of the1980s:

> Economic policies became more irrational through the 'gigantism' of excessive capacities in oil refining, petrochemistry and steel production based on raw material imports . . . that were not recouped through the value of exports. (Turnock, 2007, p. 33)

In electricity coal was favored, especially lignite. Some regional use was made of natural gas while oil was prioritized for petrochemicals (ibid., p. 59). At the time of the fall of communism, electricity distribution in

Romania was very poor, with 50 percent losses reported (ibid., p. 107). Blackouts were a significant feature of Romanian life in the 1980s.

Since the end of the 1990s, Romania has been a net exporter of electricity with an overcapacity in transmission (Diaconu et al., 2008). Roughly two-thirds of Romanian electricity is carried on the national grid operated by Transelectrica, which operates to good standards of reliability (ibid.).

Ceaușescu's desire for autarky was realized in the national vision for nuclear energy. On nuclear matters, as with much else, Romania sought to increase its distance from the Soviet Union in the 1960s (Turnock, 2007, p. 59). Romania chose to partner with Atomic Energy Canada Ltd and develop a fleet of CANDU-6 natural uranium fueled heavy water cooled and moderated reactors at Cernavoda in the south-east of the country roughly 50 km from the Black Sea port of Constanța. The choice of CANDU-6 nuclear power technology suited the development of a wholly indigenous nuclear fuel cycle. The chosen approach involved uranium mining at a range of sites around the country. Romania mines uranium through the activities of Uranium National Company s.a. (UNC) at Crucea and Botusana in the north of the country.[5] Together these mines comprise UNC's Suceava center. At present these activities are undergoing modernization. Uranium milling and processing is undertaken at UNC's Feldioara Branch in the center of the country. This facility yields sinterable UO_2 powder. It is worth stressing that the CANDU-6 nuclear power system does not require enriched fuel and so enrichment activities do not form part of the Romanian nuclear fuel cycle. Sintered UO_2 fuel pellets are produced in Pitesti, 80 km north-west of Bucharest, at the Fabrica de Combustibil Nuclear (FCN) (Nuclear Fuel Factory) part of Nuclearelectrica, a majority government-owned nuclear energy company.[6] Also in Pitesti is the headquarters of the Sucursala Cercetari Nucleare (SCN) (Institute for Nuclear Research).[7] SCN has a large range of research and production facilities including research reactors and hot cells. The Pitesti facilities of Nuclearelectrica and SCN produce qualified CANDU-6 fuel ready for use in the power stations at Cernavoda.

As part of the Ceaușescu's vision of self-reliance, Romania also developed perhaps the most technologically demanding aspect of a CANDU-6 fuel cycle: the production of heavy water for reactor moderation and cooling. This activity is undertaken by the Romag-Prod facility in the south-west of the country. The Romag-Prod facility is a key part of the Regia Autonoma Pentru Activitati Nucleare (RAAN) (Romanian Authority for Nuclear Activities) of the Ministry of Economy and Finance.[8]

Heavy water is barely consumed during the operation of CANDU-6 reactors. Once produced in sufficient quantities for each power station, little or no additional heavy water will ever be required. It is expected that Romag-Prod will soon have produced enough heavy water for the four CANDU-6 plants expected to comprise the completed Cernavoda project. Once this task is done this aging and expensive-to-operate infrastructure will be closed and decommissioned (Bucur, 2008). While most of the heavy water has already been produced for plants Cernavoda 3 and 4 (plants 1 and 2 are already operating) it might be preferable to shut the Romag-Prod facility early and to obtain the balance of heavy water required internationally (Sandulescu, 2008).

While the CANDU-6 fuel cycle is relatively simple, in that it does not require uranium enrichment, it suffers from the production of relatively large volumes of spent fuel waste. At present no final decision concerning long-term spent fuel management has been made. In particular, no final decision has been made concerning the site and specification of a radioactive waste repository. This situation is typical within Europe, with only a very few countries having made concrete progress on this issue. Reprocessing is not on the Romanian agenda although it is worth noting the Romania could, in future, enter into a reprocessing-based fuel cycle via the use of international reprocessing contracts, without needing to invest in domestic infrastructures (ibid.).

Concerning Ceaușescu's conception of the nuclear fuel cycle it is worth noting reports that prior to 1990 Romania did undertake some, at the time undeclared, research into plutonium separation, producing minute quantities of this nuclear weapons proliferation sensitive material. Romania, however, never developed a nuclear weapon and it is a signatory of the Nuclear Non-Proliferation Treaty (FAS, 2008).

At the heart of Romania's nuclear energy activities is the Cernavoda power plant complex operated by Nuclearelectrica. In the 1980s it was planned that there would be five plants, Cernavoda 1–5, to be built concurrently. Unit 5 was something of an afterthought rumored to have been forced onto the agenda by Ceaușescu himself despite the site being poorly suited for a fifth plant. While the locations for units 1–4 form a neat line beside the Danube River, the site of the fifth (part-built) plant is slightly out of line because of insufficient solid limestone foundation at that end of the Cernavoda site. To compensate for this geological difficulty very large amounts of concrete were injected to form a solid foundation for the Cernavoda 5 plant. Of the four plants part-built at the time of the 1989 revolution, the fifth plant was the least complete (at only roughly 4 percent). Given these circumstances it is now expected that Cernavoda 5 will never be finished and the Cernavoda site when complete will comprise just four CANDU-6 reactors (Mihai, 2008).

During the 1990s Romania faced significant economic challenges and the decision was made to progress the Cernavoda project in a phased way, completing the Cernavoda 1 station in 1996. It is not the purpose of this chapter to consider the financing of nuclear energy projects, and it is hoped that it will be possible to describe Romanian experience in this regard in a separate paper. Suffice it to say that the loans associated with unit 1 were repaid by 2006 and this existing asset was useful in collateralizing the costs of unit 2, completed in 2007. Loans relating to unit 2 are scheduled to be repaid by 2020 (Bucur, 2008).

In 2008 the topical issue in Romanian nuclear energy policy relates to the completion of mothballed part-built units Cernavoda 3 (17 percent complete) and Cernavoda 4 (15 percent complete) (ibid.). There is significant public and private sector interest in financing these plants and again this will be discussed in a later paper. At this stage it is sufficient to state that there is no shortage of investment funding available to complete these two units (ibid.; Sandulescu, 2008). The process has not been without some turbulence, with government late in the day increasing its stake to a controlling interest of 51 percent, much to the consternation of the private sector investors who were keen to have larger stakes in the enterprise than will be possible in the 51 percent state-owned model.

Nuclearelectrica s.a. is 90.28 percent owned by the Romanian government and it reports to the Societatea Nationala Nuclearelectrica (SNN) (National Nuclear Corporation) headquartered in Bucharest. The Cernavoda project has involved numerous collaborations with international companies, most notably Atomic Energy Canada Ltd. (AECL), designers of the CANDU-6 power plant. The Cernavoda 1 and 2 projects were delivered by AECL in collaboration with the Italian engineering firm Ansaldo. During the construction period the management team comprised representatives from SNN, AECL and Ansaldo (Mihai, 2008).

Once complete in 2015, the four-reactor Cernavoda complex will produce 2,600 MWe of baseload nuclear electricity (WNA-Romania, 2008), sufficient for roughly 40 percent of Romania's electricity needs. While in 2008 the Romanian electricity market remains 70 percent regulated and 30 percent liberalized, by 2015 it is planned that the entire market will have been liberalized. As such, it seems probable that Cernavoda 3 and 4 will spend their entire operating lives in entirely liberalized electricity markets (Bucur, 2008).

In addition to the Cernavoda nuclear power plant (NPP), south-eastern Romania and the Danube delta has a wider significance in the Romanian electricity system. The region, including offshore sites in the Black Sea, has a large (3 GWe) renewable wind energy potential (Leahu, 2008). The growth in renewables and nuclear power in this corner of Romania prompts investment for grid reinforcement in this region including 400 kV lines for the Cernavoda area (Sandulescu, 2008).

Notions of self-sufficiency remain powerful in Romanian energy politics but the Romanian government's conception of how to achieve such aims is radically different from the energy independence vision of the Ceaușescu era. The government's view is that European and wider energy markets are beneficial for energy security and not a threat to it. Sufficiency is compatible with trade and the intention is that imports of primary fuels can be balanced by the export of electricity. While an exact balance will be difficult to achieve, a net balance is a policy goal for the country (ibid.). Such a strategy is well suited to Romania's position in South-East Europe. The region has a long history of international electricity trade. For many years Bulgaria was a net electricity exporter for the region. With investment in generation and grid reinforcement, particularly in Cernavoda and the Danube Delta, Romania will be well positioned to be South-Eastern Europe's electricity hub (ibid.). It is noteworthy that Bulgaria's position as a regional power exporter was badly weakened by the imposed EU accession requirement to close down Kozloduy 3 and 4 Russian-designed VVER-230 pressurized water reactor plants near the Danube River border with Romania (WNA-Bulgaria, 2008).

Generally, energy policy for EU member states comprises a balance of economic, environmental and security of supply concerns. While all these factors are of great importance to Romania, the issue of greatest influence in the question of nuclear power is its possible contribution to electricity security of supply (Sandulescu, 2008). Romania faces the prospect of a dependence on Russian gas. Only 10 percent of Romania's electricity is generated from gas and the country's plan is to maintain, or possibly reduce, that proportion. Reductions will be difficult and may even be undesirable given that gas-fired electricity is associated with district heating and co-generation (ibid.). In 2008, 70 percent of Romanian natural gas demand is still satisfied from domestic sources. However a previous government leased those assets for 10 years to a foreign company, OMV of Austria, and this means that today profits associated with high fossil fuel prices are leaving the country.

The combined influences of rising fossil fuel prices, increased natural gas dependence on Russia and the challenge of global climate change, and an accepting regional and national public attitude to nuclear energy all lead Romania towards an expansion of nuclear power beyond the Cernavoda project. Preparatory studies for new build on a new site have been undertaken concerning location, site geology, seismology, cooling water supply and electricity network capacity (ibid.). This is to be followed by a feasibility study to provide clear answers concerning power plant scale and the technology choice. It is expected that the preferred technology will be a new and evolved technology (ibid.). It is unlikely to be the generation-II technology CANDU-6.

As regards EU energy policy goals, Romania is in a comfortable position, emerging from the burden sharing of the EU's '20:20:20 by 2020' goals.[9] While, for instance, the United Kingdom must increase its share of renewables in total energy from 3 to 15 percent, Romania's task is easier, needing only to increase from 19 to 24 percent. Furthermore, while the UK must decrease its greenhouse gas emissions by 16 percent, Romania has the right to increase its emissions by 19 percent. Given that these EU targets may be achieved by trading between EU member states it seems highly likely that there will be wealth transfers from the EU-15 to Romania as states in Western Europe struggle to achieve their binding quotas. Such wealth transfers are not necessarily undesirable given Romania's need for modernization and infrastructure renewal in order to reach EU norms.

4.2 The Baltic States

Until the completion of the 350 MW Estlink[10] electricity interconnector between Estonia and Finland in December 2006, the Baltic states formed an 'electricity exclave' or 'island' disconnected from the rest of the European Union and unable to benefit from European electricity market integration in the bulk of the EU. Within the Baltic region very significant electricity trading occurs. In 2007, the country with the largest electricity generation capacity, Lithuania,[11] traded the following amounts of electricity with neighboring states:

- Lithuania exported 3.2 TWh to Latvia and imported 1.4 TWh;
- Lithuania exported 1.1 TWh to Russia (Kaliningrad); and
- Lithuania exported 2.0 TWh to Belarus and imported 3.6 TWh (Lietuvos Energija, 2007, p.12).

The Baltic states remain connected with the old USSR power grid, both directly to Russia and via Belarus. Reflecting changing geopolitical alignments these states now seek greater westward connection, particularly to Poland and hence to Western Europe.

From 1940 until 1991[12] the Baltic states of Latvia, Lithuania and Estonia were constituent republics of the Soviet Union. As such, and in contrast to the Romanian experience, they fell completely under the centralized industrial policy of the communist government in Moscow. These command-and-control policies favored large-scale industrial investments capable of serving the needs of the Soviet Union as a whole, with little sympathy for local circumstances. In electricity such planning left the newly independent Baltic states with a difficult legacy which still causes concern. Central to that legacy are the Ignalina nuclear power reactors in Visaginas, eastern Lithuania near the border with Belarus. At the time of independence the power station comprised two RBMK-1,500 units each with a capacity of 1,360 MWe. The two units, Ignalina 1 and Ignalina 2, came on-line in 1983 and 1987, respectively. The RBMK design is a light-water-cooled graphite-moderated boiling water reactor, a type made infamous by the Chernobyl disaster of April 1986.

The Chernobyl disaster prompted the government of the then Lithuanian Republic to ask the government of the USSR to abandon the construction of the planned unit 3 RBMK plant. This request was accepted and construction of unit 3 was completely abandoned in 1989 (INPP, 2008).

In the 1990s there was much concern in Western Europe regarding the presence of Chernobyl-type (RBMK) reactors in the EU and the status of the Ignalina plant became 'one of the main issues in the Lithuanian accession negotiations' (Euro.Lt, 2008)). The fourth protocol of Lithuania's treaty of accession to the European Union[13] states in Article 1:

> Acknowledging the readiness of the Union to provide adequate additional Community assistance to the efforts by Lithuania to decommission the Ignalina Nuclear Power Plant and highlighting this expression of solidarity, Lithuania commits to the closure of Unit 1 of the Ignalina Nuclear Power Plant before 2005 and of Unit 2 of this plant by 31 December 2009 at the latest and to the subsequent decommissioning of these units. (Eur-Lex, 2003a)

Article 4 of the same protocol is, however, noteworthy:

> Without any prejudice to the provisions of Article 1, the general safeguard clause referred to in Article 37 of the Act of Accession shall apply until 31 December 2012 if energy supply is disrupted in Lithuania.

Article 37 of the Act of Accession states:

> 1. If . . . difficulties arise which are serious and liable to persist in any sector of the economy or which could bring about serious deterioration in the economic situation of a given area, a new Member State may apply for authorisation to take protective measures in order to rectify the situation and adjust the sector concerned to the economy of the common market.
> 2. Upon request by the state concerned the Commission shall, by emergency procedure, determine the protective measures that it considers necessary, specifying the conditions and modalities in which they are to be put into effect. . . .
> 3. The measures authorised under paragraph 2 may involve derogations from the rules of the EC Treaty and from this Act to such an extent and for such periods as are strictly necessary in order to attain the objectives referred to in

paragraph 1. Priority shall be given to measures as will least disturb the functioning of the common market. (Eur-Lex, 2003b)

The terms of Lithuania's EU accession are noteworthy in two important respects. First, Article 4 of the Fourth Protocol provides Lithuania with more than 10 years of protection under Article 37 of the Act of Accession rather than the usual three, for the specific issue of energy supply disruption. Article 37 permits, at the Commission's discretion, substantial measures up to, and including, derogations from the EC Treaty and the Act of Accession. However, and most importantly, such powers cannot permit a life extension for Ignalina 2 beyond December 31, 2009. That is, such measures are not in the gift of the Commission and would require a higher-level reform of the Treaty and Act of Accession itself, and as such would require the agreement of all EU member states.

Life extension for the Ignalina 2 plant is a matter of great concern in 2008 as it is widely believed that closure by the end of 2009 will place Lithuania, and perhaps even the wider Baltic region in a position of dependence on Russia for electricity security. Given the worsening relations between Russia and the west in the 2004–08 period and the role of energy in these geopolitical tensions, such a position of energy dependence causes much nervousness.

The 2007 Lithuanian National Energy Strategy document notes:

> There exist serious problems in the field of energy security, which it would be highly complicated or nearly impossible for Lithuania to deal with on its own. Key problems include the long-term reliability of natural gas supply, construction of the prospective new nuclear power plant and integration of the electricity system into EU systems. Implementation of these strategic tasks could be facilitated only by close cooperation with other Baltic countries – Estonia, Latvia and Poland. (Miskinis et al., 2008, p. 9)

The perceived jeopardy is not open-ended but merely would exist until sufficient alternative electrical generation capacity had been installed. Given the now imminent approach of the Ignalina 2 closure deadline, it seems impossible that sufficient capacity can be installed by the closure date and as such some period of jeopardy seems inevitable. It might be argued that this situation would not have arisen if the Baltic states, knowing for many years of Lithuania's Ignalina closure obligation, had made proper compensatory arrangements much earlier. In response to this, however, it might be argued that the current reality of a frosty EU–Russia relationship could not have been expected only a few years ago.

Lithuania's Prime Minister Gediminas Kirkilas is quoted as having said: 'We shall follow our commitments and we shall close the plant, but we would extend its operations for the particular period when we do not have other capacities' (EUBusiness, 2008). He went on to caution that Lithuania could be 'completely dependent on a well-known neighbouring country (Russia), for either gas or energy imports'. He added: 'Our plan is very simple – to hand [all the arguments] to the European Commission, which is responsible for energy security of all the member states and we hope it will be taken into account'.

Lithuania's Vice-Minister of the Economy, Vytautas Nauduzas, points out that the situation faced by Lithuania at the end of the first decade of the twenty-first century was not anticipated at the time of the accession negotiations. The types of energy crisis that Lithuania could face represent a sort of *force-majeure* (Nauduzas, 2008).

In April 2008 the Lithuanian government attempted to open negotiations with the European Commission (Lithuania, 2008), against a background of popular support organized by Lithuanian trades unions (Rosatom, 2008).

Lithuania's suggestions of a life extension for the Ignalina 2 reactor have been met with official silence from the Commission, and Energy Commissioner Andris Piebalgs's lack of a statement has been described as 'stonewalling' (Collier, 2008). It is clear that while some EU member states may be sympathetic to Lithuania's proposal, a number of states would appear to be implacably opposed. As the shutdown deadline approaches, there are hints that Lithuania is resigning itself to compliance with its accession treaty obligations. With Lithuania planning for a new NPP for 2020 onwards it seems probable that the country will face 10 years without domestic nuclear electricity.[14] Jurgis Vilemas of the Lithuanian Energy Institute notes that the country will benefit from new investments in combined heat and power (CHP), new combined cycle gas turbines (450 MWe in Lithuania and Latvia) and new investments in renewable energy prompted by the EU 2020 targets (Vilemas, 2008). It is important to remember that nuclear energy is only one aspect of Lithuanian infrastructure in need of updating. Arguably the most substantial challenge will be the €13 billion refurbishment of 30,000 blocks of flats in which standards of energy efficiency are currently poor (Nauduzas, 2008).

Lithuania with its large and amortized NPP enjoys very low domestic electricity prices. These prices must surely rise. First as the country covers the generation gap that will arise with the closure of Ignalina-2 and second to cover the costs of a replacement nuclear power station. That station will incur costs substantially higher than were incurred in building the original RBMK reactors in the days of the Soviet Union. The 2007 Lithuanian National Energy Strategy considered future price scenarios in some detail:

In 2005, around 70% of the total domestic electricity production was generated by the Ignalina NPP (about 21% – by thermal power plants). In 2005, the average electricity generation cost was about 8.44 Lithuanian cent/KWh (taking into account the public interest component), and the average electricity price for the final consumer – about 23 Lithuanian cent/kWh. Taking into account the decommissioning of Unit 2 of the Ignalina NPP and the forecast rise in the natural gas price, the average electricity generation cost in 2010 could stand at 16 Lithuanian cent/kWh, and the price for the final consumer could go up by 39% to 32 Lithuanian cent/kWh. A price should remain at similar levels until the planned construction of a new nuclear power plant in 2015. In the current and coming periods, the electricity price should also depend on the establishment of new electricity interconnections with Western European and Scandinavian countries and the level of electricity prices in these markets, as well as on the scope of the use of renewable energy resources in Lithuania. The evaluation of all the circumstances allows making forecasts that the electricity price in Lithuania should be somewhat lower than that in the markets of Western European or Scandinavian countries. (Miskinis et al., 2008, pp. 34 and 35)

Lithuania has well-developed plans for the construction of a new 3,200 MWe nuclear power station to be located adjacent to the existing Ignalina RBMK installation (Nauduzas, 2008). As the majority state-owned electricity transmission operator Lietuvos Energija reports:

> On 28 June 2007, the Seimas [Parliament] of the Republic of Lithuania passed the Law on the Nuclear Power Plant, the validity of which was promulgated by the President on 4 July 2007. By order of this document the Seimas gave its approval for construction of a new nuclear power plant and designated Lietuvos Energija, which had expressed a private initiative to invest in the project, to act as the national investor. (Lietuvos Energija, 2007, p. 25)

This new capacity together with a new 1,000 MW power bridge to Poland entering into service in 2016 or 2017 and plans for a similar link to Sweden should ensure Lithuania's electricity security in the 2020s (Nauduzas, 2008). The 2007 National Energy Strategy warns in a list of threats, however:

> 3) if the necessary competitive electricity-generating sources are not constructed and the reliability measures of the energy supply network, especially system interconnections with Poland and Sweden, are not implemented in proper time, the decommissioning of the Ignalina NPP and dismantling of reactors thereof, could pose a grave threat to the stable supply of electricity, while increased energy prices could become a heavy burden for consumers and the country's economy. (Miskinis et al., 2008, p. 17)

Despite this warning, electricity supply security is not inevitably compromised in the event of the closure of the Ignalina 2 plant because of the existence of the fossil-fueled 1,800 MW (2006) Lithuanian Power Plant:

The total installed electricity-generating capacity (nuclear and non-nuclear) amounts to nearly 5000 MW and exceeds the present domestic needs of Lithuania by more than two times, while the main source of electricity in the country is the Ignalina NPP which generates cheaper electricity than thermal power plants using fossil fuel. After the decommissioning of of Unit 2 of the Ignalina NPP at the end of 2009, the current generating capacities, including small capacity CHP plants that are planned to be constructed, will be sufficient to meet the national demand until 2013 in all cases of the growth in national economic needs and supply with systemic services necessary for the functioning of the system, but the Lithuanian Power Plant and the existing CHP plants with the lowest electricity generating cost during the heating season should be modernised. After the decommissioning of Unit 2 of the Ignalina NPP, the Lithuanian Power Plant will become the major electricity generating source until the construction of a new nuclear power plant, hence, it is required to carry out the necessary testing and adjustments of the power plant equipment and to ensure its reliable operation with a capacity of at least 1,500 MW from the beginning of 2010. (Ibid., p. 36)

In addition, the National Energy Strategy notes:

With the final shutdown of Unit 2 of the Ignalina NPP at the end of 2009 and without constructing a new nuclear power plant, demand for primary energy resources would increase only by approximately 25% during the period until 2025 according to the basic scenario, however total demand for fossil fuel would increase almost 1.7 times within 20 years, that is from 6 million toe in 2005 to 10.5 million toe in 2025. Natural gas demand would double – from 2.4 million toe to 4.8 million toe in 2025, and the share of natural gas in the national balance of primary energy resources would increase form 28.4% to 45% during the forecasting period. The forecasts predict that the share of indigenous (excluding indigenous oil) and renewable energy resources in the total balance of primary energy resources would grow by up to 20% in 2025, while the share of petroleum products, including orimulsion, would constitute about 35%. Having constructed a new nuclear power plant, primary energy demand would be higher due to poorer energy conversion properties of the nuclear power plant but demand for natural gas and petroleum products would decline and the diversity of primary energy resources would increase. In this case, the share of natural gas in the fuel balance could remain almost steady, that is close to 30%. (Ibid., p. 33)

However, despite these sources of comfort regarding the ability of Lithuania to cover its energy needs, the national Energy Strategy cautions: 'In the event of failure to construct necessary interconnections in time, it may be required to co-ordinate the reservation of large capacity units in the joint power system of Russia' (ibid., p. 38).

The new Ignalina 3 plant will be an entirely new plant based on western Generation 3 technology. Lietuvos Energija reports that

Between November and December [2007], meetings were held with the suppliers of modern technologies for nuclear reactors: General Electric–Hitachi

(GEH), AREVA NP, Westinghouse Electric Company LLC and Atomic Energy of Canada Ltd (AECL) with the aim of gaining knowledge about reactor technologies available on the global market, and in preparation for the tender on procurement of technology for the new nuclear power plant. (Lietuvos Energija, 2007, p. 26)

Ignalina 3 will be a 'commercial' plant based upon an innovative international approach. In 2007 Lithuania, Latvia and Estonia agreed to the construction of a new NPP to serve customers across the Baltics, and their main electricity companies contracted to collaborate on the project. Poland joined the consortium one year later. The three biggest Lithuanian energy companies that have agreed to invest with the Lithuanian state will have a 51 percent share. It is perhaps relevant that Lithuania has the most liberalized electricity industry of the four partners. The planned power bridges to Sweden and Poland will further increase the commercial attractiveness of the new NPP project (Kirkilas, 2007). The original 2006 cost estimate for the new NPP was €1,600/kWe, a figure that now (2008) seems to have been impossibly optimistic. A more realistic figure would probably be €3,000/kWe (Vilemas, 2008). Vilemas has highlighted four interesting aspects of the Ignalina 3 project. Here we present his ideas as interpreted by this author:

- Economic assessments of the new NPP project suggest that in order for the project to be successful the plant must operate with a very high load factor of approximately 8,000 hours per year.
- Placing a very large NPP in a small electricity system raises issues of the system capacity margin. The system must be able to draw upon alternative electricity sources up to at least the generation capacity of the large NPP. This need for a large capacity margin can undermine the case for the new NPP.
- It is not yet clear how waste spent fuel would be handled for a project established to serve the needs of one country. Should the four partner countries agree to share the waste burden? Or perhaps the waste burden should fall to the country hosting the plant – Lithuania.
- In contrast to the Romanian experience, none of the four partner countries associated with the Ignalina 3 project has ever built an NPP before, nor have they engaged the services of a western reactor vendor in such an endeavor. The Ignalina 1 and 2 plants, as products of the Soviet Union, were built by engineers from across the USSR. That said, Lithuania is endowed with significant nuclear operational, engineering, management competence and a world-class nuclear research center, the Lithuanian Energy Institute. These

capacities might be regarded as a legacy of the rigorous Soviet approach to the physical sciences and engineering, despite the fact that prior to independence Lithuania had no independent regulatory institutions.

The 2007 Lithuanian Energy Strategy observes that the pressures of Lithuanian electricity security in the coming years are such that the new NPP must enter service by '2015 at the latest' (Miskinis et al., 2008, p. 41). As the reactor type has not yet been selected, and some financial issues still remain unresolved, this author is skeptical that this requirement will indeed be met in time, in which case there is presumably the risk of a serious electricity security threat looming for Lithuania in the second half of the coming decade.

The opportunity to gain experience of NPP construction is arguably one of the main motivations for Poland's participation in the new NPP plan. This project allows Poland to build engineering knowledge and capacity through which it can create an option for a later nuclear energy program in Poland.

5 CONCLUSIONS

Of relevance to future EU policy is not just the data concerning issues of import dependency and fuel mix diversity but also the politics of energy security which are driven as much by perceptions and emotion as by evidence and data. This chapter is deliberately not empirical in its approach, rather it seeks to distil factors of importance concerning the interplay of nuclear energy policy in and enlargement of the European Union.

One simple message is that readers must not forget that much of the European nuclear renaissance will occur in the 'New Europe'. Experiences gained there will be every bit as valid as those gained in Finland, France and the UK and in some cases may act as better pointers to the future than the more often studied Western examples.

We accept that there is no chance of a single European voice on new nuclear build. Rather there is a consensus that the generation mix for electricity is a matter for each member state individually. That will represent the framework for nuclear energy developments for some time to come. Nevertheless the Lithuanian new NPP project presents a very powerful example to those smaller Western European countries contemplating the construction of a Generation III nuclear power plant of sizeable output (for example, 1,100–1,700 MWe). Might, for instance, the experience of Lithuania and the Baltics have lessons to offer the Netherlands and the other Benelux states? Concerning the Netherlands example: in September 2006 the environment and economics ministers submitted a paper to the Dutch parliament entitled 'Conditions for New Nuclear Power Plants' (WNA-Netherlands, 2008). The Netherlands has since resolved that any new reactor must be a Generation III model with levels of safety being equivalent to those of AREVA's European pressurized water reactor (EPR), and that any such plant should be constructed at a coastal site with operations planned for 2016 at the latest. Furthermore, in March 2008 the main advisory body of the Dutch government on national and international social and economic policy – the Social and Economic Council (SER) – said that the government should 'consider expanding nuclear energy in two years when it is due to evaluate its climate policies'. This scenario prompts the question of whether an EPR reactor of approximately 1,700 MWe can realistically be regarded as a project serving just Dutch consumers or whether the involvement of neighboring member states should be made more formal and explicit as the Lithuanians have done in the case of the plans for Ignalina 3.

As progress is made on building a single European electricity market international electricity systems are being created (for example, Nordpool in Scandinavia and SEMO in Ireland). These developments will eventually run up against the notion that nuclear energy is a national matter subject only to sensitivity to neighbors concerning environmental and safety risks. Increasingly European neighbors will become aware of the benefits of nuclear power investments in neighboring EU member states and in this way it is hoped that there can a further Europeanization of policy for nuclear energy to complement top-down moves from the Directorates General Research (RTD) and Transport and Energy (TREN) of the European Commission. The Eastern European experience reminds us that it is possible to complete an NPP project straighforwardly (Cernavoda 2, Romania) and for different countries to come together to address their common concerns through a single NPP project (Ignalina 3, Lithuania). The experiences of Romania and Lithuania show us that the nuclear renaissance is clearly achievable – it may even be achievable in Western Europe.

NOTES

1. The author is most grateful to all those who have provided insights and advice including: Ionel Bucur, Cristian Busu, Olivia Comsa, Raphael Heffron, Christian Kirchsteiger, Cristiana Leahu, Vaclovas Miskinis, Alexandru Sandelescu, Derek Taylor, Simon Taylor, Sami Tulonen and Jurgis Vilemas. The opinions expressed in this chapter are not necessarily shared by those who have provided assistance and all responsibility for

errors and omissions rests with the author. The author is most grateful to the European Commission Sixth Framework programme project CESSA. The author also acknowledges the assistance of the ESRC Electricity Policy Research Group.
2. Small research reactors are neglected. Some European countries, for example, Portugal (see, for example, the Sacavem reactor – http://www.itn.pt/uk/uk_main.htm) and Greece (see: http://ipta.demokritos.gr/Documents/MOISSIS.pdf) have operated such reactors while never having operated a nuclear power station.
3. See, for instance, UK Royal Commission on Environmental Pollution, sixth report, 1976, 'Nuclear power and the environment', 'The Flowers Report'.
4. With a brief hiatus in the spring of 2007 while impeachment proceedings went to a national referendum in which the proposal was rejected by the people.
5. See: http://www.cnu.ro/en/about.html.
6. See: http://www.nuclearelectrica.ro/?.
7. See: http://www.nuclear.ro/index_en.html.
8. See: http://www.raan.ro/en/index.html.
9. See: http://news.bbc.co.uk/1/hi/world/europe/7296564.stm.
10. See: http://www.nordicenergylink.com/index.php?id=29.
11. Juska and Miskinis (2007, p. 15, table 9: Data of the Baltic States 2005–2006: Gross production 2006 (GWh): Estonia: 9,731, Latvia: 4,891, Lithuania: 12,482, of which 8,651 from Ignalina 2 NPP).
12. On September 6, 1991 the Soviet Union recognized the independence of the three Baltic states. Only a few weeks later the Soviet Union itself ceased to exist, with the dissolution of the Supreme Soviet on December 26, 1991.
13. Lithuania's accession to the European Union was implemented via the 2003 Treaty for the Accession of the Czech Republic, Estonia, Cyprus, Latvia, Lithuania, Hungary, Malta, Poland, Slovenia and Slovakia. See: http://eurlex.europa.eu/LexUriServ/LexUriServ.do?uri=OJ:L:2003:236:0017:0032:EN:PDF.
14. The year 2020 would appear to be a plausible estimate.

REFERENCES

Bucur, I. (Cernavoda NPP Director) (2008): Personal communication, May 27.
Collier, M. (2008): 'Vilnius wages fight with EU to keep nuke', *Business Week*, March 18.
de Esteban, F. (Deputy Director-General, Directorate for Energy and Transport, European Commission) (2002): 'The future of nuclear energy in the European Union', speech, Brussels, May 23.
Diaconu, O., G. Operescu and R. Pittman (2008): 'Electricity reform in Romania', Working Paper 08-11, Centre for Competition Policy, University of East Anglia.
EUBusiness (2008): 'Lithuanian PM seeks EU deal to boost Soviet-era reactor life-span', February 5, available at: http://www.eubusiness.com/news-eu/1202217433.57/ (accessed August 2008).
Eur-Lex (2003a): *Official Journal of the European Union*, Protocols to the 2003 Treaty of Accession, available at: http://eur-lex.europa.eu/en/treaties/dat/12003T/htm/L2003236EN.093100.htm (accessed August 2008).
Eur-Lex (2003b): *Official Journal of the European Union*, Act Concerning the Conditions of Accession 2003, available at: http://eur-lex.europa.eu/LexUriServ/LexUriServ.do?uri=OJ:L:2003:236:0033:0049:EN:PDF (accessed August 2008).

Euro.Lt – Lithuania in the European Union website, administered by the Office of the Government of Lithuania, available at: http://www.euro.lt/en/lithuanias-membership-in-the-eu/ignalina-npp/ (accessed August 2008).

European Energy Forum (2006): 'Green Paper on a European Strategy for Sustainable, Competitive and Secure, Energy', ID:COM(2006)105, March 8.

FAS (2008): 'Romania Special Weapons', Federation of American Scientists, quoting *Ziua* newspaper (April 5, 2005), available at: http://www.fas.org/nuke/guide/romania/index.html (accessed August 2008).

INPP – Ignalina Nuclear Power Plant, website page – construction history, available at: http://www.iae.lt/inpp_en.asp?lang=1&subsub=6 (accessed August 2008).

Ireland, Irish Statute Book, Electricity Regulation Act 1999, section 18 item 6, available at: http://www.irishstatutebook.ie/1999/en/act/pub/0023/index.html.

Juska, A.P. and V. Miskinis (2007): 'Energy in Lithuania 2007', Kaunas: Lithuanian Energy Institute, LT-44403.

Kirkilas, G. (Lithuanian Prime Minister) (2007): Speaking at Royal United Services Institute, London, May 21.

Leahu, C. (Adviser, Energy Policy Department, Ministry of Economy and Finance, Romania) (2008): Personal communication, May 28.

Lietuvos Energija AB (2007): *2007 Annual Report*, Vilnius, LT-09310.

Lithuania Government Website, 'Lithuania in the European Union', available at: http://www.euro.lt/en/news/lithuanias-membership-in-the-eu/news/2970/ (accessed September 2008).

Mihai, A. (Cernavoda 2 NPP operator) (2008): Personal communication, May 27.

Miskinis, V., A. Galinis and J. Vilemas (eds) (2008), *National Energy Strategy 2007*, Lithuanian Energy Institute and Lithuanian Ministry of Economy, Vilnius: Zara.

Nauduzas, V. (2008): Speaking at 18th Economic Forum, Krynica, Poland, September 11.

Rosatom, website: Press Centre of Nuclear Energy and Industry, available at: http://www.rosatom.ru/en/news/8747_05.03.2008 (accessed September 2008).

Sandulescu, A. (General Director, Energy Policy Department, Ministry of Economy and Finance, Romania) (2008): Personal communication, May 28.

Tolnay, A. (2002): 'Ceauşescu's Journey to the East', 4th Annual Kokkalis Graduate Student Workshop, Kennedy School of Government, Harvard University, February, available at: http://www.hks.harvard.edu/kokkalis/GSW4/TolnayPAPER.PDF (accessed August 2008).

Turnock, D. (2007), *Aspects of Romania's Economic History with Particular Reference to Transition for EU Accession*, Aldershot, UK: Ashgate.

Vilemas, J. (2008): Presentation at 18th Economic Forum, Krynica, Poland, September 11.

WNA-Bulgaria, Country Briefing Bulgaria, World Nuclear Association, available at: http://www.world-nuclear.org/info/inf87.html (accessed August 2008).

WNA-Netherlands, Country Briefing The Netherlands, World Nuclear Association, available at: http://www.world-nuclear.org/info/inf107.html (accessed September 2008).

WNA-Romania, Country Briefing Romania, World Nuclear Association, available at: http://www.world-nuclear.org/info/inf93.html (accessed August 2008).

WNA-Sweden, Country Briefing Sweden, World Nuclear Association, available at: http://www.world-nuclear.org/info/inf42.html (accessed September 2008).

PART III

Hydrogen

9. Supply security and hydrogen
Julián Barquín and Ignacio Pérez-Arriaga

1 INTRODUCTION

Today's energy systems are based on the use of fossil fuels, which presently provide for about 80 percent of the total world energy needs (IEA, 2007a). As a consequence, the energy system is both unsustainable and prone to security of supply concerns. Indeed, fossil fuel combustion is thought to be the main cause of climate change (IPPC, 2007). Furthermore, Europe is highly dependent on imported fossil fuels, and in particular on oil and natural gas, from sources not always reliable. The ultimate amount of reserves is also uncertain, although substantial depletion of oil is expected before mid-century (IEA, 2007b).

Most likely, any solution will require the use of a wide portfolio of technological and policy options. Hydrogen technologies have been proposed as a part of this portfolio, potentially addressing two important issues: storage of energy from intermittent energy sources and provision of an alternative fuel for transportation (Sherif et al., 2005).

This chapter is mainly based on the CESSA conference 'Prospects for a European Hydrogen Economy' presentations and discussions. The conference was held in Madrid, April 14–15, 2008. However, the authors are exclusively responsible for its content. The rest of the chapter is organized as follows. The next section discusses the reasons that support the development of hydrogen technologies. Sections 3, 4 and 5 focus, respectively, on hydrogen production, infrastructure development and final use. Section 6 deals with research and development issues. Finally, Section 7 concludes.

2 WHY A HYDROGEN ECONOMY?

As stated above, the two main reasons for a hydrogen economy are linked to energy storage and to its possible use as a transportation fuel.

Regarding energy storage, the starting point is the realization of the new conditions that are likely to prevail in the future energy systems, and specifically in electric power systems (Jamasb et al., 2006). From both an environmental and a security of supply viewpoint, energy production from indigenous renewable sources is highly recommended (see Pérez-Arriaga and Barquín, 2005; EC, 2007). However, most renewable energy is actually renewable electricity (biomass is at present exceptional in this regard, as it can also be processed in solid or liquid fuels; as well as to some extent low-temperature solar thermal applications for heating), and most of it is 'intermittent', that is, not available on command but subject to uncontrollable conditions (time-of-day, cloud cover, wind, and so on). Large-scale economic electricity storage cannot be addressed with present technologies, which poses difficulties for deployment of intermittent renewable electricity in Europe as generation must balance demand at any moment. Use of electrical energy storage facilities is a must if Europe intends to obtain renewable electricity in quotas similar to those of fossil fuels today, without investing huge amounts of capital in back-up facilities to cover the gaps when renewable generation is not available. Other energy storage technologies that could fill this role (such as compressed air, enhanced hydro facilities, flow cells, plug-in hybrid cars or inertial energy storage devices) also deserve attention (Barton and Infield, 2004).

Hydrogen could also be a substitute for hydrocarbon-based liquid fuels in transportation uses. Arguably, transportation poses the most difficult challenge in the process of de-carbonizing the world economy and freeing Europe from the need of importing most of its primary energy. Other alternatives are biofuels, which could be limited from the availability of land that may be needed for other purposes (Kløverpris et al., 2008), and electricity, which appears to have a major potential as plug-in hybrid cars may constitute a new paradigm for future road transportation, especially if improved batteries are developed soon (Simpson, 2006).

A weakness of the hydrogen path in the design of a more sustainable future energy model is the need for further significant technological development. This is the cause of much uncertainty regarding the future of this technology. In order to assess the potential of hydrogen as a major future energy carrier, as electricity is at present, it is useful to examine separately the processes of production, transportation, distribution and final use of hydrogen.

3 HYDROGEN PRODUCTION

Currently, the main use of hydrogen is in the chemical industry, and the dominant technology for producing it is natural gas steam reforming. This procedure, like any other that is based on fossil fuels, results in the emission of carbon dioxide. Therefore, sustainable large-scale hydrogen production from fossil fuels will require CCS (carbon capture and storage). Furthermore, if hydrogen were produced from natural gas, European security of supply problems could be seriously increased. These reasons, on top of economic considerations, indicate that hydrogen production from coal should be regarded as a better long-term alternative. On the other hand, some 'cleaner coal' technologies, such as gasification, are actually based on hydrogen production. Generally speaking, production technologies of 'hydrogen from fossil fuels' are well advanced, although significant improvements are still required for generalized industrial deployment (García Peña, 2008).

Water electrolysis is a well-known, commercially available technology for very pure hydrogen production (Ivy, 2004). It would have the additional advantage of allowing decentralized operation, therefore easing the infrastructure building effort as compared to production from fossil fuels, biomass or nuclear energy. However, significant reductions in the costs of electrolysis equipment and improvements in efficiency would be required in order for electrolysis to become commercially viable unless electricity price, when compared with natural gas or coal prices, ends up being significantly less than expected (EC, 2006). Promising new research developments could drastically change this perspective but, in any case, there would be a significant time lapse before commercial use could be achieved.

Nuclear energy could provide an electricity source for hydrogen production (Verfondern, 2007) if general concerns about security, environmental impact and cost are met in the wider arena of electricity generation by nuclear plants. There are also proposals for direct hydrogen production by high-temperature water thermolysis. More speculatively, fusion nuclear reactors are pulsed machines that could be harnessed for hydrogen production in the latter case (Nuttall et al., 2005), although this pathway requires the marriage of two still untested technologies.

4 HYDROGEN INFRASTRUCTURE

If hydrogen is to become an energy vector on the same footing as electricity today, vast amounts of investment in building infrastructure will be needed (HyWays, 2007; Wietschel et al., this volume, ch. 11), especially if it is produced in centralized facilities either from fossil fuels or by water thermolysis in nuclear plants. Note that the amount of energy that would have to be carried is expected to be greater than the amount that is currently transported in natural gas pipelines. Moreover, the required technology is likely to be more complex than for natural gas transport today.

Hydrogen transportation requires special pipelines and transportation in liquefied or pressurized hydrogen vessels, possibly by train or truck. If the decision to make a transition to a hydrogen economy is adopted at some point in the future, a careful plan will be needed, possibly including an initial phase characterized by underutilized infrastructure. Co-opting the present natural gas infrastructure remains an open issue as it will probably be unsuitable (Adams et al., 2005).

In some scenarios, an initial phase of onsite hydrogen production is envisaged. This phase could become permanent in sparsely populated or remote areas. In other areas, it could be followed by supply via trucks carrying liquefied hydrogen, also to be used as a competing alternative to onsite production for remote locations. Liquefied hydrogen is also a relevant option for energy storage from stranded renewable energy sources. In the final stage, the delivery in high-demand areas would be mainly done by pipeline. Distribution pipelines could supply large fueling stations in dense areas. Compressed gas hydrogen trucks would be used during the transition from the liquefied hydrogen to the pipeline transportation phase, but it could be also an option for local distribution of produced hydrogen for less-dense areas.

A high hydrogen penetration rate can be critical in order to justify the huge infrastructure investment required, as well as the huge involved R&D cost. Therefore, subsidies intended to boost early hydrogen demand could also be established.

5 HYDROGEN END USES

Given the huge costs involved in building and operating the hydrogen production and distribution infrastructure, a major question is what end use of hydrogen could justify the cost that would be incurred in its massive production and distribution?

Stationary hydrogen use is currently limited to niche applications of fuel cells, where reliable electricity generation with very low local environmental impact is required. These applications have been made possible and, in turn, have stimulated significant advances in fuel cell technologies. They could also play a role in the design and operation of flexible electricity grids, although at present the costs are still high.

Use of hydrogen in transportation is viewed by many as the real justification of a hydrogen economy. At present, applications are limited to demonstration projects and niche applications, in which cost is a secondary concern. However, if hydrogen vehicles for short-range, low-velocity applications (wheelchairs, bikes, urban buses, and so on) are adopted, practical experience can be obtained and later applied if more extended applications are addressed. Widespread adoption hinges on technological breakthroughs that allow cheaper fuel cells and, more importantly, improved ways to store hydrogen in vehicles in order for them to boast ranges commensurate with those of present gasoline cars. At present, low-temperature fuel cells suitable for vehicle use cost around €8,000/kW. It is estimated that the cost should be decreased by two orders of magnitude in order to be competitive. If, however, an early transition towards a hydrogen-based transportation system were decided (possibly based on hydrogen production from fossil fuels), hydrogen internal combustion engines could be transitorily deployed. Dual fuel (hydrogen and gasoline) engines, already developed by car manufacturers, boast much lower levels of organic compounds and other contaminant emissions. They could also become part of a longer-term strategy to bridge the transition period. In any case, efficiency could be boosted by designing hydrogen cars as hybrid vehicles.

Use of hydrogen in automobiles highlights the need for and difficulties of developing an infrastructure that makes them as attractive as gasoline cars to the users. However, the competition of (oil-based) plug-in hybrid electric vehicles (PHEVs) will be fierce and, at the present moment, the odds seem to favor the PHEV versus hydrogen-based fuel-cell electric vehicles (FCEVs). It is becoming increasingly clear that the future car for sustainable transportation should probably be the PHEV, with some support from an internal combustion engine fueled by biofuels, or even gasoline or diesel, until electric batteries are able to provide energy for long distances and can be recharged in a reasonable time (Chan and Wong, 2004). Hydrogen-fed fuel cells could also play a role, if the deployment of the required infrastructure can be economically justified because of the high price of hydrocarbon-based fuels.

Another issue is the storage problem that could take place in electric power systems if there is a very high level of penetration of intermittent renewable sources (Kintner-Meyer et al., 2007), whose surplus may exceed the storage capability that the PHEV car batteries might provide. If storage of surplus renewable electricity in car batteries is not sufficient, then it might make sense to produce hydrogen by electrolysis, which could be used in stationary applications or, if development of hydrogen distribution infrastructure could be competitive with liquid hydrocarbon-based fuels (diesel, gasoline, biofuels), in cars.

6 RESEARCH AND DEVELOPMENT

Even if some hydrogen technologies have been well known for a long time, we feel that a significant effort in R&D is still needed, and that plans for deployment in the near future are premature. Therefore, an increased R&D effort is advised, especially when it is taken into account that most hydrogen technologies could play a significant role in the future energy system even if this one is not based on hydrogen. For instance, hydrogen from coal technologies is important in connection with cleaner coal, hydrogen fuel-cell cars are nothing more than electric cars in which the batteries have been substituted by a fuel cell and a hydrogen tank, and nuclear fusion is a worthwhile objective even if it is only used to generate electricity. There are a host of other hydrogen technologies, such as direct thermo-solar hydrogen generation or photo-biological hydrogen production, which are not only fascinating but potentially relevant, and whose development could provide important breakthroughs in other energy technologies as byproducts. However, they are still in their infancy, and more research is required.

Therefore, R&D activities cover a huge range of activities, which strongly advises a joint effort by member states, instead of the mostly autonomous and arguably poorly coordinated activities that are currently pursued (Stoft and Dopazo, this volume, ch. 13).[1] Comparison with US efforts, coordinated by the Department of Energy, makes the case even more compelling. On the other hand, US experience also warns on the dangers of subsidizing premature development of immature technologies. Comparison with other industries that also show network effects (that is, the internet or even the dawn of the present oil-based transportation system) rather suggests that the way forward is a combination of basic research publicly funded and private development possibly based on its initial phases in niche (that is, urban public transportation) and 'related' (that is, hydrogen as a subproduct of coal gasification) markets.

We also believe that there is widespread support for the notion that the required research efforts should not detract from other energy research areas, but rather complement them. European research effort in energy is clearly insufficient, and it should be increased considerably.

7 CONCLUSION

Hydrogen technologies are mostly still in their infancy. They promise a 'second' energy vector, in addition to electricity, that could allow renewable, nuclear and cleaner coal energy sources to be harnessed for a variety of purposes, most notably transportation. Reduced dependence on oil and natural gas has obvious and mostly positive implications regarding both sustainability and security of supply concerns. On the other hand, a significant amount of research is still needed, and short-term deployment is likely to be premature. As is the case with other technologies, hydrogen is a long-term bet that nevertheless deserves careful consideration.

NOTE

1. For example, for solar power generation, see http://www.technologyreview.com/Energy/21155/.

REFERENCES

Adams, T.M., R. Sindelar, G. Rawls and P.-S. Lam (2005): 'Evaluation of natural gas pipeline materials for hydrogen/mixed hydrogen-natural gas service', November, available at: http://www.hydrogen.energy.gov/pdfs/progress05/v_a_4_adams.pdf (accessed March 2008).

Barton, J.P. and D.G. Infield (2004): 'Energy storage and its use with intermittent renewable energy', *IEEE Transactions on Energy Conversion*, Vol. 19, No. 2, pp. 441–8.

Chan, C.C., and Y.S. Wong (2004): 'The state of the art of electric vehicles technology', *The 4th International Power Electronics and Motion Control Conference (IPEMC 2004)*, Vol. 1, pp. 46–57, Xi'an, China, August 14–16, 2004.

EC (2006): *World Energy Technology Outlook 2050: WETO H_2*, Brussels.

EC (2007): 'Towards a European Strategy for the Security of Supply', Green Paper, COM(2007)769 Final, Brussels.

García Peña, F. (2008): 'Technological and economic aspects of hydrogen production from fossil fuels', Presentation at CESSA conference 'Prospects for a European Hydrogen Economy', Madrid, April 14–15.

HyWays (2007): 'The European hydrogen roadmap', Project Report, available at: www.hyways.de (accessed October 2008).

IEA (2007a): *Key World Energy Statistics 2007*, Paris: OECD.

IEA (2007b): *World Energy Outlook 2007*, Paris: OECD.

IPCC (2007): 'Fourth Assessment Report', Intergovernmental Panel on Climate Change, UN.

Ivy, J. (2004): 'Summary of electrolytic hydrogen production', Report NREL/MP-560-36734, National Renewable Energy Laboratory, Golden, USA, available at: http://www.nrel.gov/docs/fy04osti/36734.pdf (accessed March 2008).

Jamasb, T., W.J. Nuttall and M.G. Pollitt (eds) (2006): *Future Electricity Technologies and Systems*, Cambridge: Cambridge University Press.

Kintner-Meyer, M., K. Schneider and R. Pratt (2007): 'Impacts assessment of plug-in hybrid vehicles on electric utilities and regional US power grids. Part 1: technical analysis', available at: http://www.pnl.gov/energy/eed/etd/pdfs/phev_feasibility_analysis_combined.pdf (accessed March 2008).

Kløverpris J., H. Wenzel, M. Banse, L. Milà i Canals and A. Reenberg (2008): 'Conference and workshop on modelling global land use implications in the environmental assessment of biofuels', *International Journal of Life Cycle Assessment*, Vol. 13, No. 3, pp. 178–83.

Nuttall, W.J., B. Glowacki and R. Clarke (2005): 'A trip to "Fusion Island"', *The Engineer*, October 31, pp. 16–18.
Pérez-Arriaga, I. and J. Barquín (2005): 'Towards a sustainable European energy model: investment for sustainability', Paper prepared for the SESSA project.
Sherif, S.A., F. Barbir and T.N. Veziroglu (2005): 'Towards a hydrogen economy', *The Electricity Journal*, Vol. 18, No. 6, pp. 62–76.
Simpson, A. (2006): 'Cost–benefit analysis of plug-in hybrid electric vehicle technology', National Renewable Energy Laboratory conference report CP-540-40485, Golden, CO.
Verfondern, K. (2007): 'Nuclear energy for hydrogen production', Institut für Sicherheitsforschung und Reaktortechnik (IEF-6), Jülich.

10. Hydrogen from renewables
Dries Haeseldonckx and William D'haeseleer

1 INTRODUCTION

At the start of the twenty-first century, we face significant energy challenges. An important aspect of sustainable development is 'de-fossilizing' our future energy economy. This mitigation from fossil fuels towards more sustainable energy technologies is driven by several factors:

- The need for a drastic reduction of CO_2 emissions, that is, 20 percent reduction of greenhouse gases (GHGs) by 2020 as proposed by the EU Commission (2007).
- The worldwide energy-dependence issue; fossil fuels alone will not suffice, moreover differentiation in primary energy sources will improve the security of supply.
- The exhaustibility of fossil sources; at current consumption and production levels, the world's proven reserves of oil, natural gas and coal are expected to be 'depleted' in 42, 64 and 155 years, respectively (IEA, 2006). Although these 'years left' are moving targets, prices will rise substantially when oil fields become more depleted.
- The needs of developing economies; it might be fair to leave the 'easy sources' for them.

The substitution of fossil fuels by renewable energy sources is a mitigation strategy that is advocated by non-governmental organizations, some research institutes and groups of stakeholders, and can be found in concrete policy goals. The future use of renewables is now accepted by almost every energy analyst; only the level of penetration is currently a matter for discussion. The main modern renewable energy technologies that produce electricity are small hydropower, solar photovoltaics, concentrating solar power, biomass, geothermal power and wind energy (IEA, 2003). Nevertheless, despite being clean and abundant, most of these sources face another major problem: intermittency. Due to the fluctuating power delivery from most renewables, a good integration in the electric grid is needed, which brings with it costly adaptations. Furthermore, electricity storage in large quantities remains a major unsolved problem. Indirect storage via pump/turbine hydro stations can offer a solution, but only when it is geographically possible. Another possible 'solution' for electricity storage is the future use of plug-in hybrid cars, but the 'success' of that route remains to be seen.

This is where hydrogen (H_2) enters the scene. Because hydrogen can (in principle) serve as a storage medium for (excess) electricity, the interest for hydrogen first originated from the renewable advocates. Later on, hydrogen was adopted by the coal sector, since H_2 might go hand in hand with the development of carbon capture and storage (CCS). Next, the nuclear sector also jumped on the 'bandwagon'. Of course, there is the possibility of producing hydrogen by means of high-temperature reactors (Gen IV), but – being clean and emission free – the combination with hydrogen might also help the nuclear revival.

In all cases, hydrogen fulfills its role as an energy carrier or storage medium, having some unique characteristics (Veziroglu and Barbir, 1992):

- it can be converted into other forms of energy in more ways and more efficiently, by means of fuel cells, than any other fuel;
- it can be stored as liquid, gas or embedded in solids;
- it can be transported over large distances using pipelines, tankers or rail truck;
- it can be converted into electricity and heat in the absence of local emissions. This is particularly of interest for transport applications; and
- it may allow the integration of renewable, intermittent energy sources.

This chapter further focuses on hydrogen from renewables, the complementarity of hydrogen and electricity, the corresponding environmental impact and some general infrastructural and economic considerations. A more detailed technological overview of renewable hydrogen production can be found in Turner et al. (2008). As highlighted by Sherif et al. (2005), both hydrogen and electricity complement renewable energy sources particularly well, by presenting them to the end user in a convenient form and at a convenient time.

2 TECHNOLOGY OVERVIEW

In theory, renewable energy sources can easily meet the world's *energy* demand. In practice, however, there is a large difference between the theoretical, technological and economic potential as the available resources are strongly determined by geographical and climatological aspects. Even if the economic potential should be fully developed, there still remains an instantaneous *power* issue: intermittency must be overcome by means of storage or grid/demand-side management.

This section discusses the different renewable energy sources and hydrogen production technologies. An overview of the main characteristics and technological developments will be given.

2.1 Renewable Energy Sources

Wind energy
Wind power is the conversion of wind energy into a useful form, such as electricity, using wind turbines. As can be seen in Figure 10.1, installed global capacity has risen from 1.9 GW_e in 1990 to nearly 95 GW_e at the end of 2007. Wind power is the fastest growing renewable energy sector with annual growth rates averaging around 30 percent, resulting in an expected installed capacity of around 400 GW_e by 2013. Today, wind generators produce approximately 1 percent of global electrical energy output (WWEA, 2008).

There are two types of wind turbines used today based on the direction of the rotating shaft: horizontal- and vertical-axis wind turbines. The size of these turbines varies widely. Over the past 20 years the generating capacities of individual units have grown from 25 kW to about 2.5 MW.

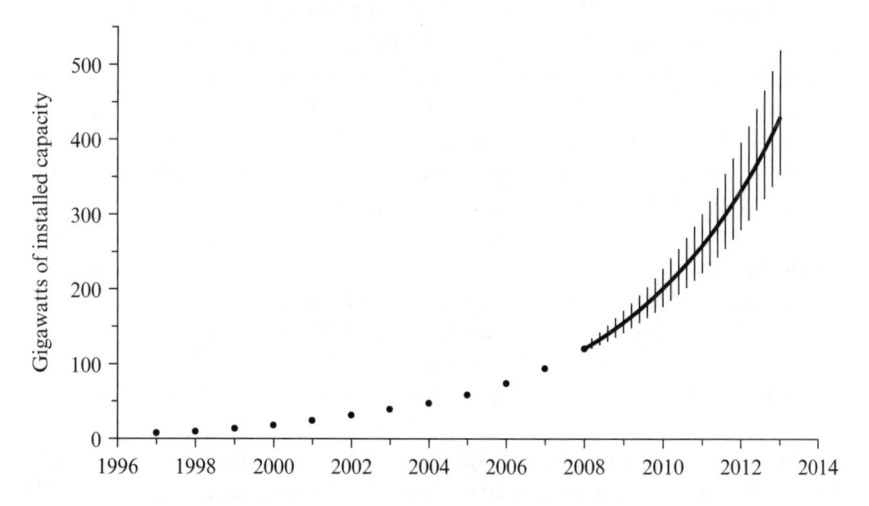

Source: WWEA (2008).

Figure 10.1 Worldwide installed wind energy capacity

As a result of better designs, prototype turbines are now exceeding capacities of 5 MW.

Electricity generated from wind power can be highly variable at several different timescales: on a second by second basis, as well as hourly, daily, and even seasonally. In order to correctly assess the relevance of intermittent power sources in the expansion of the power system, the concept of capacity credit is commonly used. It expresses the amount of installed conventional power that can be avoided or replaced by intermittent power sources. This capacity credit is the fraction of the installed renewable power for which no 'double investment' is needed. For example, 1,000 MW of installed wind power with a capacity credit of 30 percent, can avoid a 300 MW investment in conventional dispatchable power (Voorspools and D'haeseleer, 2006). In good locations with a high mean wind velocity and with an efficient wind turbine, capacity credits may reach 33–38 percent (Sherif et al., 2005).

Moreover, this intermittency and the non-dispatchable nature of wind-energy production can raise costs for regulation and could require demand-side management, load shedding or storage solutions such as hydrogen at high penetration levels. Studies have shown that the grid is able to absorb most of the wind power produced, as long as wind power is less than 20 percent of the maximum load (ibid.).

As far as costs are concerned, capital cost of 'early 2000' wind turbines was about €1,000/kW$_e$ (EUSUSTEL, 2007); since 2006, prices have risen to €1,000–€1,500, because of higher world market prices for raw materials and primary sources. It has generally been expected that the cost of wind turbines would follow a downward trend as larger multi-megawatt turbines are mass produced. However, since fewer facilities can produce large modern turbines and their towers and foundations, constraints develop in the supply of wind turbines possibly resulting in higher costs (GWEC, 2008). In areas with good wind resources, electricity generation costs are generally as low as €0.04–€0.06/kWh$_e$ (Sherif et al., 2005).

Solar energy

Each year, approximately 90 petawatts of sunlight reach the Earth's surface, being almost 6,000 times more than the 15 terawatts of average power currently consumed by humans. Furthermore, solar electric generation has the highest power density (global mean of 170 W/m^2) among renewable energies.

The most widespread use of solar radiation is the conversion into electricity by means of photovoltaic (PV) cells. Simply stated, in a PV cell photons from sunlight knock electrons into a higher state of energy, creating electricity. Growth in installment of PV cells has been 30 percent over the past decade, but the baseline is small. Figure 10.2 shows the evolution

Figure 10.2   Worldwide installed PV capacity

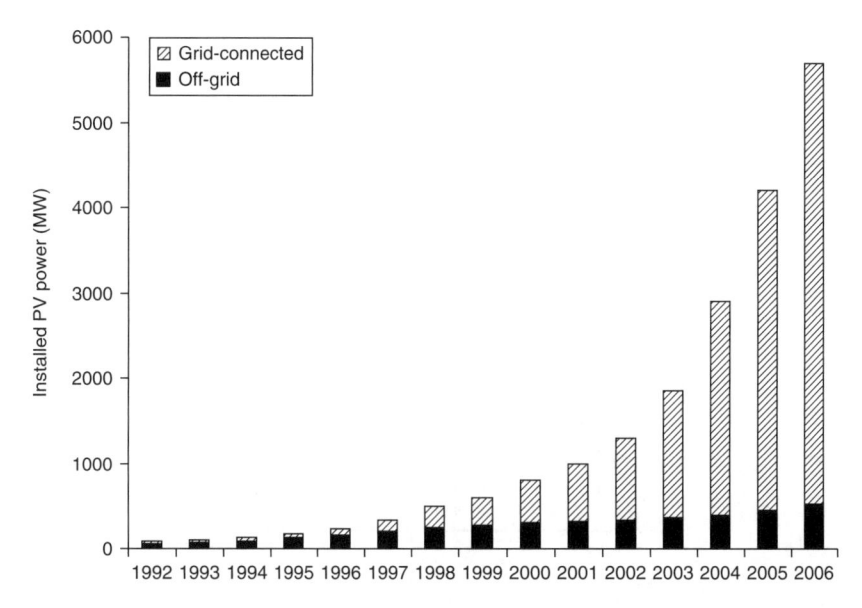

Source: Adapted from IEA (2007).

of globally installed PV power, resulting in a capacity of almost 6 GW at the end of 2006 (IEA, 2007).

The most important issue with solar panels is capital cost, mainly due to the use of silicon wafers. However, recent developments offer alternatives to these standard crystalline silicon modules including casting wafers instead of sawing, thin film and continuous printing processes. This should enhance the conversion efficiency, which is necessary for lowering the balance-of-system costs. In addition, due to economies of scale, costs are expected to drop in the years to come. On the other hand, capacity credits of PV systems merely amount to approximately 10 percent, which has a detrimental effect on the price per kWh_e. Installation costs for an entire system are situated between €3,000 and €10,000/kW, while solar electricity prices fluctuate between €0.2 and €0.4/kWh_e depending on the system and geographic circumstances (Lipman et al., 2006; EUSUSTEL, 2007; Zoulias and Lymberopoulos, 2007; Solarbuzz, 2008).

Besides PV cells, solar radiation can also be used in concentrating solar thermal systems or for the photo-electrochemical decomposition of water. Concentrating solar systems use lenses or mirrors and tracking systems to focus a large area of sunlight into a small beam. The concentrated light

Figure 10.3 Working principle of photo-electrochemical decomposition of water

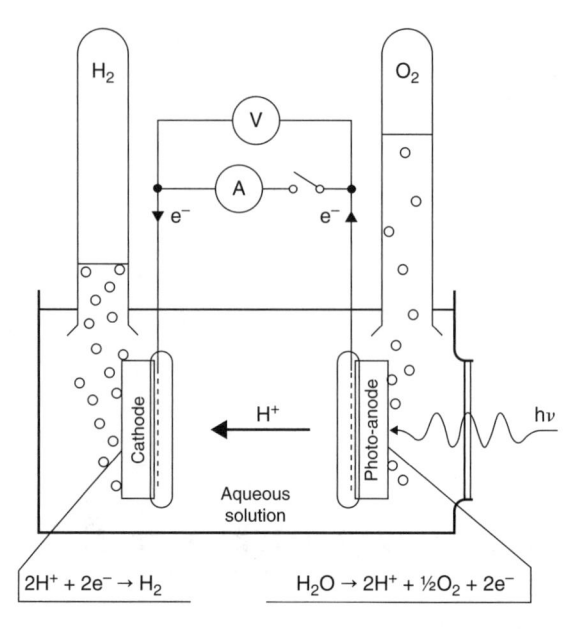

Source: Nowotny et al. (2005).

is then used to heat a working fluid for conventional power systems or Stirling engines, or to thermally dissociate water into hydrogen and oxygen. Although a wide range of concentrating technologies exists, the most developed are the solar trough, parabolic dish and solar power tower. Worldwide only a few prototype systems are currently installed and it is estimated that levelized energy costs of €0.04/kWh$_e$ are achievable by 2020 (Taggart, 2008).

The principle of photo-electrochemical decomposition of water is shown in Figure 10.3. The development of this technology requires new photo-sensitive materials to be used as photo-electrodes for electrochemical devices converting solar energy into chemical energy. According to the US Department of Energy, this technique will become commercially viable when the efficiency of the conversion of solar energy into chemical energy is greater than 10 percent. Currently, this technology is still under development (Nowotny et al., 2005).

Hydropower

Hydropower is power that is derived from the energy of moving water and can be divided into different categories:

- waterwheels, used for hundreds of years to power mills and machinery;
- hydroelectricity, usually referring to hydroelectric dams;
- damless hydro, which captures the kinetic energy in rivers, streams and oceans;
- tidal power, which captures energy from the tides in a horizontal direction;
- tidal stream power, which captures energy from the tides in a vertical direction; and
- wave power, which uses the energy in waves.

Of these different technologies, hydroelectricity is by far the most developed, providing 10.7 EJ of primary electricity, or 16.5 percent of global electricity production in 2005 (Moriarty and Honnery, 2007). Most hydroelectric power comes from the potential energy of dammed water driving a water turbine and generator. In this case the energy extracted from the water depends on the volume and on the difference in height (called the head) between the source and the water's outflow. The amount of potential energy in water is proportional to the head. Depending on the head and the flow rate of the water, two different types of turbines can be used: impulse and reaction turbines.

Impulse turbines change the direction of flow of the water stream. Water impinges on the turbine's blades which reverses the flow of water. The resulting change in momentum (impulse) causes the turbine to spin. The diverted water flow is left with diminished energy. Prior to hitting the turbine blades, the water's energy is converted to kinetic energy by a nozzle and focused on the turbine. No pressure change occurs at the turbine blades and the turbine does not require housing for operation. Examples are the Pelton, Turgo and crossflow turbine.

Reaction turbines are acted on by water, which changes pressure as it moves through the turbine. Reaction turbines must be encased to contain the water pressure (or suction). Or they must be fully submerged in the water flow. The most important types are the propeller and Francis turbines (Voets et al., 2001; DOE, 2004).

The highest efficiencies are reached by the Francis and propeller turbines, up to 90 percent, but they decrease fast as the flow rate decreases. The other turbines show better 'partial-load' characteristics and achieve efficiencies between 75 and 90 percent. An overview of all the different turbine efficiencies is given in Figure 10.4.

Hydroelectricity is a well-developed technology, leaving the remaining hydro potential in the industrializing countries, mainly in tropical or subtropical regions. At many potential sites, tropical forests would be

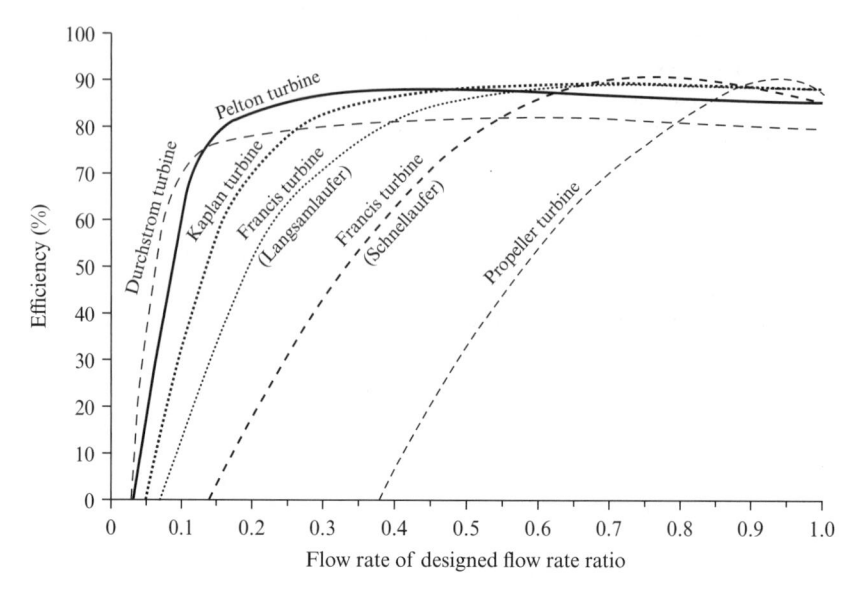

Figure 10.4   Efficiencies of different hydropower turbines

inundated. As trees die, they release their carbon into the atmosphere, which could have a negative impact on the sustainable character of future hydro projects. Other technologies such as tidal (stream) power and wave power also have a large potential, but with no installed commercial plants, a detailed technology overview is currently impossible (Moriarty and Honnery, 2007).

Geothermal energy
Geothermal energy is useful energy generated by thermal energy stored beneath the Earth's surface. It can provide both electrical power and direct heat on a continuous basis over the life of the installation, making it a non-intermittent renewable energy source. In 2004, geothermal electricity production was about 0.2 EJ, with a somewhat larger production of thermal energy used directly as heat (ibid., 2007). The global potential of geothermal sources might amount to 140 GW of electrical generation capacity (GEA, 2008).

Geothermal energy can be used for electricity production, for direct use purposes, and for home heating efficiency (through geothermal heat pumps). To develop electricity from geothermal resources, wells are drilled into the natural hot water or steam, known as a geothermal reservoir. The reservoir collects many meters below the groundwater table. Wells bring the geothermal liquid to the surface, where it is converted at a power plant into electricity. An economically competitive geothermal power plant can cost as low as €2,800 per kilowatt installed. The levelized generation costs of new plants vary between €0.04 and €0.08 per kWh$_e$ (GEA, 2008).

Biomass energy
Biomass is one of the most abundant renewable resources. It is formed by fixing carbon dioxide in the atmosphere during the process of plant photosynthesis and, therefore, it is carbon neutral when grown sustainably. Currently, biomass accounts for about 12 percent of today's world energy supply (Koroneos et al., 2008; Saxena et al., 2009). A variety of biomass resources can be used to convert to energy. According to Ni et al. (2006), they can be divided into four general categories:

- energy crops, such as wood, industrial and agricultural crops;
- agricultural residues and waste;
- forestry waste and residues; and
- industrial and municipal waste, for example, municipal solid waste and sewage sludge.

Modern biomass energy can be derived either from wastes/residues, or from dedicated energy plantations of short-rotation trees or grasses. Municipal wastes are likely to be only a minor energy source, given the growing interest in recycling. Agricultural wastes are very large, but often have existing uses. In assessing the future potential of biomass, the yield per hectare and the energy content of the crop are two important factors. These vary widely, which leads to a potential range from 33 to 1,135 EJ annually (Moriarty and Honnery, 2007).

Biomass can be converted into other forms of energy either by thermochemical or by biological processes. Thermochemical processes include combustion, pyrolysis, liquefaction and gasification, whereas the five biological processes are direct biophotolysis, indirect biophotolysis, biological water-gas shift reaction, photo-fermentation and dark fermentation. In the combustion process, biomass is converted into heat, mechanical power or electricity by burning it in air. The energy efficiency of this process varies between 10 and 30 percent and the pollutant emissions are the byproducts, making the combustion process not suitable for sustainable development. In biomass liquefaction, biomass is heated to approximately 600 K in water under a pressure of 5–20 MPa in the absence of air. The main disadvantages of liquefaction are difficulty in achieving the operation conditions and the low energy efficiency (Ni et al., 2006). All other conversion processes are more suitable for hydrogen production and will be discussed in Subsection 2.2.

2.2 Hydrogen Production Technologies

Production of hydrogen can be divided into three main categories: electrolytic, thermochemical and biological hydrogen production. This subsection gives an overview of the main characteristics of the different production technologies that are relevant for renewable hydrogen production. Therefore, technologies such as steam reforming of hydrocarbons, gasification of coal and nuclear hydrogen production will not be considered.

Electrolytic hydrogen production
In water electrolysis, electricity is passed through a conducting aqueous electrolyte, breaking down water into its constituent elements, hydrogen and oxygen via the following reaction:

$$H_2O \rightarrow 2H_2 + O_2.$$

This process is schematically represented in Figure 10.5, whereas Figure 10.6 gives an overview of an electrolysis unit including all peripheral equipment needed.

Three types of industrial electrolysis units are currently being produced. Two involve an aqueous solution of potassium hydroxide (KOH), which is used because of its high conductivity, and are referred to as alkaline electrolysers. These units can be either unipolar or bipolar. The unipolar electrolyser resembles a tank and has electrodes connected in parallel. A membrane is placed between the cathode and anode, which separate the hydrogen and oxygen as the gases are produced, but allows the transfer of ions. The bipolar design resembles a filter press. Electrolysis cells are connected in series, and hydrogen is produced on one side of the cell, oxygen on the other. Again, a membrane separates the electrodes.

The third type of electrolysis unit is a solid polymer electrolyte (SPE) electrolyser. These systems are also referred to as PEM (proton exchange membrane) electrolysers. In this unit, the electrode is a solid ion conducting membrane as opposed to the aqueous solution in alkaline electrolysers. The membrane allows an H^+ ion to transfer from the anode side of the membrane to the cathode side, where it forms hydrogen. The SPE membrane also serves to separate the hydrogen and oxygen gases, as oxygen is produced at the anode on one side of the membrane and hydrogen is produced on the opposite side of the membrane. One of the main advantages

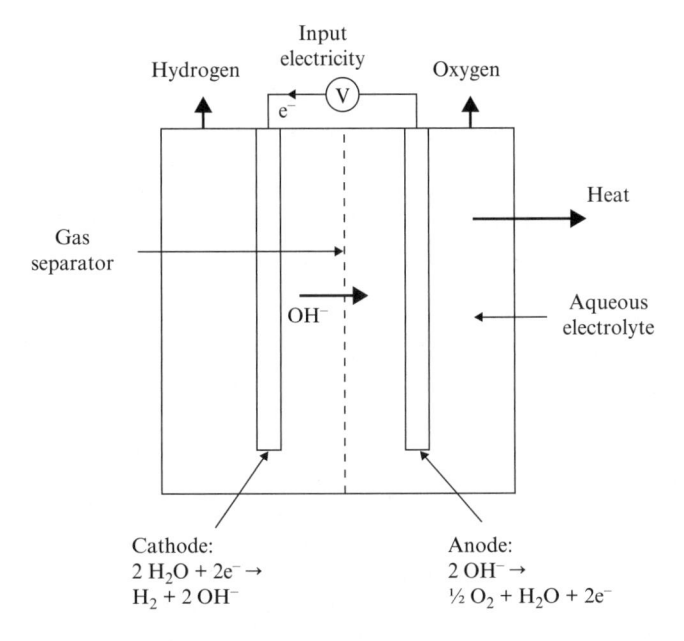

Figure 10.5   Principle of electrolytic hydrogen production

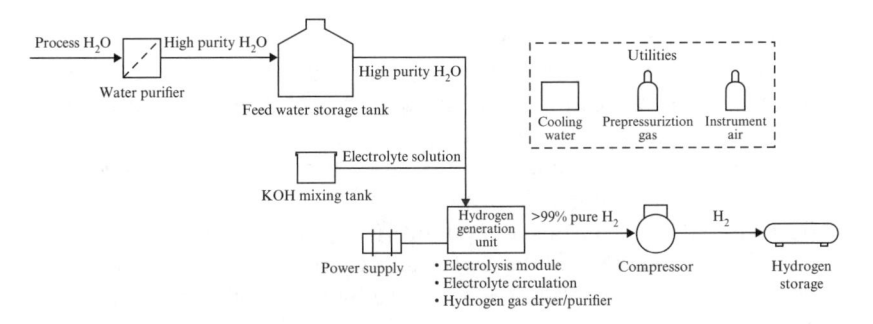

Cathode:
$2 H_2O + 2e^- \rightarrow H_2 + 2 OH^-$

Anode:
$2 OH^- \rightarrow \frac{1}{2} O_2 + H_2O + 2e^-$

Source: Ogden (1999).

Figure 10.6   Electrolysis plant with peripheral equipment

of this technology is that it allows working under high pressure, which has a positive effect on the energy efficiency.

Recently, experimental designs for electrolysers have been developed using solid-oxide electrolytes and operating temperatures at 1,000–1,200 K, or using steam to enhance the process. High-temperature electrolysis systems promise higher efficiency for converting electricity to hydrogen, because some of the work to split water is done by heat, but material requirements are more stringent (Ogden, 1999; Padro and Putsche, 1999; Ivy, 2004).

Water electrolysis can be used to produce hydrogen over a wide range of scales from a few kilowatts to hundreds of megawatts. The energy efficiency of electrolysers is defined as the higher heating value (gross energy) of hydrogen divided by the energy consumed by the electrolysis system per kilogram of hydrogen produced. Energy efficiencies range from 56 to 75 percent, which corresponds with energy requirements of 70.1 kWh$_e$/kg H$_2$ and 53.4 kWh$_e$/kg H$_2$, respectively. An efficiency goal for electrolysers in the future has been reported to be in the 50 kWh$_e$/kg H$_2$ range, or a system efficiency of 78 percent. Note that this efficiency refers to the entire system, peripheral equipment included, and not only to the electrolyser itself.

The investment cost of commercial facilities nowadays amounts to approximately €1,000–5,000 per kW, whereby larger units are generally the cheapest. Yearly maintenance costs fluctuate around 2–3 percent of the investment cost. The production cost of electrolytic hydrogen is strongly and linearly dependent on the cost of electricity. Electrolytic systems are generally competitive, with steam reforming of natural gas only where low-cost electric power is available (for example, industrial customers). The lifetime of electrolyser cell stacks is approximately 15 years (Ogden, 1999; Padro and Putsche, 1999; Adamson, 2004; Ivy, 2004; Greiner et al., 2007).

Thermochemical hydrogen production
As only hydrogen from renewables is considered in this chapter, the overview of thermochemical hydrogen production is limited to pyrolysis and gasification of biomass and the thermal dissociation of water.

Pyrolysis is the heating of biomass at a temperature of 650–800 K at 0.1–0.5 MPa in the absence of air to convert biomass into liquid oils, solid charcoal and gaseous compounds (Ni et al., 2006). Pyrolysis can be classified into slow and fast pyrolysis, whereby only the latter is considered for hydrogen production according to the following reaction scheme:

$$\text{Biomass} + \text{heat} \rightarrow H_2 + CO + CH_4 + \text{other products.}$$

Next, methane and carbon monoxide can be further converted into hydrogen by means of steam reforming or the water-gas shift reaction. The most important pyrolysis reactor types are ablative, fluidized bed, circulating fluidized bed and entrained flow reactors. Hydrogen yields can vary substantially with biomass type, facility size and process conditions but, according to Saxena et al. (2008), biomass-to-hydrogen energy conversion efficiencies can amount to 60 percent, while the estimated hydrogen production cost of biomass pyrolysis is in the range of €8–15/GJ (Padro and Putsche, 1999).

Gasification is the conversion of biomass into a combustible gas mixture by the partial oxidation of biomass at high temperatures, typically in the range of 1,100–1,200 K. Hydrogen can be produced from the gasification gaseous products through the same procedure of steam reforming and water-gas shift reaction. As the products of gasification are mainly gases, this process is more favorable for hydrogen production than pyrolysis. The most important problems of biomass gasification are the formation of tar and ash, which can be overcome by a proper reactor design, appropriate process conditions and the use of additives or catalysts (McKendry, 2002; Ni et al., 2006).

Numerous reactor types exist, among which are fixed bed, moving bed, fluidized bed, circulating fluidized bed and downdraft reactors. Researchers are still optimizing hydrogen production by using various types of biomass at various operating conditions in different reactors. Using a fluidized bed gasifier along with suitable catalysts, it is possible to achieve hydrogen production about 60 volumen percent. In recent years, new gasification methods such as hydrogen production by reaction integrated novel gasification (HyPr-RING) and supercritical water gasification are being developed and tested, resulting in high hydrogen yields and economic competitiveness with other hydrogen production methods (Ni et al., 2006). Finally, biomass can also be gasified in an integrated gasification combined cycle (IGCC). This technology is particularly interesting due to its numerous configurations and possibilities. Biomass can be converted into hydrogen as well as electricity and, furthermore, the IGCC system can easily be adapted to capture the resulting CO_2 streams (Starr et al., 2007; Huang et al., 2008; Klimantos et al., 2009).

At high temperatures, above about 1,800 K, water vapor begins to dissociate into a mixture of H_2, O_2, H_2O, O, H and OH. The extent of dissociation increases with increasing temperature and decreasing pressure (H-ION, 2008). The *thermal dissociation of water* is very similar to the electrolysis process. However, thermal power from concentrated solar radiation can be used without any mechanical or electrical input and without the aid of any catalyst to achieve water dissociation. The main disadvantage of this technology is the stringent material requirements. The materials used in the reactor must be capable of withstanding the thermal cycling and shock brought about by the intermittent nature of solar diurnal and weather cycles.

Biological hydrogen production
The five most important technologies for biological hydrogen production are direct biophotolysis, indirect biophotolysis, biological water-gas shift reaction, photo-fermentation and dark fermentation (Ni et al., 2006).

Direct biophotolysis uses the same processes found in plants and algal photosynthesis, but adapts them for the generation of hydrogen gas instead of carbon containing biomass. In this process solar energy is directly converted to hydrogen via this reaction:

$$2H_2O + \text{'light energy'} \rightarrow 2H_2 + O_2.$$

Two photosynthetic systems are responsible for the photosynthesis process: photosystem I producing reductant for CO_2 reduction and photosystem II splitting water and evolving hydrogen. In this coupled process, two photons are used for each electron removed from water and used in CO_2 reduction or hydrogen formation with the presence of hydrogenase (see Figure 10.7). Whereas green plants lack hydrogenase, microalgae (for example, green algae) have hydrogenase enzymes and can produce hydrogen under certain conditions. As hydrogenase is sensitive to oxygen, this remains a key problem keeping the efficiency of direct biophotolysis rather low (10 percent overall solar conversion efficiency) (Das and Veziroglu,

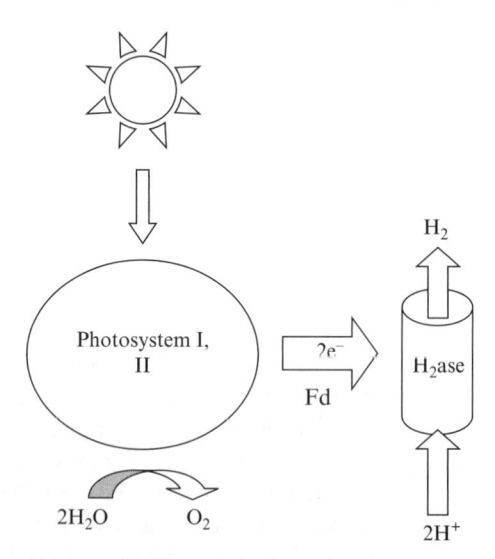

Source: Hallenbeck and Benemann (2002).

Figure 10.7 Direct biophotolysis

2001; Hallenbeck and Benemann, 2002; Ni et al., 2006; Vijayaraghavan and Soom, 2006).

In *indirect biophotolysis*, problems of sensitivity of the hydrogen evolving process[1] are potentially circumvented by separating temporally and/or spatially oxygen evolution and hydrogen evolution (Hallenbeck and Benemann, 2002). Cyanobacteria have the unique characteristics of using CO_2 in the air as a carbon source and solar energy as an energy source. The cells take up CO_2 first to produce cellular substances, which are subsequently used for hydrogen production. In a typical indirect biophotolysis, Cyanobacteria are used to produce hydrogen via the following reactions (Levin et al., 2004):

$$12H_2O + 6CO_2 + \text{'light energy'} \rightarrow C_6H_{12}O_6 + 6O_2;$$

$$C_6H_{12}O_6 + 12H_2O + \text{'light energy'} \rightarrow 12H_2 + 6CO_2.$$

Certain photoheterotrophic bacteria, such as Rubrivivax gelatinosus, are capable of performing a *biological water-gas shift reaction* at ambient temperature and atmospheric pressure. These bacteria can survive in the dark by using CO as the sole carbon source to generate adenosine triphosphate (ATP) coupling the oxidation of CO to the reduction of H^+ to H_2 (Ni et al., 2006):

$$CO + H_2O \rightarrow CO_2 + H_2.$$

In *photofermentation*, photosynthetic bacteria produce molecular hydrogen catalysed by nitrogenase using solar energy and organic acids or biomass. Despite reports of impressive hydrogen production yields, there are several drawbacks to this type of system; first, the use of nitrogenase enzyme with its inherent high energy demand; second, the low solar energy conversion efficiencies; and last, the requirement for elaborate anaerobic photobioreactors covering large areas. In conclusion, the rates and efficiencies of hydrogen production by photofermentation fall far short of even plausible economic feasibility (Hallenbeck and Benemann, 2002; Ni et al., 2006; Manish and Banerjee, 2008).

Hydrogen can be produced by anaerobic bacteria, grown in the dark on carbohydrate-rich substrates. Unlike direct and indirect biophotolysis processes, which produce only hydrogen, *dark-fermentation* processes produce a mixed biogas containing primarily hydrogen and carbon dioxide, but which may also contain lesser amounts of methane, carbon monoxide and hydrogen sulfide. Hydrogen production by these bacteria is highly dependent on the process conditions such as pH, hydraulic retention time and gas partial pressure. Due to the fact that solar radiation is not a requirement, hydrogen production by dark fermentation does not demand much land and is not affected by weather conditions, improving the feasibility of the technology (Hallenbeck and Benemann, 2002; Ni et al., 2006; Vijayaraghavan and Soom, 2006; Manish and Banerjee, 2008).

3 ENVIRONMENTAL IMPACT

Although renewable energy sources are often presented as clean, a more correct definition is that they are cleaner than those based on fossil fuel conversion. Indeed, production and delivery of energy and raw materials, components manufacturing, transport, installation, maintenance, disassembly and disposal will all determine the sustainability of renewable production units. Therefore, in order to correctly evaluate the environmental performance of renewable energy technologies, a life-cycle assessment (LCA) approach is needed. The impact categories of an LCA generally include energy resource consumption, non-energy resource consumption, and emission of GHGs, eutrophication and acidification. This section gives an overview of GHG emissions, expressed as g CO_2-eq per functional unit, that is, 1 kWh_e. A more detailed analysis of other impact categories can be found in Pehnt (2006).

For wind turbines most of the GHG emissions arise at the turbine production (rotor, tower, nacelle) and plant construction (foundation). The emissions related to the construction of the foundation of the power plant can vary widely, as offshore wind turbines require significantly higher amounts of steel and cement than an onshore counterpart. Typically, larger turbines have lower life-cycle GHG emissions than smaller ones. GHG emissions generally lie between 9 and 19 g CO_2-eq/kWh_e for offshore wind turbines, and between 8 and 30 g CO_2-eq/kWh_e for onshore units, with outliers up to 100 g CO_2-eq/kWh_e (Pehnt, 2006; Weisser, 2007; Evans et al., 2009; Martinez et al., 2009).

For photovoltaics the lion's share of GHG emissions is the result of electricity use during manufacturing. There is a wide variation in the results due to different factors, such as the quantity and grade of silicon, module efficiency and lifetime, as well as irradiation conditions. Recently calculated values of GHG emissions of different photovoltaic systems (mono-crystalline, polycrystalline, amorphous, and so on) lie between 20 and 140 g CO_2-eq/kWh_e (Pehnt, 2006; Fthenakis and Kim, 2007; Pacca et al., 2007; Uchiyama, 2007; Weisser, 2007; Evans et al., 2009). Solar thermal technologies such as parabolic trough emit around 15 g CO_2-eq/kWh_e (Pehnt, 2006).

As with wind and solar technologies, most of the GHG emissions of hydroelectric power plants arise during production and construction of the plant. In general, cooler climates, lower biomass intensities and dams with higher power densities (ratio of the capacity of the dam to the area flooded) have lower emissions per kWh$_e$. The type of terrain flooded in dam construction significantly impacts GHG emissions. For example, flooded biomass decays aerobically – producing carbon dioxide – and anaerobically, producing carbon dioxide and methane. This can lead to the emission of large quantities of greenhouse gases that exceed GHG emissions of any other technology. Therefore, GHG emissions of hydroelectric plants can range from 1 to more than 1,000 g CO_2-eq/kWh$_e$ (Pehnt, 2006; Weisser, 2007; Evans et al., 2009).

For biomass plants, GHG emissions vary with the biomass properties, the energy intensity of the fuel cycle, the plant technology and its efficiency. Life-cycle analyses of different biomass technologies report emissions between 10 and 100 g CO_2-eq/kWh$_e$ (Pehnt, 2006; Weisser, 2007; Koroneos et al., 2008).

As far as geothermal energy is concerned, geothermal plants can release CO_2 during operation. The emissions vary greatly from plant to plant and the reported range is from 0 to 400 g CO_2-eq/kWh$_e$ (Pehnt, 2006; Moriarty and Honnery, 2007).

4 GENERAL DISCUSSION AND REMARKS

As promising as the combination of renewable energy sources and hydrogen might look for building a sustainable energy future, this interaction is not always self-evident. In these concluding remarks, some issues are raised that could stand in the way the development of the renewable-hydrogen future.

First, there is the competition between hydrogen and electricity, which is determined by the respective overall energy efficiency between renewable source and end use. Whereas the energy losses in an efficient electrical grid are about 10 percent, these losses amount to 75–80 percent when hydrogen is used as a storage and transport medium (Bossel, 2005). Of course, in an 'electron' economy solutions for energy storage also have to be found, which will similarly result in a decrease of overall energy efficiency. Nevertheless, to some extent the inefficient electricity–hydrogen–electricity pathway seems incompatible with a sustainable energy future.

Second, as renewable hydrogen is seen as a storage medium, facilitating the integration of renewable energy technologies, this means that hydrogen will only be produced when there is an excess of renewable electricity. Since it is generally accepted that the existing electricity grid can readily absorb most of the fluctuating renewable energy produced, as long as the installed capacity is less than 20 percent of the maximum load, the penetration of renewable energy sources needs to be considerable before hydrogen can effectively be produced (Sherif et al., 2005). Recent research in the Netherlands has shown that the production of hydrogen from wind becomes technically and economically viable with about 8 GW or more wind energy capacity installed, which is about 30 percent of the entire Dutch power-generation capacity (Schenk et al., 2007). Moreover, the resulting very low and irregular production of hydrogen has a detrimental effect on the efficiency of electrolysers. That is, the efficiency of an electrolyser is inversely proportional to the cell potential, which in turn is determined by the current density, and that in turn directly corresponds to the rate of hydrogen production per unit of electrode active area (Sherif et al., 2005).

Third, non-intermittent renewable energy sources may also not be available for hydrogen production. Biomass, for example, can also be used to produce biofuels for transport or already has existing uses. According to Moriarty and Honnery (2007), agricultural and silvicultural systems are often needed to meet the world's demand for food, fiber and forestry products. A large fraction of crop, animal and forestry wastes may need to be left in place for soil fertility maintenance, and municipal wastes are likely to be only a minor energy source, given the growing interest in recycling. Also hydroelectricity may not be available for hydrogen production, as the high GHG emissions of some hydroelectric plants do not even make it a sustainable technology.

Fourth, even given that there is an opportunity to produce hydrogen from renewables, albeit at very low and irregular production rates, an infrastructure for hydrogen transport and storage needs to be developed in order to fulfill every customer's demand. When renewable hydrogen production is low, this infrastructure development will turn out to be very costly. In fact, this comes down to the famous chicken-and-egg problem again: who will pay for this large infrastructure development cost, when hydrogen production – and accordingly demand – are still negligible?

However, this hydrogen infrastructure problem could be overcome by also taking into account other hydrogen production technologies, such as reforming of natural gas or gasification of coal, and applying them to ensure a substantial, continuous hydrogen production rate. Nevertheless, the future use of these technologies for hydrogen production seems very unlikely as long as CCS does not become a widely adopted, commercial technology, supplemented with the appropriate infrastructure needs.

Another way to circumvent hydrogen infrastructure issues could be to make use of mixtures of hydrogen and natural gas. Although some problems still need further research, this approach certainly seems possible at the level of distribution and end use without the need for a newly developed infrastructure (IEA GHG, 2003; Haeseldonckx and D'haeseleer, 2007).

In conclusion, from a theoretical and technical point of view, hydrogen and renewable energy sources seem to match perfectly. Hydrogen's storability and transportability offer a solution to the intermittent character of some renewable energy sources. In addition, the use of renewable energy sources perfectly meets the sustainability criterion, which implies a 'de-fossilization' of our future energy economy. However, when looking closely at the technical, economic and ecological aspects of hydrogen from renewables, this pathway turns out to be far from self-evident. Rather than putting all our hopes on the 'sustainable' marriage between hydrogen and renewables, it seems wise not to exclude other hydrogen production technologies, to incorporate less-obvious pathways for the development of a hydrogen infrastructure and to carefully investigate the competition and interaction between hydrogen and other energy carriers.

NOTE

1. The hydrogen evolving process is the production of hydrogen by means of the hydrogenase enzyme reaction. This hydrogenase activity is extremely oxygen sensitive.

REFERENCES

Adamson, K. (2004): 'Hydrogen from renewables – the hundred year commitment', *Energy Policy*, Vol. 32, No. 10, pp. 1231–42.

Bossel, U. (2005): 'Does a hydrogen economy make sense?', European Fuel Cell Forum, available at: http://www.efcf.com/reports/E13.pdf (accessed April 2008).

Das, D. and T.N. Veziroglu (2001): 'Hydrogen production by biological processes: a survey of literature', *International Journal of Hydrogen Energy*, Vol. 26, No. 1, pp. 13–28.

DOE (2004): 'Energy Efficiency and Renewable Energy, Wind and Hydropower Technologies Program', US Department of Energy, August, available at: http://www.eere.energy.gov/windandhydro/ (accessed April 2008).

European Commission (2007): 'Renewable energy roadmap – renewable energies in the 21st century: building a more sustainable future', Communication from the Commission to the Council and the European Parliament, COM(2006) 848 Final, Brussels: European Commission.

EUSUSTEL (2007): 'European Sustainable Electricity; Comprehensive Analysis of Future European Demand and Generation of European Electricity and its Security of Supply', Final Technical Report, no. 006602, February, available at: http://www.eusustel.be (accessed April 2008).

Evans, A., V. Strezov and T.J. Evans (2009): 'Assessment of sustainability indicators for renewable energy technologies', *Renewable and Sustainable Energy Reviews*, Vol. 13, No. 5, pp. 1082–8.

Fthenakis, V. and H.C. Kim (2007): 'Greenhouse-gas emissions from solar electric and nuclear power: a life-cycle study', *Energy Policy*, Vol. 35, No. 4, pp. 2549–57.

GEA (Geothermal Energy Association) (2008): available at: http://www.geo-energy.org/ (accessed November 2008).

Greiner, C.J., M. Korpas and A.T. Holen (2007): 'A Norwegian case study on the production of hydrogen from wind power', *International Journal of Hydrogen Energy*, Vol. 32, No. 10–11, pp. 1500–507.

GWEC (2008): 'Global Wind 2007 Report', Global Wind Energy Council, May, available at: http://www.gwec.net (accessed November 2008).

H-ION (H-ION Solar Inc.) (2008): available at: http://www.hionsolar.com/ (accessed November 2008).

Haeseldonckx, D. and W. D'haeseleer (2007): 'The use of the natural-gas pipeline infrastructure for hydrogen transport in a changing market structure', *International Journal of Hydrogen Energy*, Vol. 32, No. 10–11, pp. 1381–6.

Hallenbeck, P.C. and J.R. Benemann (2002): 'Biological hydrogen production; fundamentals and limiting processes', *International Journal of Hydrogen Energy*, Vol. 27, No. 11–12, pp. 1185–93.

Huang, Y., S. Rezvani, D. McIlveen-Wright, A. Minchener and N. Hewitt (2008): 'Techno-economic study of CO_2 capture and storage in coal fired oxygen fed entrained flow IGCC power plants', *Fuel Processing Technology*, Vol. 89, No. 9, pp. 916–25.

IEA (2003): *Renewables for Power Generation: Status and Prospects*, Paris: OECD.

IEA (2006): *World Energy Outlook 2006*, Paris: OECD.

IEA (2007): 'Photovoltaic Power Systems Programme', June, available at: http://www.iea-pvps.org (accessed November 2008).

IEA GHG (2003): 'Reduction of CO_2 emissions by adding hydrogen to natural gas', IEA Greenhouse Gas R&D Programme, Report No. PH4/24, Cheltenham, UK, October.

Ivy J. (2004): 'Summary of electrolytic hydrogen production, Milestone completion report', US Department of Energy, National Renewable Energy Laboratory, report no. NREL/MP-560-35948, April.

Klimantos, P., N. Koukouzas, A. Katsiadakis and E. Kakaras (2009): 'Air-blown biomass gasification combined cycles (BGCC): system analysis and economic assessment', *Energy*, Vol. 34, No. 5, pp. 708–14.

Koroneos, C., A. Dompros and G. Roumbas (2008): 'Hydrogen production via biomass gasification – a life cycle assessment approach', *Chemical Engineering and Processing: Process Intensification*, Vol. 47, No. 8, pp. 1261–8.

Levin, D., L. Pitt and M. Love (2004): 'Biohydrogen production: prospects and limitations to practical application', *International Journal of Hydrogen Energy*, Vol. 29, No. 2, pp. 173–85.

Lipman, T.E., J.L. Edwards and C. Brooks (2006): 'Renewable hydrogen: technology review and policy recommendations for state-level sustainable energy futures', Institute of Transportation Studies, University of California, Davis, UCD-ITS-RR-06-06.

Manish, S. and R. Banerjee (2008): 'Comparison of biohydrogen production processes', *International Journal of Hydrogen Energy*, Vol. 33, No. 1, pp. 279–86.

Martinez, E., F. Sanz, S. Pellegrini, E. Jiménez and J. Blanco (2009): 'Life cycle assessment of a multi-megawatt wind turbine', *Renewable Energy*, Vol. 34, No. 3, pp. 667–73.

McKendry, P. (2002): 'Energy production from biomass (part 2): Conversion technologies', *Bioresource Technology*, Vol. 83, No. 1, pp. 47–54.

Moriarty, P. and D. Honnery (2007): 'Intermittent renewable energy: the only future source of hydrogen?', *International Journal of Hydrogen Energy*, Vol. 32, No. 12, pp. 1616–24.

Ni, M., D.Y.C. Leung, M.K.H. Leung and K. Sumathy (2006): 'An overview of hydrogen production from biomass', *Fuel Processing Technology*, Vol. 87, No. 5, pp. 461–72.

Nowotny, J., C.C. Sorrell, L.R. Sheppard and T. Bak (2005): 'Solar-hydrogen: environmentally safe fuel for the future', *International Journal of Hydrogen Energy*, Vol. 30, No. 5, pp. 521–44.

Ogden, J.M. (1999): 'Prospects for building a hydrogen energy infrastructure', *Annual Review of Energy and the Environment*, Vol. 24, pp. 227–9.

Pacca, S., D. Sivaraman and G.A. Keoleian (2007): 'Parameters affecting the life cycle performance of PV technologies and systems', *Energy Policy*, Vol. 35, No. 6, pp. 3316–26.

Padro, C. and V. Putsche (1999): 'Survey of the Economics of Hydrogen Technologies', US Department of Energy, National Renewable Energy Laboratory, report no. NREL/TP-570-27079, September.

Pehnt, M. (2006): 'Dynamic life cycle assessment (LCA) of renewable energy technologies', *Renewable Energy*, Vol. 31, pp. 55–71.

Saxena, R.C., D.K. Adhikari and H.B. Goyal (2009): 'Biomass-based energy fuel through biochemical routes: a review', *Renewable and Sustainable Energy Reviews*, Vol. 13, No. 1, pp. 167–78.

Saxena, R.C., D. Seal, S. Kumar and H.B. Goyal (2008): 'Thermo-chemical routes for hydrogen-rich gas from biomass: a review', *Renewable and Sustainable Energy Reviews*, Vol. 12, No. 7, pp. 1909–27.

Schenk, N., H.C. Moll, J. Potting and R.M.J. Benders (2007): 'Wind energy, electricity, and hydrogen in the Netherlands', *Energy*, Vol. 32, No. 10, pp. 1960–71.

Sherif, S., F. Barbir and T.N. Veziroglu (2005): 'Wind energy and the hydrogen economy – review of the technology', *Solar Energy*, Vol. 78, No. 5, pp. 647–60.

Solarbuzz (2008): available at: http://www.solarbuzz.com/ (accessed November 2008).

Starr, F., E. Tzimas and S. Peteves (2007): 'Critical factors in the design, operation and economics of coal gasification plants: the case of the flexible co-production of hydrogen and electricity', *International Journal of Hydrogen Energy*, Vol. 32, No. 10–11, pp. 1477–85.

Taggart, S. (2008): 'Hot stuff: CSP and the power tower', *Renewable Energy Focus*, Vol. 9, No. 3, pp. 51–4.

Turner, J., G. Sverdrup, M.K. Mann, P.-C. Maness, B. Kroposki, M. Ghirardi, R.J. Evans and D. Blake (2008): 'Renewable hydrogen production', *International Journal of Energy Research*, Vol. 32, No. 5, pp. 379–407.

Uchiyama, Y. (2007): 'Life cycle assessment of renewable energy generation technologies', *IEEJ Transactions on Electrical and Electronic Engineering*, Vol. 2, No. 1, pp. 44–8.
Veziroglu, T.N. and F. Barbir (1992): 'Hydrogen: the wonder fuel', *International Journal of Hydrogen Energy*, Vol. 17, No. 6, pp. 391–404.
Vijayaraghavan, K. and M.S.M. Soom (2006): 'Trends in bio-hydrogen generation – a review', *Environmental Sciences*, Vol. 3, No. 4, pp. 255–71.
Voets, P., T. Sels, R. Belmans and W. D'haeseleer (2001): 'Kleinschalige Waterkracht' (Small-scale hydro power) HERN-REG project report, EI/EDV/HERN-(REG)/10.00/FIN, 'Kleinschalige aanwending van hernieuwbare bronnen in Vlaanderen en wetenschappelijke ondersteuning van een beleid voor rationeel energiegebruik (op distributieniveau)' (Small-scale application of renewable resources in Flanders and scientific support of a policy for rational energy use (at the distribution level)), K.U. Leuven Energy Institute, Leuven (in Dutch).
Voorspools, K. and W. D'haeseleer (2006): 'Impact assessment of using wind power', *Solar Energy*, Vol. 80, No. 9, pp. 1165–78.
Weisser, D. (2007): 'A guide to life-cycle greenhouse gas (GHG) emissions from electric supply technologies', *Energy*, Vol. 32, No. 9, pp. 1543–59.
WWEA (World Wind Energy Association) (2008): available at: http://www.wwindea.org (accessed November 2008).
Zoulias, E. and N. Lymberopoulos (2007): 'Techno-economic analysis of the integration of hydrogen energy technologies in renewable energy-based stand-alone power systems', *Renewable Energy*, Vol. 32, No. 4, pp. 680–96.

11. Build-up of a hydrogen infrastructure in Europe

Martin Wietschel, Philipp Seydel and Christoph Stiller

1 INTRODUCTION

The potential benefits of a hydrogen economy are recognized to differing degrees by national governments and supranational institutions, although the pathways and timeframes to achieve such a transition are highly contended. The development of hydrogen-powered fuel cell vehicles that are economically and technically competitive with conventional vehicles is a crucial prerequisite for the successful introduction of hydrogen as an automotive fuel. In addition, there are various other factors that are vital for a successful transition to a hydrogen economy, in particular, the build-up of an infrastructure for supplying hydrogen. Developing a hydrogen infrastructure requires the selection of user centers, deciding on a mix of production technologies, the siting and sizing of production plants, the selection of transport options and the location and sizing of refueling stations. Integrating all this into an existing energy system constitutes a challenging task for the introduction of hydrogen as an energy carrier. The implementation of an operational infrastructure will require considerable investments over several decades and involves a high investment risk regarding the future of hydrogen demand. In addition, the supply of hydrogen needs to be integrated into the energy system as a whole, as its production will affect the entire conventional energy system.

This chapter presents regional hydrogen demand and supply build-up scenarios over time which were created for Europe by considering the available resources as well as national policies and stakeholder interests. The purpose is to evaluate different infrastructure options in economic terms and to derive recommendations for introducing hydrogen as a transport fuel in the next decades. This chapter is based on the EU project HyWays.[1] The objective of HyWays, an integrated project co-funded by research institutes, industry and the European Commission under the

6th Framework Programme, is to develop a validated and well-accepted roadmap for introducing hydrogen into the energy system in Europe. The HyWays project combines technology databases and socio-, techno-economic analyses to evaluate selected stakeholder scenarios for future sustainable hydrogen energy systems. These scenarios are based on member states' visions of the introduction of hydrogen technologies which were developed after extensive interaction between science and stakeholders at over 50 workshops. For each country, the theoretical, optimum economic choice is calculated and evaluated by the member states on an iterative basis. A multinational approach covering, at that time, 80 percent of the EU land area and over 70 percent of the population ensures wide diversity in terms of feedstocks, regional- and infrastructure-related conditions and preferences.

The chapter is structured as follows: Section 2 discusses the deployment of hydrogen end-use applications. Section 3 presents major assumptions and the methodology. Section 4 shows the results of the analysis. Section 5 summarizes the conclusions derived from the infrastructure analysis.

2 DEPLOYMENT OF HYDROGEN END-USE APPLICATIONS

Hydrogen can be used for mobile, stationary, and portable applications. Because of the amount of energy derivable for mobile and stationary applications, an analysis of specific applications follows.

The development of an infrastructure for mobile applications has been the objective of much research. Today's energy and transport system, which is based mainly on fossil fuels, can in no way be evaluated as sustainable. In the light of the projected increase in global energy demand, concerns over energy supply security, climate change, local air pollution and increasing prices of energy services are having a growing impact on policy making throughout the world. Hydrogen in combination with fuel cell vehicles is one of the mostly frequently discussed solutions for a more sustainable transport system. Figure 11.1 shows the market penetration of hydrogen passenger cars until 2050 in the different scenarios as developed by the HyWays project for the EU (HyWays, 2007). The figure also includes the most optimistic world hydrogen penetration scenario developed by the International Energy Agency which also includes light/medium trucks (IEA, 2005).

For mobile applications, the dominant fuel cell type is the proton exchange membrane (PEM) fuel cell, which only functions with pure hydrogen. The situation is different for stationary (high temperature) fuel

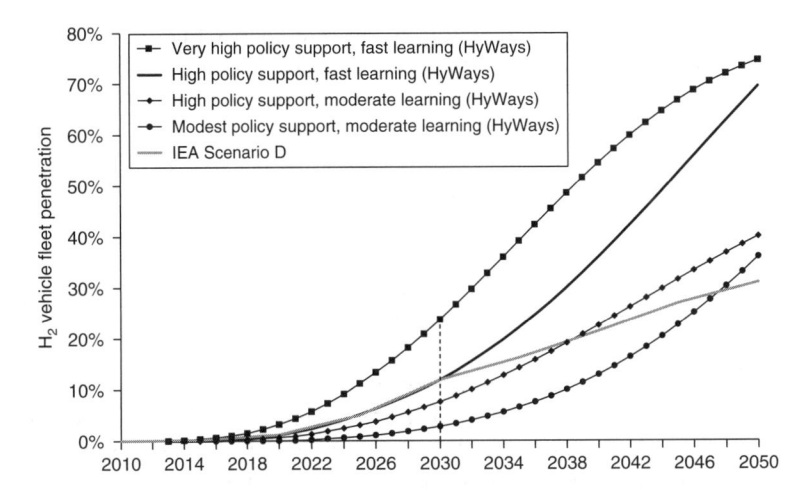

Figure 11.1 Hydrogen vehicle penetration rates

Legend:
- Very high policy support, fast learning (HyWays)
- High policy support, fast learning (HyWays)
- High policy support, moderate learning (HyWays)
- Modest policy support, moderate learning (HyWays)
- IEA Scenario D

Y-axis: H_2 vehicle fleet penetration (0%–80%)
X-axis: 2010–2050

Sources: IEA (2005); HyWays (2007).

cells – and hence distributed heat and power generation – because they can also use, for example, natural gas directly from the gas mains; converting the gas to hydrogen would only reduce their overall efficiency. Unless existing natural gas pipelines can be used to transport and distribute hydrogen, a dedicated hydrogen pipeline will also be required to supply hydrogen to the residential and commercial sectors.

Unlike transport applications, for which biomass-based fuels and CO_2-free or CO_2- lean electricity are the only sustainable alternatives to hydrogen, there are many alternatives to hydrogen when supplying sustainable heat and power to the residential and commercial sectors. These include the use of electricity produced centrally or locally from renewable resources, and the use of 'renewable heat' produced locally by means of solar collectors or heat pumps, or supplied centrally through a biomass-fired district heating system. In addition, the heat demand in houses and buildings can be greatly reduced through improved insulation.

However, there are still some arguments in favor of hydrogen for stationary applications. Hydrogen can be used as a medium for energy storage to remedy the mismatch between energy demand and supply in a renewable electricity system based mainly on intermittent resources such as wind energy. Hydrogen produced locally or centrally during periods of excess electricity can provide back-up power via local combined heat and power (CHP) plants or central power units in periods of limited supply. In the short term, energy storage will play a vital role in the integration of large shares of locally available renewable energy into island energy systems and other stand-alone and weak grid situations. On the other hand, islands and remote areas only make up a small part of the energy demand in the EU member states.

Based on the above arguments, the following description of infrastructure build-up focuses on mobile applications.

3 INFRASTRUCTURE ANALYSIS METHODOLOGY AND KEY ASSUMPTIONS

The economics and aspects of hydrogen infrastructure development have been studied by several groups for Europe (HyNet, 2004; Tzimas et al., 2007) and on a country level (Ball et al., 2006). The methodology of the HyWays infrastructure analysis goes further by studying the 10 countries participating in HyWays (Finland, France, Germany, Greece, Italy, the Netherlands, Norway, Poland, Spain and the UK) individually based on country-specific inputs formulated by a large group of stakeholders. The results give insights into each specific country as well as, aggregated, into a large part of Europe.

Figure 11.2 shows an overview of how the different infrastructure analysis subtasks are organized. Inputs from other parts of the project or from the partners and stakeholders are shown on the left, and subtasks on the right. The tasks can be divided into demand and supply sides, where the hydrogen supply of each region is required to match the corresponding demand, but does not impose any feedback on demand.

Four snapshots T1–T4 are used for the time discretization. These phases are defined by the number of cars that will be on European roads. A connection to the calendar year can be established using hydrogen vehicle market penetration curves elaborated by the HyWays consortium (see HyWays). The infrastructure analysis focuses on the early phase of hydrogen deployment with a relatively low penetration of hydrogen vehicles (up to approximately 8 percent), because regional aspects are the most crucial in this phase.

The spatial resolution for the analyses is based on the NUTS3 classification (NUTS),[2] resulting in around 1,000 areas in the 10 HyWays countries.

In the first snapshot (T1), the fueling stations (FSs) for local hydrogen traffic are situated only in 'early-user centers'. In each country, 4–6 areas or agglomerations were selected based on a qualitative evaluation of a list of regional indicators: local pollution, cars per household, size of cars, stationary use possibility, availability of experts, existing demo-projects,

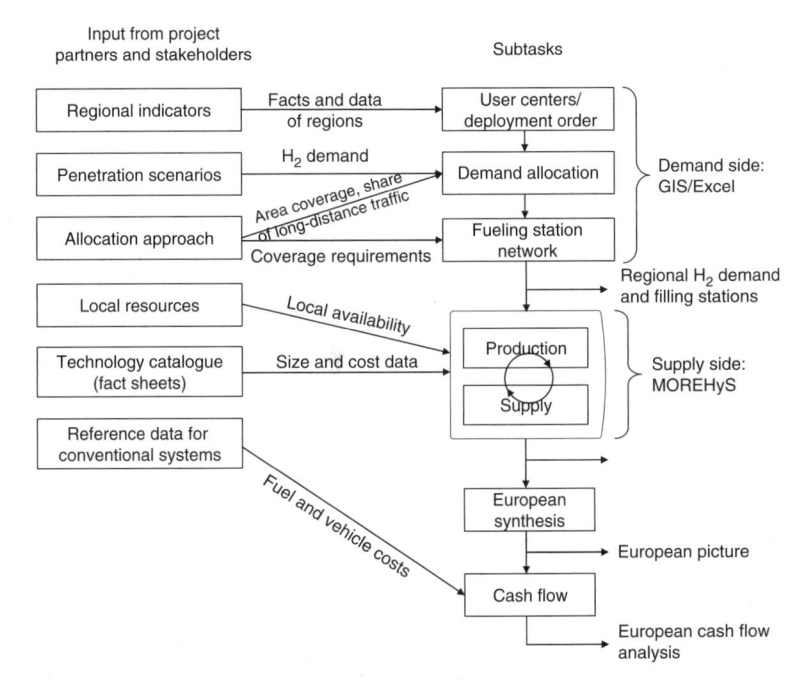

Figure 11.2   Overview of the methodology

advantageous hydrogen production portfolio (renewable energy, byproduct), customer base, political commitment and stakeholder consensus. For long-distance traffic in T1, a few 'early corridors' were selected which mainly serve to connect these early-user centers and allow commuting in their vicinity.

The further regional rollout of hydrogen FSs for local traffic in the later snapshots (T2–T4) is determined by ranking the areas based on weighted socio-economic indicators (catchment population, purchasing power, cars per person). Three different FS capacities are considered (small, medium-sized, and large, with 1, 4, and 10 dispensers, respectively). The number and size share of FSs required in each area is determined according to the calculated local traffic hydrogen demand, but user accessibility must also be guaranteed by a minimum number of FSs in one area and a certain degree of overcapacity in order to compensate fluctuations in FS usage. A common assumption is that 10–30 percent of all conventional FSs must dispense the alternative fuel to ensure broad user acceptance (Nicholas et al., 2005). The long-distance traffic FSs are calculated accordingly, assuming 80 km between two FSs in T2 and T3 on average, and 60 km in T4 (multi-lane roads have one FS on each side).

The production and supply side is analysed mainly using the MOREHyS model (Ball, 2006; Seydel, 2008). MOREHyS is a technology-based (bottom-up), mixed-integer, linear optimization model. The objective function used for the optimization, which is carried out sequentially, is yearly cost minimization for the whole country and the complete supply chain (production to dispensing) in each snapshot.

MOREHyS is used separately for each country. A country is divided into areas, and all capacities and demand are described at this level. Hydrogen demand areas are defined on the basis of NUTS areas. Due to computing limitations, areas with similar indicators are merged. In total, 20–26 regions are distinguished per country and a distinction is made between urban and rural regions. Both types play an important role in the build-up of a hydrogen infrastructure.

A large amount of input data, assumptions and projections were employed to complete the described studies (Figure 11.3). To achieve coherence within the project, assumptions from within HyWays were used wherever possible, and some results were adopted from other models (for example, the projected hydrogen demand from the energy market model Markal). The forecasts for fossil energy costs were taken from the WETO-H_2 study (European Commission, 2006)). Technology costs and performance data are mainly based on the data of the EUCAR–CONCAWE–JRC study (2006), along with reviews and updates by the project partners. Country-specific data such as anticipated feedstocks for hydrogen production were obtained from discussions with the national stakeholders and literature research. For the full set of assumptions and input data used, see the HyWays website, where detailed reports are available.

In the base-case scenario, the shares of hydrogen production feedstocks were set according to stakeholder perceptions, and it was assumed that 20 percent of hydrogen must be dispensed in liquid form (LH_2) in all countries except Norway (100 percent CGH_2). Other scenarios vary with regard to the share of population supplied, when the long-distance road network takes off, market penetration rates over time, unbound feedstock shares, higher bounds for renewables, or whether LH_2 is required.

4 RESULTS

4.1 European Member State Visions of Hydrogen Sources

The vision about how hydrogen could be introduced to the energy system played a major role in HyWays. Over 50 member state workshops were conducted with key stakeholders. In addition, market scenarios for

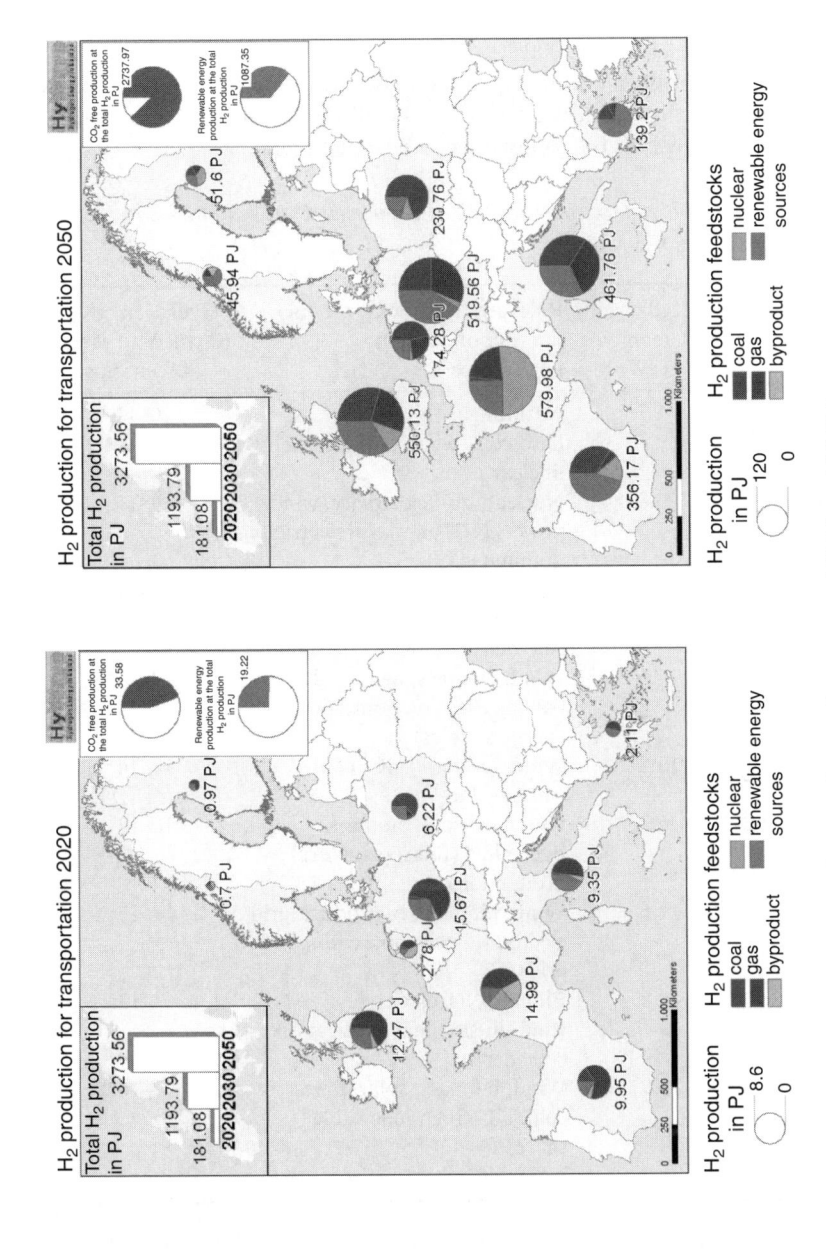

Figure 11.3   Feedstocks for H_2 production for road transport (2020 and 2050)

hydrogen end-use applications were discussed. These were provided by the HyWays partners and outcomes of model analysis and led to further refinement of the member states' visions. Each country outlined its own preferences (see HyWays, 2007).

Table 11.1 shows the source characteristics of the analyzed countries as well as the hydrogen production cost at the filling stations and total hydrogen demand. Overall, the following can be concluded:

Table 11.1 *Characteristics for each of the 10 HyWays countries in the 2025–30 timeframe*

Country	Hydrogen demand in (GWh)	Relevant feedstocks, roughly in order of declining importance	Hydrogen costs (€-ct/kWh), range of scenarios
Finland	1.7	Natural gas (NG), biomass, hard coal, wind, grid electricity, nuclear	10–11
France	25.8	Nuclear, grid electricity, wind power, NG, biomass (electricity dominated)	9–11
Germany	26.1	Hard coal, biomass, wind, byproduct, NG, grid electricity	8–11
Greece	4.6	Wind, biomass, lignite, NG (strong focus on domestic energy sources)	9–16*
Italy	17.8	Wind, biomass, NG, coal, waste, solar	10–14*
Netherlands	6.2	NG, hard coal, biomass, byproduct (focus on central production)	10–13
Norway	1.6	Wind, biomass, byproduct, grid electricity, NG (no existing NG grid)	11–12
Poland	9.6	Biomass, hard coal, lignite, NG, wind (in-situ coal gasification considered)	8–13
Spain	14.9	Wind, biomass, solar, hard coal, NG (high renewable share)	12–16*
UK	21.1	NG, coal, wind, nuclear, waste	10–13

Note: * The high maximum prices mainly result from scenarios with a high share of renewables (particularly wind).

- There is a good distribution of usage of primary energy types across Europe.
- The feedstock use is mainly migrating from fossil fuels in 2020 (about 33 percent) to about 80 percent renewable, CO_2-lean or CO_2-free supply in 2050.

4.2 User Centers and Hydrogen Demand Rollout

Table 11.2 summarizes the variation in the hydrogen demand scenarios. The time snapshots on which the infrastructure analysis is based are related to the number of hydrogen vehicles on European roads. A discrete infrastructure scenario consists of a penetration scenario (when will this number of vehicles be on the roads?), a local use scenario (what percentage

Table 11.2 Demand scenario summary

Phase	T1	T2	T3	T4
Hydrogen vehicles in EU-27	10	500	4 mill.	16 mill.
Penetration scenarios (see Figure 11.1)	Year these vehicle numbers will be realized			
Very high policy support, high learning	2012	2015	2019	2024
High policy support, high learning	2014	2017	2021	2027
Modest policy support, modest learning	2017	2021	2028	2036
Local-use scenarios	% of population in areas with hydrogen fueling stations			
Distributed users	26*	32*	52*	85
Concentrated users	26*	75	90	100
Long-distance road scenarios	Long-distance road network supplied (10 MS)			
Early road network	25,000 km**	70,000 km***	70,000 km***	70,000 km***
Late road network	0 km	25,000 km**	25,000 km**	70,000 km***

Notes:
*   The percentages vary between the countries depending on their structure and choice of early-user centers (for example, Greece: 33% of population live in Athens).
** Early H_2 transit roads.
*** All transit roads.

of the population will have access to hydrogen to drive these vehicles?) and a long-distance road scenario (when will early users be able to commute between the user centers?).

Figure 11.4 shows the early-user centers of all HyWays countries elaborated by the country representatives and the stakeholders based on a list of quantitative and qualitative indicators (see above), and the early hydrogen corridors.

Most countries focus on densely populated areas for the early adoption of hydrogen due to the obvious infrastructure advantages arising from a high density (shorter distribution distances, more users reached per FS). Besides size, indicators such as the availability of hydrogen and technology experts, political commitment, existing demonstration projects and, to some extent, the availability of resources, played a major role in selecting the early-user centers. Some countries also include remote areas in the early-user center portfolio, namely Navarra (Spain), Kyklades (Greece), Koszalinski (Poland), the Shetland Islands (UK), and the Åland archipelago (Finland). This is mainly because stranded renewable energy resources can be tapped, and the need for a transit road network is lower due to the remoteness of these regions leading to a stronger focus on the local use of road vehicles.

The resulting early transit road network focuses on connecting early-user centers within the HyWays countries, but also on international links. Furthermore, the motorways around early-user centers with high population densities should be equipped with hydrogen FSs to facilitate daily commuting in these regions.

Figure 11.4 shows the subsequent regional deployment of hydrogen use in the later snapshots (the lighter the shading, the later the region is supplied with hydrogen) for the 'concentrated-user' scenario, together with the full network of hydrogen corridors. This rollout was estimated by the above described chronological deployment order which was determined based on purchasing power, catchment population and cars per person in the regions, each with weights decided by the national stakeholders. Accordingly, in the later phases, the existing user centers are extended and simultaneously new user centers are developed until almost the entire area of the countries is covered in the last snapshot.

4.3 Fueling Stations

Figures 11.5 and 11.6 provide a geographic picture of the FS distribution for the first (T1) and last (T4) considered snapshot for two scenarios. No distinction is made here between highway and local FSs. The geographic position of the stations is only roughly indicated in terms of the

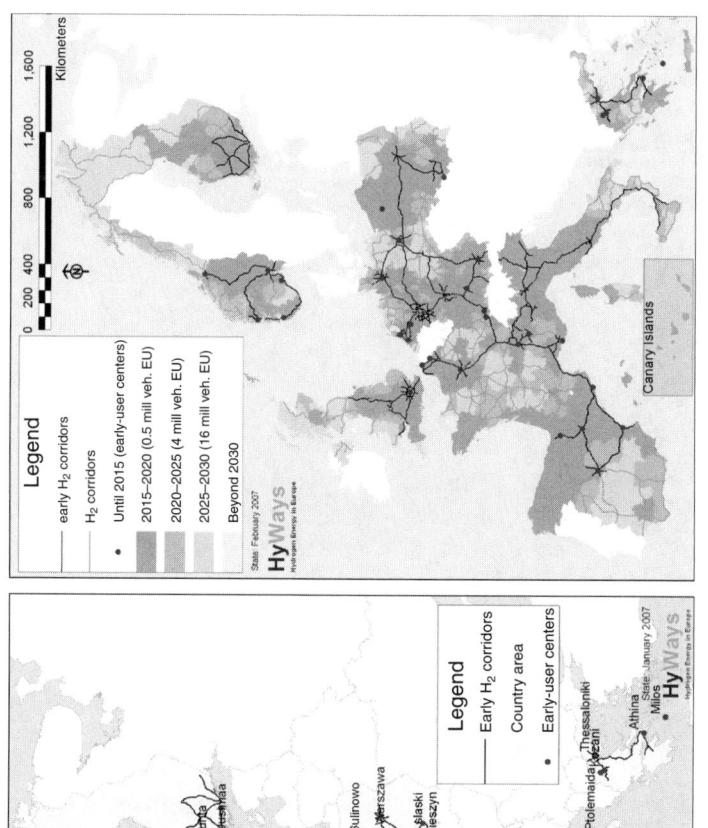

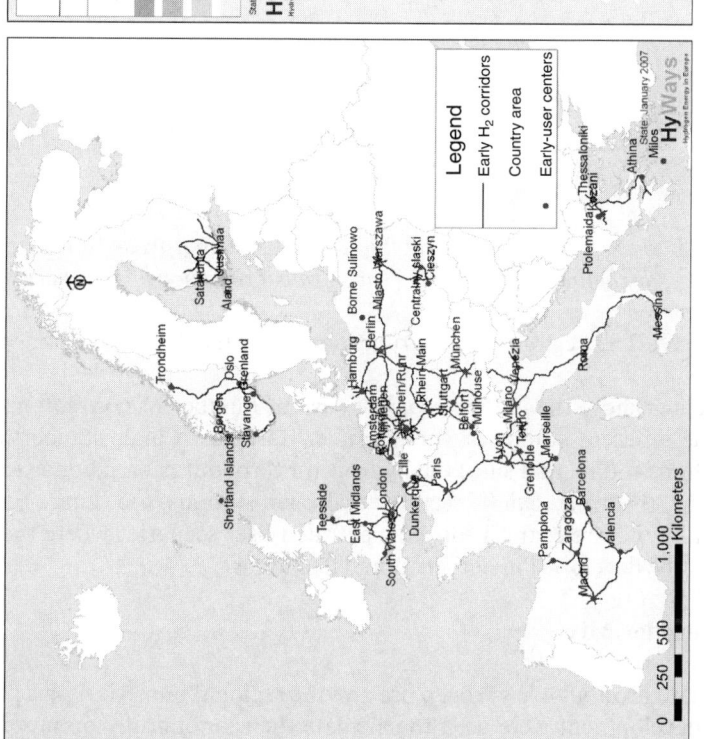

Figure 11.4 Early-user centers for hydrogen in the 10 HyWays countries (left) and time of first supply of all regions (right)

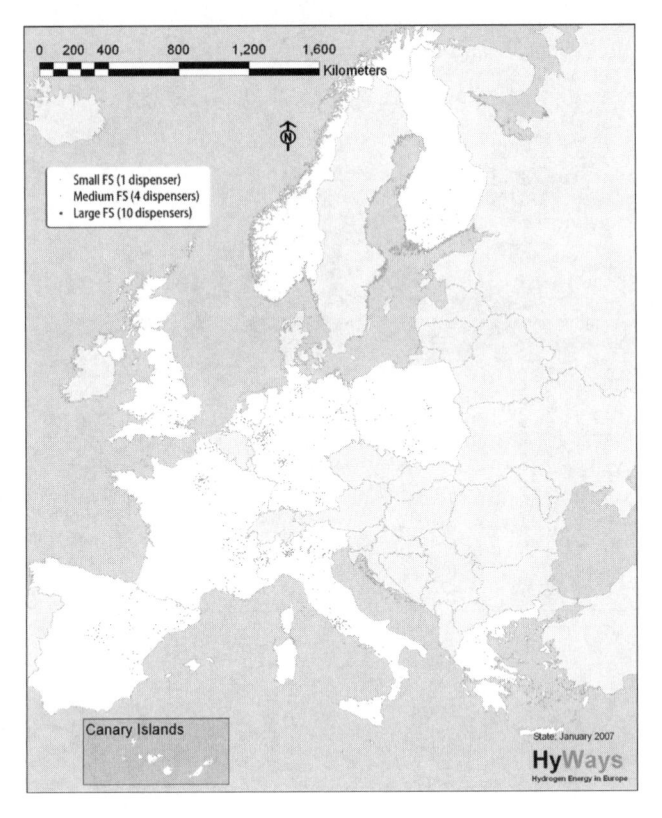

Figure 11.5 Fueling stations during the early-user center phase (T1) – including highway FSs (early transit road network scenario)

respective NUTS3 region. The distribution within the NUTS3 regions is arbitrary.

The figures show the influence of the current number of conventional FSs in the different countries: for example, Italy has a high number of small stations today, and since the hydrogen FS rollout is based on existing stations, this trend can be seen for hydrogen stations, too. Large FSs are basically only relevant in the concentrated-user scenario, where they would first be developed in densely populated areas.

4.4 Production Mix

The estimated rollout of hydrogen use and the regional numeric hydrogen demand development were used to calculate the economically optimized production and supply infrastructure in the 10 HyWays countries, while

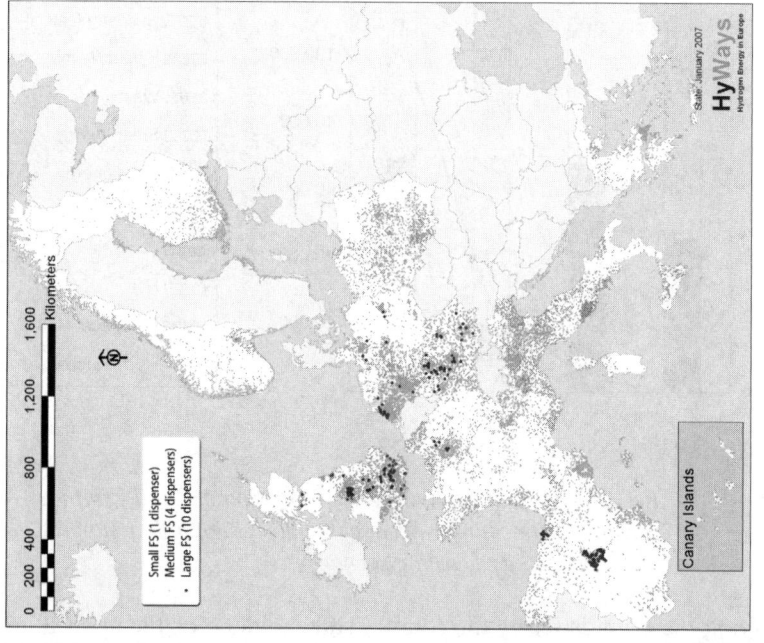

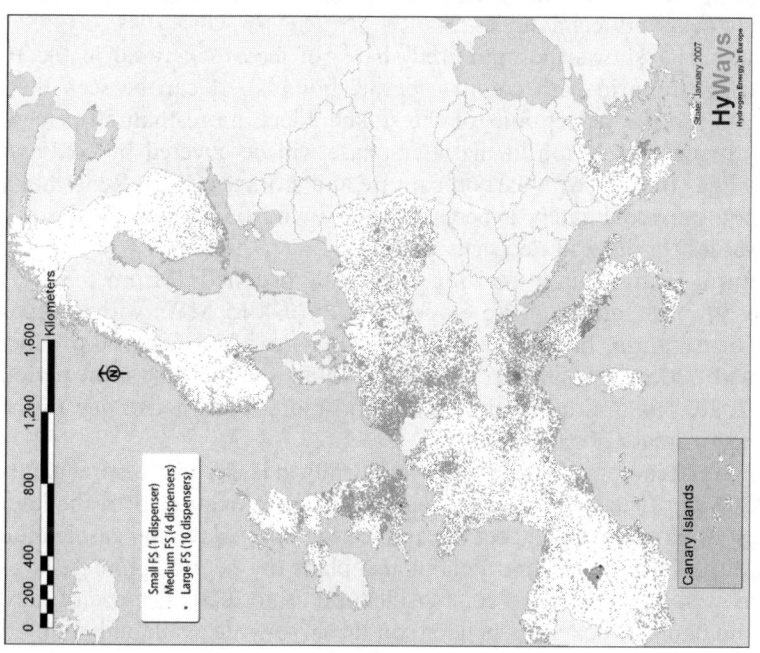

Figure 11.6 Fueling stations in the last phase (T4) – distributed- (left) and concentrated- (right) user scenarios

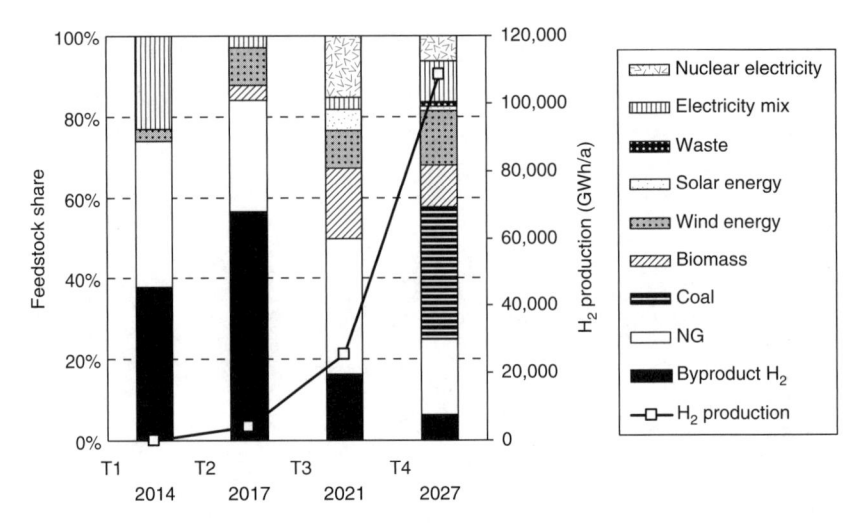

Figure 11.7 Feedstocks used for hydrogen production in T1–T4 in the scenario with country-specific bounds (high policy support, high-technology learning scenario)

satisfying the bounds set by the country stakeholders for the application of feedstocks and technologies.

Figure 11.7 shows the optimized shares of feedstocks used in the 10 countries (scenario with country-specific bounds). It can be seen that, according to the perceptions of the stakeholders, more than 50 percent of hydrogen production in the later phase will be covered by coal and natural gas (mostly with carbon capture and storage: CCS). Renewables contribute approximately 25 percent, mainly wind and biomass, plus some renewables via the grid electricity pathway.

Natural gas is mainly used for small-scale production from the early phase onwards, either onsite or decentrally (0.4–15 MW) with gaseous truck distribution. In later phases, it also plays a role for central production with CCS (>300 MW). However, looking at the long-term period after 2030, this option becomes more and more unattractive due to the assumed increase of gas prices.

Hard coal can only be gasified economically in large-scale central plants (~800 MW). Due to low initial demand these are restricted to the later phases. Most countries expect the application of CCS to coal gasification from 2020–25 onwards, and Poland also plans to operate lignite gasification as well as in-situ hard coal gasification after 2020. The use of hard coal and lignite is heavily dependent on the large-scale availability of CCS technology.

Biomass gasification is the cheapest renewable hydrogen supply option; however, it has restricted potential and end-use competition with biofuels, and other sectors must also be taken into account. Biomass gasification is applied by the model in decentral (~50 MW) and central plants (>300 MW) in later periods.

Wind energy is mainly utilized with green certificates and onsite electrolysis (0.4–2.7 MW), or direct hydrogen production and transportation. To what extent the grid is capable of this depends on grid characteristics and the electricity production mix. Many countries do expect hydrogen to be produced from wind, even if the specific costs cannot compete with most other options. However, further research is needed (for example, into storage of hydrogen in underground caverns) because integrating wind electricity adequately into an energy system is a complex issue.

Onsite electrolysis from grid-mix electricity is expected to some extent in countries where excess electricity may become available (for example, in Norway due to the phasing out of heavy industry, in France due to the high nuclear power share). However, it is questionable whether excess electricity will be used in the hydrogen market or the European electricity market. Given that hydrogen will come into its own in a world with high conventional energy prices, it might be misleading to assume the availability of cheap excess electricity for hydrogen production. This has an impact on liquefaction, too, because cost competitiveness here is sensitive to electricity prices. Choosing onsite electrolysis from grid-mix is, however, also linked to the interest in using lean-CO_2 electricity in the respective countries, as long as CCS is not available.

Nuclear power plants dedicated to hydrogen production are envisaged by the stakeholders in France, Finland, Spain, Poland and the UK during later periods. The last three countries will also rely on nuclear thermocycles (such as the sulphur–iodine cycle, ~1 GW), while Finland and France envisage standard central water electrolysis and high-temperature electrolysis (~2.4 MW per module), respectively. This is a stakeholder view and not necessarily consistent with government positions (for example, in Spain no new nuclear power plants are being considered by the government).

Solar energy for hydrogen production is currently only predicted in Spain and Italy, but restricted to a lower share in later phases. Thermochemical cycles (sulphur–iodine, mixed ferrites) are also envisaged here.

Hydrogen occurring as a byproduct of the chemical industry which is already used thermally today in conventional plants (various capacities) is seen as a cheap option, especially for the initial phase, because it can be substituted by natural gas. However, investments in purification might be needed and have to be evaluated case by case. This type of hydrogen is mainly used where user centers are nearby. It will also be present to a certain extent during later phases, but with a lower share due to its limited potential.

4.5 Role of Transport Options

Figure 11.8 shows the shares of transport and distribution options over time as optimized by the model. Here, central production means that inter-regional transport is necessary, while decentral production is located within the same region as the consumption and only requires distribution to the fueling stations. Centrally produced hydrogen that is transported to the target region via pipeline can be distributed from the pipeline terminal via distribution pipeline or CGH_2 truck. For CGH_2 and LH_2, a distinction is made between trucks transporting within the same region (that is, from decentral production or a reception terminal, that is, from a long-distance pipeline) and long-distance trucks from central plants.

Figure 11.9 shows a static sensitivity analysis for a region with 100 fueling stations. The transport costs have been calculated to show the influence of the different parameters based on certain assumptions about transport distance and cars per FS. In the sensitivity analysis shown, the assumptions were 10 km dedicated distribution pipeline for each FS and 60 €/MWh electricity costs for liquefaction. Onsite technologies are considered if transportation costs exceed 9 ct/kWh of hydrogen and if utilization is not too low (no fewer than 100 cars per FS).

From the accumulated results and more detailed scenario evaluations, the following role of the supply options can be derived:

- Hydrogen production at the FS (onsite) from natural gas or electricity is considered over the whole period studied in areas where the demand is too low for more centralized schemes. Practical problems may hinder the application of onsite technology. For instance, in densely populated areas, the space requirement can be disadvantageous and onsite steam methane reforming might not be an option for FSs with very low initial utilization due to its limited part-load ability and very high specific investments compared to central plants with higher utilization. Due to the fact that low initial FS utilization is assumed in this study, and that natural gas and electricity become more expensive in later phases, onsite production has a relatively low share here. However, this is strongly sensitive to the ingoing assumptions and it can be stated that, in most imaginable scenarios, onsite production will be indispensable in certain locations.

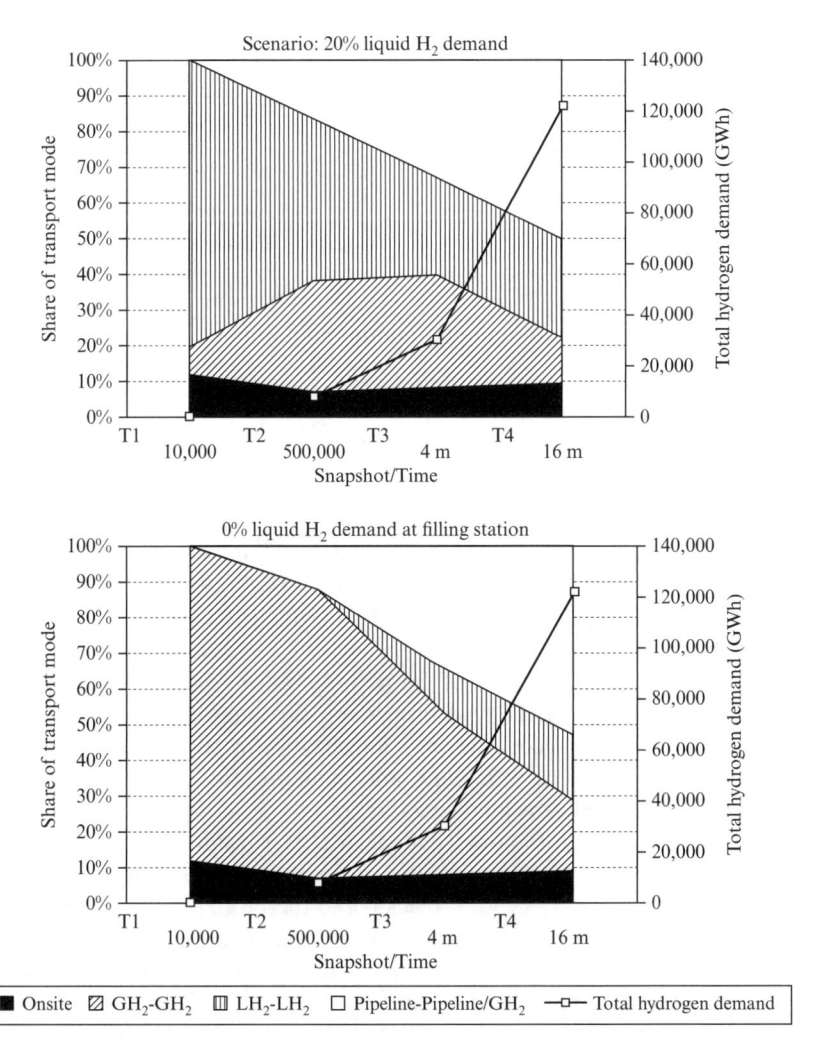

Figure 11.8 Share of hydrogen delivery options over time for the 10 countries (scenarios with 20 and 0 percent liquid hydrogen demand)

- Tube trailer trucks delivering compressed gaseous hydrogen (CGH_2) have high variable costs (also per distance) due to the small volume of hydrogen they hold, but they are flexible and have comparatively low fixed costs. Under the current assumptions they appear to be advantageous for the distribution of decentrally produced hydrogen within regions of intermediate density. The highest share can be

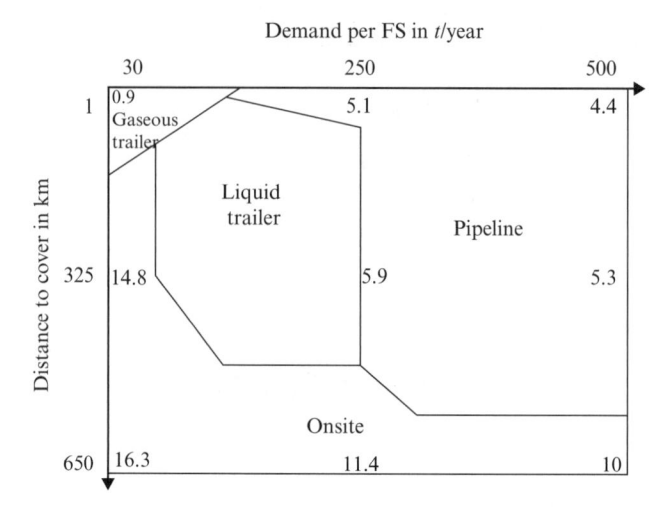

```
                    Demand per FS in t/year
          30              250              500
     1  ┌─0.9──────────────5.1──────────────4.4─┐
        │ Gaseous                               │
        │ trailer                               │
        │         Liquid                        │
        │         trailer                       │
        │                        Pipeline       │
    325 │14.8             5.9              5.3  │
        │                                       │
        │                                       │
        │              Onsite                   │
    650 │16.3             11.4              10  │
        └───────────────────────────────────────┘
   Distance to cover in km
```

Figure 11.9 Sensitivity of transport cost (€ct/kWh) and mode depending on the number of vehicles per FS (not refuelings per day) and transport distance in an area with 100 fueling stations

observed in the middle of the period studied, making CGH_2 trucks a technology for the transition to pipeline delivery. Their flexibility is also advantageous where FS utilization is too low to use onsite production in a technically and economically reasonable way, if delivery distance is less than 100 km.

- Truck delivery of liquid hydrogen (LH_2) results in lower variable costs per km (LH_2 trucks can store approximately eight times as much hydrogen as CGH_2 trucks) and therefore longer distances can be covered between production and use location. One drawback, however, is the required liquefaction process which has high investment and variable costs (due to high energy demand). In the very early phase, the capacity of existing liquefiers may be sufficient. Due to the 20 percent LH_2 demand assumption, LH_2 delivery dominates the early phase when most FSs receive only LH_2 and evaporate some of this to dispense CGH_2. Because onsite and small-scale liquefaction are not considered to be economically viable, vehicles with LH_2 storage can only be supplied with LH_2 trucks. These are mostly used to supply small and medium-sized fueling stations, both outlying ones (for example, along motorways) and in centers where neither onsite production nor CGH_2 trucks are preferable options due to space restrictions and high delivery frequencies, respectively. The decision as to whether hydrogen should be distributed as a liquid or

a gas is very sensitive to the liquefaction energy costs, the transport distance and volume involved.
- Pipeline delivery implies high investment, but negligible variable costs. The investment is in proportion to the distance involved; while the influence of capacity is lower (that is, halving the pipeline capacity will only bring about 9 percent cost reduction for the same pressure drop). Therefore pipelines are most economic over relatively short distances and with high turnover. Distribution pipelines are envisaged for the long-term supply of medium and large FSs in user centers as well as around production sites. For the transportation of hydrogen from a central plant to neighboring regions, pipelines are used to a large extent in later phases. Pipeline distribution is mainly an option for densely populated areas and larger FSs. This indicates that they will become more attractive as hydrogen penetration advances. A positive side-effect is that their intrinsic storage capacity may facilitate the use of intermittent renewable energy sources.
- The dual supply of LH_2 and CGH_2 is an option for later phases. In early phases and for small FSs where 20 percent of the hydrogen is dispensed as a liquid (the rest gaseous), the most economic option is to deliver all the hydrogen in liquid form and then vaporize the part to be dispensed as a gas (by applying cryogenic compression). For medium and large FSs in later phases, dual hydrogen supply may also make sense, that is, with gaseous hydrogen from either onsite or pipeline supply. A positive side-effect is that liquid hydrogen vaporization could serve as a back-up for CGH_2 at little extra cost.

It must be pointed out that Figure 11.8 shows the model results of a scenario which is exposed to many sensitivities (for example, transport distances, FS turnover/cars served, demand for LH_2, energy prices, density of FSs in a region) and should not be regarded as an 'ultimate strategy'. The results show that each of the transport options may play a role under specific conditions. The distance to be covered has the strongest impact on transport costs, which in turn have a much larger impact on the total supply costs of hydrogen than is the case for today's liquid fuels. The primary optimization goal should therefore be to minimize the average hydrogen transport distances through well-planned and -distributed siting of the production plants.

4.6 Costs

Figure 11.10 shows the average specific hydrogen costs (including feedstock, production, transport and refueling), and the cumulated investment

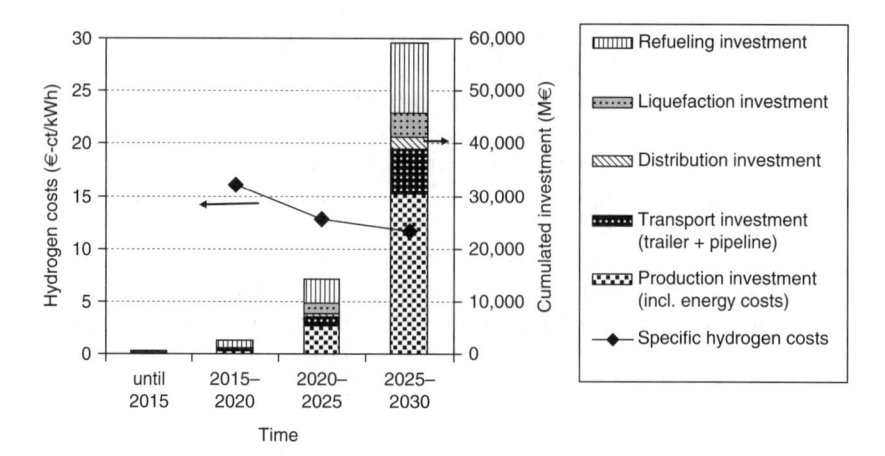

Figure 11.10 *Aggregated total hydrogen costs (base-case scenario with country-specific feedstock bounds and 20 percent LH_2 demand)*

in hydrogen infrastructure aggregated for all countries for the base-case scenario described above.

While refueling dominates infrastructure investments in the early phases, in the later phases it is superseded by production. The total investment of the 10 countries until T4 (that is, to reach a hydrogen vehicle penetration rate of approximately 8 percent) is around €60 billion. It should be noted that conventional fuels also require large investments: for example, the IEA recently assumed that a global investment of as much as US$4,300 billion will be required in the oil sector until 2030 in order to maintain current production levels. Even though a direct comparison of these numbers is not valid, they may provide a relative context to the investment needed for hydrogen infrastructure.

The early phase of hydrogen deployment (that is, approximately 10,000 hydrogen vehicles EU-wide) would show the high specific costs of hydrogen (intentionally left out in Figure 11.10). The main reason for this is the underutilization of the production and supply infrastructure due to technology-related capacity thresholds and the overcapacity of the refueling infrastructure needed for user convenience. The total hydrogen costs are very sensitive to the required number of FSs; establishing an early transit corridor network therefore leads to a drastic cost increase. Furthermore, if liquid hydrogen is to be available in all supplied areas (for example, 20 percent of the total demand), or if other large-scale production technologies are mandatory due to the bound setting of the country stakeholders, this will result in a further cost increase due to plant underutilization. Cost differences between countries can mainly be explained by the use of different feedstocks and differences in population density.

Nevertheless, the total investment for the early infrastructure on a country level is limited to €30–120 million. Assuming approximately 1,000 vehicles per country, this represents a high specific infrastructure investment per vehicle because the FS utilization is assumed to be very low. This is thought to be necessary for the initialization of hydrogen deployment and must be overcome by adequate policy measures. Substantially higher vehicle penetration rates will level out the costs to values between 11 and 16 ct/kWh hydrogen in the medium term. When comparing these numbers to today's fuel costs, the substantial reduction in consumption due to improved fuel efficiency must be taken into account.

The rollout strategy for hydrogen in the snapshots T2–T4 (concentrated users or distributed users) substantially shapes the development of the hydrogen landscape; the distributed user strategy with its early widespread use leads to a more even penetration of hydrogen because more users will have early access to hydrogen. In the concentrated-user strategy, areas supplied later will have a backlog in penetration compared to areas supplied earlier. This also affects the FSs: if a certain hydrogen demand is 'spread' over a larger area, a greater number of smaller stations will be required than if the same demand is concentrated in denser areas. As a consequence of the higher number of stations, the distributed-user strategy leads to 10–20 percent higher specific hydrogen costs in the early phases (T2), which level out to 5–10 percent later (T4).

4.7 Cash-flow Analysis

The cash-flow analysis compares the expenses for hydrogen production, supply and vehicles (as calculated by MOREHyS) with the savings gained from replacing conventional fuel and vehicles over time. The basic assumption for the cash-flow analysis is that each hydrogen vehicle replaces a conventional vehicle.

Two cumulative cash-flow graphs result:

- The *fuel cash-flow* graphs show the balance between the costs for supplying hydrogen fuel (accounted negatively; including feedstock, production and supply infrastructure and fueling stations and taking into account the higher efficiency of hydrogen vehicles) and the savings made in conventional gasoline and diesel fuel (accounted positively; projection of the price without tax at the pump).

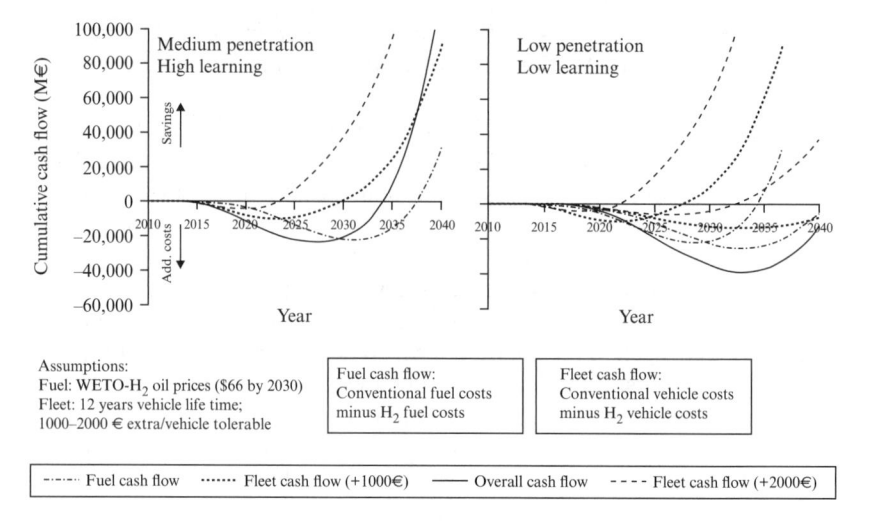

Figure 11.11 Results of the cash-flow analysis

- The *fleet cash-flow* graphs show the balance between investments in hydrogen vehicles (accounted negatively) and investment savings for conventional vehicles (accounted positively).

The cash-flow graphs are scenario specific, that is, all the assumptions made about hydrogen costs, vehicle costs, penetration rates, fuel economy and so on are reflected in the resulting graphs. To obtain a smooth curve over the time period considered, polynomial curve fitting was applied between the time snapshots (T1–T4).

Figure 11.11 shows the result of the cash-flow analysis described above. From the fuel graphs, it can be seen that hydrogen is expected to break even with conventional fuels after 2030, almost independently of the scenario assumptions made about infrastructure rollout and penetration. Of course, energy prices (especially oil) play an important role in the balance. Furthermore, it can be seen that the high specific hydrogen cost in the early phase does not cause high economic losses on a macro scale; in fact the period after 2020 is the most costly with still significantly higher costs for hydrogen than for conventional fuels despite the already high vehicle penetration.

The savings made due to hydrogen after reaching the break-even point are enormous. If the growth in hydrogen demand takes place slowly, the cost of the infrastructure build-up increases significantly (because of the long period of underutilization of capacities). For the full commercialization phase, it can be concluded that hydrogen costs at the filling station in comparison to oil-based fuels are not a relevant barrier to hydrogen if crude oil prices stay above $50/b for densely populated countries (such as Germany) and $60–70/b for less populated countries (such as Norway or Finland).

For the fleet cash flow graphs, it was assumed that either €1,000 or €2,000 per vehicle would be covered by government funding or customers willing to pay more for an environmentally friendly and quieter vehicle (see Potoglou and Kanaroglou, 2006). The fleet graphs show that the scenario assumptions here play a much more important role in reaching the economic break-even point of hydrogen vehicles; that is, the number of vehicles built and the learning rate determine the shape of the curve. The overall cash flow (including €1,000 per vehicle) for hydrogen fuel and the hydrogen vehicle fleet breaks even between 2025 and 2035.

4.8 Context of European Spatial Structure

In 1989, RECLUS, a group of French geographers led by Roger Brunet (Brunet, 2007), presented a study on the development opportunities of urban areas in the European economy. The study was meant to be a warning signal for the public authorities in Paris: since France was not connected to the central growth axis from London towards Milan, the French might fail to benefit from the European single market. The press named this core zone in Europe the 'Blue Banana'. Historians had already identified this area as the backbone of European economic development and, according to them, the Blue Banana dated back to Medieval or even Roman times: it encompasses centuries-old trade routes (the Alpine–Rhine axis) and the borders of Roman-Catholic and German-Protestant Europe. It was along this belt that the Industrial Revolution spread all over Europe from 1800 onwards.

If anything, the Blue Banana shows how long-term structures may continue to be important to the present day. The Blue Banana still differs from other European locations in demographic, economic, infrastructural and cultural–educational aspects. First, the Blue Banana is densely populated and highly urbanized. The area comprises many large or medium-sized cities (for example, London, Amsterdam, Brussels, Frankfurt, Zurich and Milan), in which 40 percent of the EU population (1996) lives.

Statistics show that the regions within the Blue Banana have higher per capita incomes and lower unemployment rates compared with the rest of Europe. In addition, this zone includes large industrial concentrations (for example, the West Midlands and the Ruhr region) as well as strongly developed service centers, particularly in the field of business services, banking and public services.

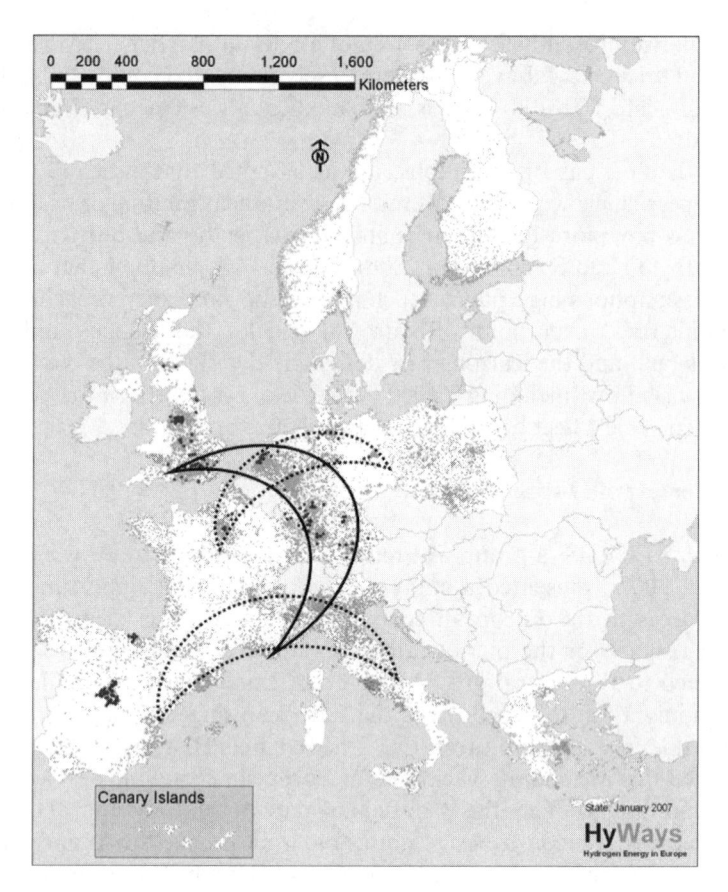

Figure 11.12   Regional use of hydrogen

The Blue Banana has a well-developed physical and telecommunications infrastructure as well as dense traffic networks. Finally, this area attracts attention within Europe because of its relatively large supply of cultural and educational facilities. Nowhere in Europe can one visit as many exhibitions, museums and conferences as in the Blue Banana, and most European universities and colleges are located here.

Figure 11.12 and Table 11.3 show the Blue Banana region and two other regions which seem to be very interesting in terms of population, population density and well-developed infrastructure. These three regions might be worthy of further research into developing an initial European hydrogen infrastructure.

Table 11.3 Population centers in the Blue Banana region

City	Region	Population (millions)
London	United Kingdom	14.0
Paris	France	11.5
Lille–Kortrijk–Tournai	France/Belgium	1.8
Flemish Diamond	Belgium	5.5
The Randstad	Netherlands	10.5
Meuse–Rhine	Belg/Neth/Germany	3.9
Rhine–Ruhr	Germany	12.0
Frankfurt Rhine-Main	Germany	5.2
Basel	Switzerland	0.7
Zurich	Switzerland	1.3
Milan	Italy	9.0
Total		75.4

5 CONCLUSIONS

Hydrogen infrastructure build-up is much more relevant for mobile applications than for stationary ones. How hydrogen supply infrastructure would develop and what this would look like depends heavily on country-specific conditions such as the available feedstocks (such as renewable energies), population density and geographic factors, and must therefore be assessed on a country-by-country basis. Nevertheless, it is possible to derive some robust strategies and cross-national communalities.

Hydrogen infrastructure rollout is seen as taking place predominantly in population centers and also to some extent in remote or less populated areas to start with. Further drivers for the selection of such seedpoints are the political commitment and importance of the regions as well as their activities and innovativeness in research and development (mainly in hydrogen and fuel cells) which ensure political support of the process and also its visibility among interested individuals. The availability of cheap, local, primary energy resources for hydrogen production is regarded as only a minor point. It can be concluded that a purely economic optimization of the regional demand allocation adapted to the supply options is not really desired in an industrial context.

There are high specific hydrogen costs in the early phase due to the required overcapacity of the supply infrastructure and high-technology investments because of the early phase of technology learning. However, the cash-flow analysis shows that the total economic impact of the early phase is minor compared to later phases due to the comparatively low turnover here. It can be concluded that gradually building up infrastructure, concentrating on agreed user centers at first, effectively eliminates the often cited chicken-and-egg problem. In order to make hydrogen an attractive fuel and facilitate its deployment among users, hydrogen supply along an early road network is required, but this also keeps the total initial investment in infrastructure comparatively small. Long-term hydrogen costs of 11–16 ct/kWh can be achieved. The technology progress of end-use technologies, and particularly the cost reductions and improved performance of fuel cells and on-board storage technologies constitute the most influential cost factors.

Assuming that 20 percent of the hydrogen demanded will be supplied in liquid form, initially, most of the hydrogen will be delivered by LH_2 trucks (with evaporation for CGH_2 demand). In later phases, pipelines will gradually take over the transportation and distribution of gaseous hydrogen. Pipelines for medium and large fueling stations may become relevant once a significant market penetration of hydrogen vehicles has been achieved, but these are mostly used for local distribution in highly populated areas and for large-scale interregional energy transport. Along with decentral, regional production, CGH_2 truck distribution is one solution for the transition phase to pipelines. In less populated and remote areas, onsite supply and LH_2 transport remain the most economic choice even in later phases.

Despite the high weighting of transport costs, more than half the hydrogen required may come from large, central production plants in combination with interregional transport in all phases. This underlines the fact that it is important to consider larger regions and the interconnections between them when targeting an economically optimized build-up of hydrogen infrastructure. Well-planned siting of the production plants is essential to minimize transport costs.

The cash-flow analysis shows that hydrogen fuel and fuel cell vehicles can become competitive with conventional fuel and vehicles (at oil prices over 60 $/b) between 2025 and 2035 under the given framework assumptions. This applies to both scenarios with high penetration and quick technology improvement and those with lower penetration and moderate technology learning, although the break-even point is delayed in these cases.

NOTES

1. For detailed information about HyWays, see: http://www.hyways.de/.
2. NUTS: Nomenclature of Territorial Units for Statistics, used by the European Union,

Regions at NUTS3 level have a population size between 150,000 and 800,000. More information at: http://ec.europa.eu/eurostat/ramon/nuts/homc_regions_en.html (accessed February 3, 2010).

REFERENCES

Ball, M. (2006): 'Integration einer Wasserstoffwirtschaft in ein nationales Energiesystem am Beispiel Deutschlands' (Integration of a hydrogen economy in a national energy system, using the example of Germany), Dissertation, VDI Fortschritt-Berichte 16, Nr. 177, VDI Verlag, Düsseldorf.

Ball, M., M. Wietschel and O. Rentz (2006): 'Integration of a hydrogen economy into the German energy system: an optimising modelling approach', *International Journal of Hydrogen Energy*, Vol. 32, No. 10–11, pp. 1355–68.

Brunet, R. (2007): available at: http://www.ersa.org/ersaconfs/ersa02/cd-rom/papers/210.pdf (accessed May 2007).

EUCAR, CONCAWE and JRC (2006): 'Well-to-Wheels Analysis of Future Automotive Fuels and Powertrains in the European Context', 2nd edn, May, available at: http://ies.jrc.ec.europa.eu./WTW.html (accessed February 3, 2010).

European Commission (2006): *World Energy Technology Outlook – WETO H_2*, Brussels.

HyNet (2004): The European Thematic Network on Hydrogen Energy, Network of Excellence under the 5th Framework Programme of the European Commission, 2001–2004, available at: http://www.hynet.info (accessed May 2008).

HyWays (2007): 'Hydrogen Energy in Europe', Integrated project under the 6th Framework Programme of the European Commission to develop the European Hydrogen Energy Roadmap (contract No. 502596), 2004-2007, available at: www.hyways.de (accessed May 2008).

IEA (2005): *Prospects for Hydrogen and Fuel Cells*, Paris: OECD.

Nicholas, M., J. Weinert and M. Miller (2005): 'Hydrogen station economics and station siting in Southern California', in *Electric Vehicle Symposium 21*, Monaco, April 2–6, 2005.

Potoglou, D. and P.S. Kanaroglou (2006): 'Household Demand and Willingness to Pay for Alternative Fuelled Vehicles', Center for Spatial Analysis, CSpA Working Paper 18, April, McMaster University, Canada, available at: http://sciwebserver.science.mcmaster.ca/cspa/papers/CSpA%20WP%20018.pdf.

Seydel, P. (2008): '*Entwicklung und Bewertung einer langfristigen regionalen Strategie zum Aufbau einer Wasserstoffinfrastruktur – auf Basis der Modellverknüpfung eines geografischen Informationssystems und eines Energiesystemmodells*' (Development and assessment of a long-term regional strategy to build up a hydrogen infrastructure – linking to a geographical information system and an energy system model), Dissertation, ETH Zurich.

Tzimas, E., P. Castello and S. Peteves (2007): 'The evolution of size and cost of a hydrogen delivery infrastructure in Europe in the medium and long term', *International Journal of Hydrogen Energy*, Vol. 32, No. 10–11, pp. 1369–80.

12. The contributions of the hydrogen transition to the goals of the EU energy and climate policy

Anders Chr. Hansen

1 SOCIETAL PRIORITIES IN ENERGY AND TRANSPORT POLICY

1.1 The Integrated Energy and Climate Policy

The goals of the integrated European energy and climate policy are to achieve security of supply, competitiveness and environmental sustainability. Advances in energy efficiency have a positive effect on all of these goals and may therefore be seen as a goal in itself as well as a means to achieve the other goals. European Union (EU) countries share these goals in their national energy policies as well as in the EU policies.

The energy used for automotive transport is particularly critical and the EU has initiated a large number of research, development, and demonstration projects focusing on the use of hydrogen and fuel cell technology in transport. These activities give rise to the more general question about the extent to which a transition from petrol and diesel to hydrogen as transport fuel will contribute to achieving the goals of the European energy and climate policy. This is the question addressed in this chapter.

1.2 Transitions to Hydrogen Fuels in Transport

The transport sector is almost entirely fueled by oil products as it has been for almost a century. The technological development in fuels, vehicles and related infrastructures are 'locked in' to the specific configuration of oil products and combustion engines. Basing transport exclusively on crude oil is, however, unsustainable (IEA, 2007). The reserves are limited and the competition for them is increasing. The environment suffers from the emissions of fossil fuel combustion. The power that it confers on whoever controls the remaining reserves is undesirable for Europe.

Against this backdrop, we are at the beginning of a transition in transport away from oil to something else, and hydrogen is one of the options.[1]

A future transition to hydrogen is a challenging process that involves government coordination at all levels and consistently though a long period of many decades. For such a sustained all-European government policy to be realistic, the transition will have to contribute significantly to the overall policy goals listed above.

The electro-motor is in many respects superior to the internal combustion engine (ICE). It can be much more energy efficient, has no tail-pipe emissions, a low noise level, and is in some respects less complex. The problem is that electricity is difficult to store on-board and thus difficult to use for automotive transport. Automotive transport carries its own energy source and the system must have a high energy content related to volume and weight (volumetric and gravimetric energy density). Hydrogen and fuel cell technology offers such a high energy density power-train.

Hydrogen is like a convertible currency. Any source of energy can be converted to hydrogen and the fuel cell converts the hydrogen to electricity. Hydrogen can even serve as fuel for an ICE or it can enrich and improve other combustible fuels. Thus, hydrogen can serve as a bridge between any other primary energy source into use as a transport fuel. It is, however, important to note that hydrogen is only an energy carrier, not an energy source. Thus, the answers to the question addressed in this chapter must depend on the primary energy sources from which the hydrogen is produced.

2 THE BEGINNING OF THE TRANSITION

2.1 The Fuel Cell Electric Vehicle (FCEV)

The fuel cell electric vehicle (FCEV) is an electric vehicle just like the battery electric vehicle (BEV), the hybrid electric vehicle (HEV) and the plug-in hybrid electric vehicle (PHEV). HEVs have been sold since 1997 and BEVs were marketed in a period in the 1990s and in the early 2000s, but with little success. Several large car producers have announced that they will reintroduce BEVs and PHEVs in 2009–11.

The drawback of the BEV is the low gravimetric energy density of batteries resulting in a relatively short range of even a heavy battery (for example, 150 km per charging, lower in a cold climate) combined with lengthy recharging and uncertain durability of batteries. There is a heated debate about whether innovation can change these properties significantly, but even BEVs with these properties can meet the requirements of a significant share of car users, particularly if the plans for building a network of battery exchange stations are realized.[2]

The market segment that requires a longer range per refilling is already offered a range of advanced energy-efficient internal combustion vehicles (ICVs) and there is little doubt that these and their descendants will gain an increasing market share. Hydrogen also applies as fuel for a combustion engine (H2IC) with few emissions. Car users who prefer electric driving, but with a range, performance, and refilling comparable to ICVs will be able to choose PHEVs, providing an extended range based on ICE technology. FCEVs will offer a fully zero emission solution, that is, a full electric mode with a range, performance, and refilling comparable to ICVs. FCEVs will definitely be more expensive than BEVs when they are introduced in large numbers at the market, but they will address another market segment.

There are still important technological breakthroughs that must be achieved for the fuel cell technology to be a realistic option for mass production. The major challenge seems to be the fuel cell itself, which today is made with heavy use of platinum as catalyst and with insufficient durability. Cheaper and more accessible catalysts and longer durability are necessary to achieve comparable costs. The ideal solution is a two-way fuel cell that can serve as an electrolyser, that is, produce hydrogen from electricity, when it does not use hydrogen to produce electricity, but this still seems a rather distant option.

Another research priority is to develop low-weight solid materials suitable for absorption and release of hydrogen with little energy loss. Breakthroughs in this technology are desirable, but not necessarily as crucial as breakthroughs in the fuel cell technology. Demonstrations of the mastered technology of hydrogen storage under high pressure suggest that it could be a workable solution until efforts to achieve low-weight solid hydrides are successful.

There are widely differing views about whether and when such breakthroughs will take place. The European Hydrogen & Fuel Cell Technology Platform (European HFP, 2007) is now the Joint Technology Initiative for Fuel Cells and Hydrogen, a public–private partnership for development and deployment of these technologies in Europe. The program anticipates introduction of FCEVs on the European market in numbers of 400,000 to 1,800,000 a year in the 2015–20 period. This will allow 'mass roll-out' in the 2020s.

From 1993 on the US Department of Energy has pursued research and development programs to develop a new generation of cars in cooperation with the major US car producers. The development in the 1990s followed a diversified portfolio of technologies including advanced ICVs, BEVs, PHEVs, HEVs, and FCEVs with the aim of presenting prototypes in 2004. In the 2000s, however, the perspective changed to a more long-term perspective of FCEVs. In 2009, government priorities changed again towards allocating more funds to the near-term options such as batteries and advanced ICE technology and concentrating the hydrogen and fuel cell efforts more on solving the most critical problems such as fuel cell cost and durability. The US Department of Energy (DOE, 2007) aims to push the technologies towards specified performance and cost targets for 2015. Achieving these targets will enable decisions on commercialization to be made by about 2020.

The California state program for advancing zero emission vehicles (ZEVs = BEVs and FCEVs) and partly zero emission vehicles (PZEVs = PHEVs, HEVs, H2ICs and so on) has been effective since 1990. It is probably the most ambitious program for advancing the use of these technologies. The zero emission program demands from car makers at the California market that they supply at least 25,000 FCEVs or 7,500 BEVs and 58,333 H2ICs (or similar) in 2012–14 to the Californian market (CARB, 2008).

Frontrunners such as Daimler and Honda plan to start production in modest series (Honda 200 vehicles over three years) in the near future, but other car manufacturers plan to engage in FCEV production in 2020 or later.

Similarly, in Japan, the car industry plans to begin an early commercialization phase in 2015 to prepare for assembly-line mass production of FCEVs at a later date (FCCJ, 2008).

It is not the purpose of this chapter to settle the debate on when commercialization of FCEVs can be expected to take place. Rather, in the following we shall assume that it will happen at some time in the 2015–25 period. This will mean that the FCEV can be produced at competitive costs, performance, and durability at some point of time in that period.

It should also be noted that the hydrogen and fuel cell technologies are also developed for other than automotive uses. Commercial opportunities for fuel cells are already identified and exploited in 'early' niche markets such as forklifts, in emergency power generator back-up, in portable equipment such as laptops, and in stationary use as combined heat and power units. Synergies between innovative progress in these different fields should advance the technology development for FCEVs.

With the timeframe assumed above, most of the contributions from the hydrogen transition to achieving the three goals will occur well after 2020, and thus it is an important assumption that the policy continues after the EU 2020 targets have been reached.

2.2 The Primary Energy Basis for Hydrogen

At the current hydrogen market, hydrogen is a chemical rather than a fuel. Refineries use increasing amounts of hydrogen for desulfurization and upgrading of heavier oil fractions. Ammonia production is another large hydrogen consumer. It is also used for numerous chemical processes involving among others metal, methanol, and plastics. Most of the hydrogen is produced by steam reforming of fossil fuels, in particular natural gas. A small fraction is supplied by electrolysis. It is produced by both technologies as a byproduct as well as an on-purpose product.

An expanding market for hydrogen as a transport fuel will change the properties of the hydrogen demand from a few large to many small consumers and from chemical industry-intensive areas to car-intensive areas. It will require a storage and transport network which differs from the current one by a finer grid of pipelines and more delivery of compressed rather than of cryogenic[3] hydrogen. Moreover, in the boundaries of expanding hydrogen delivery networks, onsite production of hydrogen can be expected to supply the transport hydrogen demand. That is, hydrogen production in small- or medium-scale natural gas-based plants or electrolysers at fuel stations.

Future technologies for hydrogen production include among others high-temperature electrolysis, gasification of biomass and hydrocarbons and separation of hydrogen from the gas, and microbiological processes. Combining coal and biomass gasification with carbon capture and storage (CCS) would enable a sustainable use of the coal reserves in countries that still possess large reserves. The energy loss, emission leaks, infrastructures, and costs associated with this technology are, however, still too uncertain to determine its future competitiveness. Research and demonstration projects planned in the EU will probably make this knowledge available before 2020.

The performance and properties of these technologies are in the nature of the case unknown and it is even unknown whether they will be practical options in 2020 or later. When they are, they will, however, have to be competitive with regard to the important properties. Consequently, the role of the transition in achieving the societal goals will be considered mainly in the light of the two main hydrogen production technologies that have already been mastered, and the primary energy sources that can be expanded in the 2015–25 perspective. These pathways are shown in Table 12.1.

The table shows that hydrogen can be produced directly from combustible fuels by gasification, partial oxidation, or steam reforming or from non-combustible resources, primarily by low- or high-temperature electrolysis (water splitting). Hydrogen could in principle be produced from electricity generated from combustible resources, but it would entail

Table 12.1   Types of primary energy feedstock transformable to hydrogen

	Renewable	Non-renewable
Non-combustible (LT/HT electrolysis)	Hydro, wind, wave, tidal, geothermal, PV, microbial	Nuclear
Combustible (gasification, steam reforming)	Biomass	Fossil: oil, gas, coal, tar sands

an unnecessary conversion loss and is thus excluded. Biomass-based fuels often make most sense as a replacement for coal in power and heat generation.

Currently, hydrogen is often produced in combination with other products. In oil refining, hydrogen is a byproduct as well as a main product used as an input to refining processes or for heating. Chlorine-alkali plants produce hydrogen as a byproduct of electrolysis, and ammonia production uses it as an input. The industrial gas industry produces hydrogen in combination with other gases.

In the future, hydrogen will probably also be produced in combination with power, heat, other gases and chemicals. Numerous other synergies can be expected to be exploited. Production of hydrogen without byproducts will most likely take place at hydrogen filling stations and in other small- and medium-scale applications, but even at this scale ongoing innovation pursues synergies and byproducts.

Thus, the following analysis is confined to assumptions of the key parameters that are certain to determine hydrogen fuel cost: efficiencies of transforming non-combustible and combustible energy sources to hydrogen and the non-energy costs of production plants and infrastructure. Fewer assumptions should make the analysis more transparent.

3   ENERGY EFFICIENCY

The most attractive feature of the fuel cell technology is probably its superior energy efficiency as a part of a hydrogen-fuel cell-electro-motor power-train, that is, in the tank-to-wheel (TtW) part of the fuel chain. On the other hand, the energy loss in hydrogen production, that is, in the well-to-tank (WtT) part of the fuel chain, is considerable. The total well-to-wheel (WtW) efficiency is potentially superior to even advanced ICVs or hybrid solutions.

Earlier studies of feasible scenarios for the introduction of hydrogen

Table 12.2 Expected efficiency advantage of FCEVs above grid-independent HEVs with advanced ICE technology in 2010 (%)

ICE technology*	WtW (2010+)	GREET (2015–20)
Port Injection Spark Ignition (PISI)	48–72	58–59
Direct Injection Compression Ignition (DICI)	50–55	45–47

Note: * PISI and DICI are basically petrol and diesel engines, respectively, but they can be adapted to various alternative fuels.

Sources: Author's calculations based on Edwards et al. (2007); Argonne National Laboratory (2008).

and fuel cell technology in automotive transport often compared the high efficiency of the fuel cell vehicle (FCEV) with rather fuel-inefficient ICE vehicles. The Alternative Fuels Contact Group (2003) assumed the fuel cell system to be 100 percent more efficient than an ICE system. The US National Academy of Science (2004) assumed a 66 percent efficiency advantage[4] of FCEVs over ICVs. Ogden et al. (2004) assumed an efficiency advantage of 79 percent and the International Energy Agency (2005) an efficiency advantage of 82 percent relative to advanced ICVs.

When FCEVs are introduced to the market, however, they will most likely compete with vehicles that are far more energy efficient. Table 12.2 shows the efficiency advantages of FCEVs over future grid-independent HEVs expected by the European WtW database (Edwards et al., 2007) and the GREET model in the US Argonne National Laboratory (2008).[5]

Table 12.2 shows that the expected fuel efficiencies of the nearest competing technologies are not that far from the expected fuel efficiency of the FCEV. The efficiency of FCEVs is outstanding, but compared to its future competitors, one should not assume an efficiency advantage of more than 50 percent.[6] This assumption is also used in the comprehensive study of feasible hydrogen roadmaps in Europe, the Hyways Project (2007).

The flipside of the high energy efficiency of all grid-dependent electric vehicles (including BEVs, HEVs and FCEVs) is the high energy loss in transformation of primary energy to power and hydrogen. Natural gas reforming and electrolysis are used today with conversion efficiencies of 60–65 percent (see Hansen, 2007b, 2007e). For transport fuel use, it will also be necessary to use energy for compression, filling and so on, today amounting to maybe 7–14 percent of the hydrogen produced.

It is difficult to predict system efficiency of hydrogen production from natural gas and electricity in 2015–25. The European HFP (2007) aims for a conversion efficiency by low-temperature electrolysis of above 70 percent in 2015 and the US DOE (2007) for 71 percent in 2017, but in system efficiency, energy use for compression and refilling also has to be taken into account. For the conversion efficiency in natural gas reforming, the US target is 75 percent in 2015 whereas there is no European target.

It can be calculated that with a 50 percent TtW efficiency advantage and a system efficiency of conventional fuels of 92 percent, the system efficiency can be as low as 62 percent before the overall WtW efficiency is lower for hydrogen and fuel cell technology than for the conventional oil product and ICE technology (Hansen, 2007c). Thus, with these assumptions a minimum system efficiency of 62 percent is required for the hydrogen and fuel cell technology to contribute to a general progress in energy efficiency.

Nevertheless, it is a challenge to achieve even this modest efficiency in hydrogen production. For instance, hydrogen losses of 5–10 percent were reported by the European hydrogen bus project (CUTE, 2008) due to the purging of system components and background leakage. With such a loss rate it can be difficult to achieve an efficiency of the overall hydrogen production, purification, compression and filling system. The causes of the losses were, however, technical problems that probably can be solved.

When hydrogen becomes a fuel rather than a chemical, it may also be a problem to expand the practice of using cryogenic hydrogen delivery and storage and at the same time maintain high system efficiency, because of the large energy loss associated with this method.

The lower limit of 62 percent system efficiency as a societal priority should, however, not be interpreted as an absolute limit. First, the increasing demand for hydrogen in the refining processes leads to lower system efficiency for conventional fuels. Second, the different energy forms involved (oil, gas, coal, electricity, heat and so on) differ in usefulness and thus in value. The value (or exergy) per GJ of one energy commodity does not necessarily equal the value per GJ of another. They also differ by environmental as well as security of supply properties. Third, conversion efficiencies can probably be developed significantly by industrial learning.

4 SECURITY OF SUPPLY

Security of energy supply to the European Union is a complex problem that involves at long list of problems of which we shall address only a few: first, the geological–economic capability of supply to respond to the increasing global demand for energy at the world markets – the response is constrained by the geology of the reserves as well as the institutional framework for their exploitation; second, the geographical distribution of the remaining reserves and the related geopolitical and market power issues; third, the resilience of the global production, transport, and transformation networks providing the technical basis for the throughput of combustible energy; and fourth, the problem is not only inadequate supply, it is also the high prices resulting from an inadequate supply and the vulnerability of the European economies to these high prices.

Hydrogen produced with or without CCS from natural gas or coal is in many scenarios expected to supply a large share of the market for hydrogen transport fuel in the future. Such a transition path will diversify the primary energy basis of transport from oil to natural gas and coal and in this sense improve the security of supply. Natural gas and coal supplies are, however, troubled by similar constraints due to geological–economic scarcity, geographical distribution of reserves, resilience of the global throughput, and vulnerability of the economies to supply failure and world market price increases.

Comparative advantage is one of the factors behind the specialization of particular economies in industries and products. It can be roughly indicated by resource availability, although they should be interpreted cautiously.

Table 12.3 shows that Europe is not generously endowed with indigenous

Table 12.3 Recent estimates of Europe's share of fossil fuel reserves

Fossil fuel	Source	Date	Share (%)
Oil	BP Statistical Review	Year-end 2007	1.3
	Oil & Gas Journal	January 1, 2008	1.1
	World Oil	Year-end 2006	1.3
Natural gas	BP Statistical Review	Year-end 2007	3.3
	CEDIGAZ	January 1, 2008	3.5
	Oil & Gas Journal	January 1, 2008	2.8
	World Oil	Year-end 2006	2.7
Coal	Recoverable Anthracite and Bituminous		2.0
	Recoverable Lignite and Subbituminous		9.1
	Total Recoverable Coal		5.5

Sources: DOE (2007); *Oil and Gas Journal* (quoted in DOE, 2007); *World Oil Magazine*, September 2007; *BP Statistical Review 2008*, London; CEDIGAZ: *The Gas Year in Review 2009*, Rueil-Malmaison.

fossil energy sources when compared to the 9 percent of the world population living in Europe. The only exception is coal with low energy content and high environmental damage in extraction (lignite and subbituminous coal).

Natural gas is expected to supply an increasing share of European energy demand in the future, but it is a non-renewable resource and the supply cannot increase forever. Much of the remaining reserves are comfortably situated around Europe in Russia, Central Asia, the Middle East, and North Africa, but Europe will not be the only market for natural gas from these sources in the future. Just like oil, the natural gas market faces increasing competition for an output that is unable to adapt to the growing demand and even more so if natural gas increasingly replaces crude oil as a primary energy source for transport. Jonathan Stern suggests (this volume, ch. 3) that this will become a constraint to European natural gas supply beyond 2020. In this case it would not be beneficial for Europe to replace crude oil as the primary energy basis for transport with natural gas, either via hydrogen or directly.

The ongoing expansion of the capacity for exporting liquefied natural gas (LNG) from the natural gas producing countries and importing it into Europe should improve the resilience of the European natural gas supply. It may even in a few years exert a downward pressure on natural gas prices independently of the oil prices. In the long run, however, it would be very optimistic to expect independence between the oil and the natural gas price (see Hansen, 2007d, 2007e). Ongoing attempts to create an independent spot market price with gas-to-gas competition have achieved short-term and local deviations in the spot market price. The long-term relation between the oil price and the natural gas spot market price does, however, seem to persist (see European Commission, 2004; Panagiotidis and Rutledge, 2007; for the US, see Villar and Joutz, 2006). This long-term covariance could even be strengthened as natural gas liquefaction and its use as transport fuel increase the substitutability with oil.

It is also important to recognize that liberalization of the European natural gas market will not change the concentration of upstream supply. Recently, the formation of a natural gas cartel comprising Russia, Iran, and Qatar as well as the alliance of natural gas suppliers in Central Asia with Gazprom signals intentions of using this latent market power. Three-quarters of the remaining global oil reserves are controlled by only nine countries, but the same countries control a similar share of the remaining natural gas reserves that are relevant to Europe (Hansen, 2007d).

The European economy is vulnerable to increasing oil and gas prices because of its significant net imports of oil and gas compared to the GDP. On the member state level, this vulnerability is considerably larger in the new member states than in the old ones (ibid.). Thus, an oil price increase implies a significant leak of national income from the economic circuit in Europe. Replacing oil and natural gas imports by energy from European sources would not only improve the overall terms of trade between Europe and the rest of the world, but also create jobs in the industries, providing the indigenous energy supply. Replacing oil by natural gas, however, would not have these effects except to the extent that natural gas based hydrogen as a transport fuel is more fuel efficient and thus marginally reduces the need for primary energy.

According to the government reports on remaining reserves, *coal* is a relatively abundant energy resource with plenty of proven and recoverable reserves, but the quality of these data is increasingly questioned by independent centers for resource assessment. The Energy Watch Group (2007) even predicts that the global coal production due to geological–economic scarcity will reach its peak in the 2020s. According to Gerling et al. (2006) only 6 percent of the global hard coal reserves and 3 percent of the brown coal reserves are situated in Europe (classifying coal differently from in Table 12.3). It is, however, questionable whether these reserves should be classified as economically recoverable reserves in the same way as, for example, South African coal is. In Europe the inexpensively exploitable coal reserves were mined long ago, and coal production has been in decline for decades as it becomes still more expensive to mine coal from still deeper and thinner seams. Much of the European coal production is only kept alive with the help of considerable government subsidies to extraction and/or coal power plants. These subsidies obviously work against the Lisbon goal of economic growth as well as the climate policy goal of reducing CO_2 emissions. They are only allowed by the EU because they are indigenous resources and in this capacity provide some security of energy supply. Access to indigenous sources of energy is a central means to achieve security of supply (Council of the European Union, 2002). When subsidies are terminated and the restructuring of the European coal industry is complete, the reserves are likely to be considerably lower.

The hard coal consumed in Europe originates predominantly from distant sources such as South Africa, Colombia, and Russia. A value-added chain analysis by Gerling et al. (2006) showed that about half the cost of hard coal[7] in Europe in 2005 comprised transport costs that vary closely with the price of oil. In addition, there is a substitution effect on the coal cost net of transport costs. It is not as strong as the oil–gas substitution effect, but it will rise if more coal is used for production of synthetic diesel, DME,[8] or other fuels. Thus, switching from oil to coal as a primary energy basis for transport fuels will only partly avoid the negative effects from the oil market.

Table 12.4   The EU share of the world's recoverable uranium resources (RAR+inferred) by extraction cost

	< US$40/kgU	< US$80/kgU	< US$130/kgU
EU share (%)	3.4	0.5	1.9

Source: IAEA (2008).

The ongoing innovation of CCS technology will probably before 2020 show which solutions we can expect to be competitive. In many future scenarios, this technology is expected to open up for large amounts of low carbon energy to the European market. However, a note of caution is warranted on this perspective, keeping the high transport cost of coal in mind. When a competitive CCS technology is ready, it will be available all over the world. Thus, it is possible that it will be more competitive to generate hydrogen from coal at the location of coal extraction (for example, South Africa, Central Asia) and then ship the hydrogen to Europe, than it will be to ship the coal and generate the hydrogen in Europe. In that case, however, CCS technology in Europe would still be useful in combustion and gasification of European combustible energy sources.

Uranium deposits occur in many countries, but as shown in Table 12.4 the economically recoverable reserves, like oil and gas reserves, are primarily located outside of Europe. The OECD Nuclear Energy Agency and the International Atomic Energy Agency (IAEA, 2008) expects that it will be possible to double the global uranium extraction at cost below US$80/kgU to approximately 120,000 tU per year in 2016. At a constant rate of extraction these resources would allow for uranium extraction well into the 2040s.

However, the binding constraint for expansion of nuclear energy in Europe is public acceptance rather than this economic–geological scarcity. Locations where the general public accepts nuclear energy production, fuel processing, and waste deposition are very scarce and have been so for decades, irrespective of economic costs. New generations of nuclear energy technology may change that as well as make use of more abundant resources such as thorium.

The European Commission (EC, 2007) considers generally renewable energy to 'contribute to security of supply by increasing the share of domestically produced energy, diversifying the fuel mix, diversifying the sources of energy imports and increasing the proportion of energy obtained from politically stable regions' (p. 14). They are, however, also subject to a combined geographical–geological–economic scarcity – many

Table 12.5 EU-27 shares of OECD+BRICS renewable energy potentials realizable by 2020 (%)

Renewable energy	EU-27 share	Renewable energy	EU-27 share
Biogas	19	Onshore wind	31
Solid biomass	20	Offshore wind	59
Renewable municipal waste	15	Total RES-E	20
Geothermal electricity	11	Biofuels (domestic)	27
Hydropower	11	Solar thermal heat	27
Solar photovoltaics	23	Geothermal heat	25
Solar thermal electricity	33	Biomass CHP heat	24
Tidal and wave energy	80	Total RES-H	25

Source: OECD (2008).

of them are land intensive as onshore wind power involves land use where several other interests are at play. Thus, the resources that realistically can be recovered depend to a high degree on public acceptance, which in turn is affected by the arrangements for redistribution of the resource rents. Moreover, many renewable resource technologies are under development and their development partly depends on the rate of their deployment. Due to these complicated aspects, it is difficult to assess the economically recoverable renewable energy resources.

Assessments of such resources differentiate between the physical (or theoretical), practical (or technical) and realizable (or economic) potential within a given timeframe. According to a recent study of the renewable energy potentials in the OECD countries and the BRICS (Brazil, Russia, India, China and South Africa), the total renewable energy potential amounts to approximately 9,000 TWh electricity, 5,700 TWh heat, and 1,700 TWh biofuels for transport in the 2020 perspective for these countries (OECD, 2008). The European shares of these potentials, shown in Table 12.5, can be rich sources of either 'classical' biofuels or hydrogen production.

The EU is inhabited by 15 percent of the total OECD and BRICS population and in this perspective the EU is especially well endowed with wind, wave/tide, and solar thermal electricity resources. Such renewable and non-combustible energy resources could form an indigenous primary energy basis for hydrogen.

Like coal, indigenous first-generation biofuels from Europe are not competitive with the products of large foreign producers. For second-generation technologies, the biomass resources shown in Table 12.5 can be rich sources of either 'classical' biofuels or hydrogen production.

Wind resources are abundant in Europe, in particular in Northern Europe and in mountain areas. The vast offshore wind resources in the Atlantic, the North Sea, and the Baltic Sea are only little exploited. Concerted investments in the necessary transmission grid and adapting the national electricity grids to smart grids will allow for more of this to enter the European market.

In the 2020s, Europe will have a much larger power supply generated by non-combustible sources that are difficult to regulate according to demand changes. Thus, considerable amounts of low-cost electricity will be available at off-peak hours and at times with good wind. Hydrogen based on such resources would represent a relatively secure supply of transport energy by virtue of its indigenous origin. A hydrogen infrastructure and smart grid configuration that enables the transformation of this low-cost electricity to high-value transport energy would increase the social benefits of developing these non-combustible power resources. It would also reduce the vulnerability of the European economies to increasing oil prices and create jobs.

5   ECO-EFFICIENCY

The EU target of reducing CO_2 emissions by 20 percent of the 1990 level in 2020 implies a similar targeted rate of progress in *eco-efficiency*, the ratio of an indicator of economic activity to an indicator of the environmental pressure, it causes. On the level of aggregate GDP for EU-27, the macroeconomic requirement derived from the GHG target is to sustain an average growth rate in eco-efficiency of 3.3 percent from 2005 to 2020. This is ambitious, too, as the GHG-efficiency growth rate achieved from 1995 to 2005 was on average 2.4 percent.

Since the start of the GHG accounts in 1990, transport activities have caused a rising share of Europe's total GHG emissions to the level of 21 percent in 2004 (EU-15). The 20 percent target is hardly achievable without reversing this trend, and it raises the question whether hydrogen and fuel cells in automotive use can contribute to this.

The immediate answer is no, for the simple reason that until 2020 there will in any case be a very small number of fuel cell vehicles on the roads. Most likely, they will be too few to make any difference in the European GHG accounts. However, climate policy does not end in 2020 and the perspective as far as the EU is concerned is to continue to reduce GHG emissions to a level that is 60–80 percent lower than the 1990 level in 2050.

To study the possible contribution to GHG emission reduction from the introduction of passenger cars with hydrogen and fuel cell technology

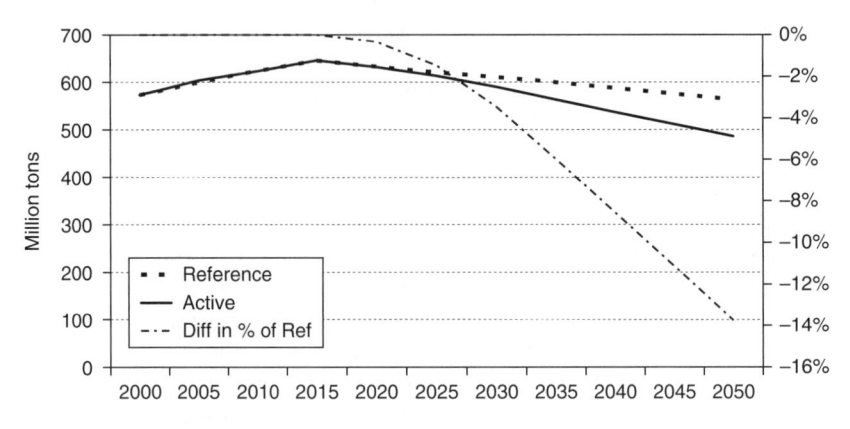

Figure 12.1 *Impact on European GHG emissions from passenger cars of replacing oil products by natural gas-based hydrogen in fuel cell cars reaching a market share of 43 percent in 2050*

on the European market, a series of scenarios were produced with the Sustainable Mobility Project model (WBCSD and IEA, 2004).They are documented in Hansen (2007a).

The scenarios introduced passenger cars with fuel cell technology on the European market from 2015 with a market share growing to 43 percent in 2050. Two different scenarios with respect to feedstock for hydrogen production were created. One scenario assumed that the hydrogen was produced on the basis of natural gas, whereas the other scenario assumed that it was produced by electrolysis from renewable or nuclear energy.

The contribution to the GHG emission reduction was very different in the two scenarios. In the natural gas-based scenario, the aggregate GHG emissions from passenger cars in Europe 2050 was reduced by 14 percent, corresponding to 5 percent of the emissions from the total transport sector. This scenario is shown in Figure 12.1.

The reference scenario emissions in the figure (the dotted bold curve, left axis) are expected to decline due to a higher market share of energy efficient cars (particularly diesel) as fuel prices increase. The active scenario shows the emissions that would result from introducing hydrogen and fuel cell cars (bold curve, left axis). The slim curve refers to the right axis and shows the deviation of the active from the reference scenario in percent.

The result shows that there will be a reduction in GHG emissions, but a rather modest one. Even if all diesel and petrol cars were replaced by fuel cell cars, half of the GHG emissions would still remain because hydrogen

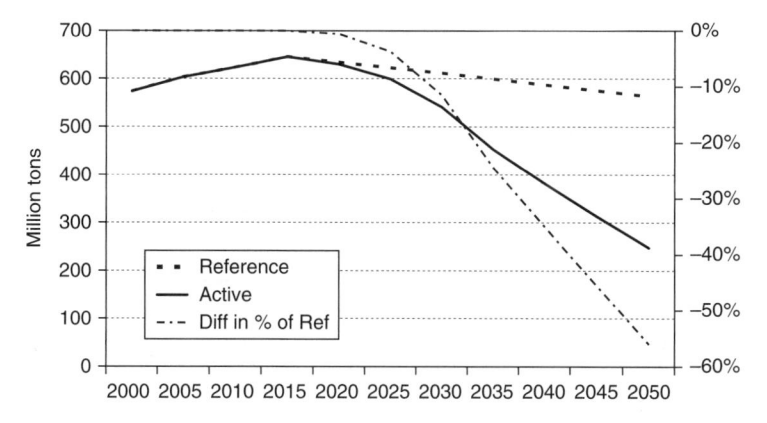

Contributions of the hydrogen transition to EU energy and climate policy 263

Source: Hansen (2007a).

Figure 12.2 Impact on European GHG emissions from passenger cars of replacing oil products by non-combustible-based hydrogen in fuel cell cars reaching a market share of 43 percent in 2050

is produced by natural gas. This is, however, not necessary. The same scenario with hydrogen produced from non-combustible sources can produce quite different results.

Figure 12.2 shows that with the production of hydrogen from CO_2 neutral feedstock the emissions from passenger car transport will be reduced by almost 60 percent.

With reference to these scenarios, European governments would have important environmental reasons to support hydrogen as a transport fuel as long as it is based on non-combustible energy, but only little reason if it is based on natural gas (or coal) without CCS. Biomass and coal-based hydrogen with CCS can reduce GHG emissions from fossil fuel combustion by up to 80 percent.

The two scenarios were also used to study the impact on local air pollutants emitted from passenger car transport such as particulate matter (PM), nitrate oxides (NOx), volatile organic compounds (VOCs), and carbon monoxide (CO). These pollutants affect human mortality and morbidity and are responsible for damage to environmental quality.

The study showed that on the aggregate level the emissions of these pollutants were already drastically reduced in the reference scenario at the time when the fuel cell cars are introduced to the market. This is a result of the EU and member state efficiency requirements, fuel and exhaust standards, and other initiatives under the CAFE program for eliminating air pollution that damages human health. The results are shown in Figure 12.3.

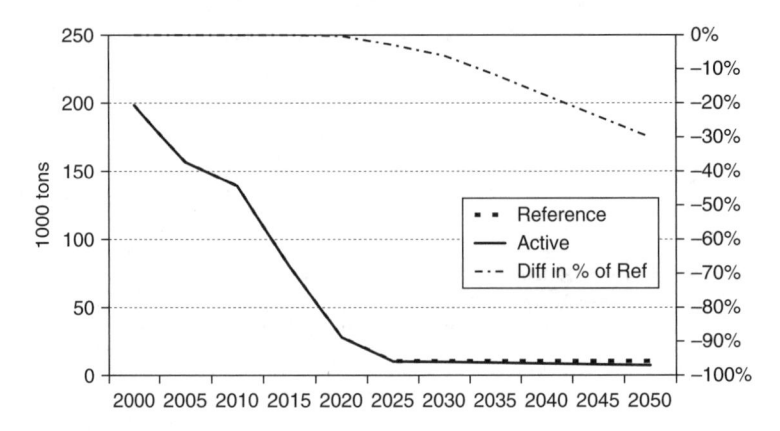

Figure 12.3 *Impact on PM emissions from passenger cars of replacing oil products by carbon-free lean-based hydrogen in fuel cell cars reaching a market share of 43 percent in 2050 in Europe*

The non-combustible-based hydrogen in the figure shows a very modest impact of using green carbon-free hydrogen for transport. This is because the CAFE policies aim at eliminating the health-damaging pollution from stationary as well as mobile sources. The reference scenario assumes that these programs will be implemented effectively and in a timely fashion, and with the desired results. These assumptions may be overly optimistic, but much of the local air pollution can be avoided by applying the exhaust, engine, and fuel standards, filters, and so on as planned according to the EU CAFE policies.

Natural gas reforming would relocate emissions from the mobile to the stationary sources, and with the same assumptions of effective elimination of local pollutants a similar result could have been achieved.

The aggregate emissions shown in Figure 12.3 are, however, not the adequate indicators for *local* pollutants. These pollutants are trapped in locked air-sheds and city air at several locations in Europe. In these locations, governments have a particular reason for continuously supporting the use of electric vehicles, whether battery or fuel cell electric, and even in some places hybrid electric. Figure 12.4 shows how these spots on the European map looked in the year 2000 and how they are expected to look in 2020 after the implementation of the EU Air Strategy and Maximum Climate Action policies.

The indicator used in the figure is the particulate-caused mortality measured as the loss of statistical life expectancy due to very fine particulates

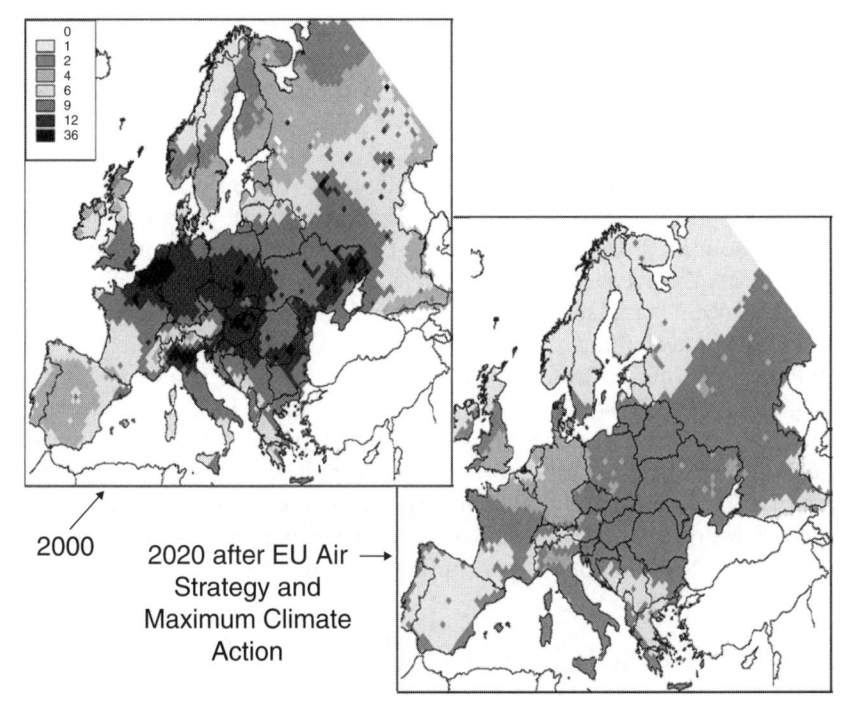

Contributions of the hydrogen transition to EU energy and climate policy 265

Source: Hansen (2007a).

Figure 12.4 Concentrations of PM 2.5 exceeding the health (mortality) limits in 2000 and 2020

(2.5 microns) from combustion. Several other local pollutants could have been mapped here, but their geographical distribution and intensity are not very different from that of the fine particulate pollution shown here.

The maps in Figure 12.4 show in accordance with the modeled scenarios above that the programs improving the fuel and exhaust standards, emission standards and so on in Europe have the potential to solve a large part of the most severe pollution problems, but not all. There is a strong case for regional policies on parking fees, parking rights, road tolls, and so on favoring the use of BEVs, HEVs, PHEVs, and FCEVs. Policies along these lines are already being planned or adopted in cities such as London and Milan.

Summing up, the high fuel efficiency of FCEVs contributes in any case to the reduction of GHG emissions as they gain a higher market share on the European car market. This contribution is, however, small relative to the contribution from higher market shares of the already existing energy-efficient cars and the stricter regulation of fuels, exhaust, and so on. In the scenarios above, when basing hydrogen on non-combustible energy sources (or GHG-free sources), even a 43 percent market share in 2050 would eliminate more than half of the GHG emissions from passenger cars in 2050.

The effect on emissions of local pollutants is a different story. The regulation of such emissions not only from car exhaust, but also from stationary sources, is capable of reducing the future pollution considerably, even before the hydrogen and fuel cell technology has matured sufficiently to introduce it on a large scale in automotive transport. Thus, hydrogen and fuel cell technology is not essential for reducing the bulk of the air pollution causing health damage. However, the results shown in Figure 12.4 indicate that electric vehicles – and among them FCEVs – with their zero tail-pipe emissions will be essential for eliminating the unacceptable health impact from air pollution in the most polluted cities and regions of Europe. This conclusion applies in particular to the European regions with the highest population (and car) density where fuel cell buses can also be expected to be an important part of the solution.

6 COST EFFICIENCY

The cost efficiency of FCEVs relative to their competitors depends on the vehicle ownership costs as well as the fuel cost per kilometer. It can be assumed that FCEVs at some time in the future can be produced at costs comparable to the costs of similar ICVs, HEVs or PHEVs. It is difficult to predict the time required for the necessary research breakthroughs to emerge and for the necessary industrial learning to take place. If it is further assumed that FCEV production at comparable costs will be possible at some time in 2015–25, the question of competitiveness becomes a matter of fuel cost per kilometer at that time.

The cost of petrol and diesel depends on the international oil price, but so does natural gas-based hydrogen as well, albeit to a different degree. Most of the earlier studies such as the Alternative Fuels Contact Group (2003), the US National Academy of Science (2004), Ogden et al. (2004), and the International Energy Agency (2005) have all envisaged transition scenarios in which hydrogen in the beginning of the transition was produced from natural gas. This is because they have assumed oil prices in the $25–60/bbl (barrel) interval. During 2008, however, leading centers for oil market analysis have elevated their expectations of future oil prices to levels beyond $110/bbl. The IEA (2008) and other international oil market analysts seem to increasingly share the view that 'the era of cheap oil is over'.[9] Thus, there are good reasons for reconsidering the vision of a hydrogen transition, which initially is based on natural gas.

The earlier studies also assume a large gap between the fuel economy of the FCEV and that of the ICV – typically an efficiency advantage of 100 percent of the FCEV over the ICV. That is, if the FCEV delivers 35 km per amount of hydrogen corresponding to 1 litre of conventional fuel, the ICV delivers 17.5 km/l. However, this is the *average* fuel economy expected for ICVs in Europe for the 2010s. FCEVs will compete with the most efficient ICVs and hybrids rather than the average car.

Hansen (2010) consequently studies the competitiveness of hydrogen as a transport fuel, assuming only 50 percent efficiency advantage of FCEVs over its competitors and within a much wider range of oil prices ($0–200/bbl). The model is based on the simple assumptions of the cost of the primary energy commodity, the energy transformation efficiency, and the non-energy costs of each link in the fuel chain as explained in Hansen (ibid.).

The study considers the refining of crude oil to petrol and diesel, steam reforming of natural gas to hydrogen, and electrolytic hydrogen production based on non-combustible power sources. Hydrogen pathways based on coal and biomass with CCS could also be an option, but the cost and efficiencies of these solutions are still ambiguous and their energy, eco-, and cost effectiveness must be comparable to the presently mastered solutions if they are introduced on a large scale.

Electricity from non-combustible sources is assumed to cost €61–90/MWh independently of the oil price. It can be expected that the future smart grid will enable the trade of electricity at hours when the price is very low. Furthermore, the generation of wind power (and in some countries nuclear power) can be expected to grow considerably, creating a large supply of low-cost off-peak and excess-generation electricity. Hydrogen is one of the electricity storage technologies that can make this electricity available for transport purposes. The storage technologies will, however, also make the electricity available for other purposes, possibly even in competition with peak-hour electricity. Thus, the currently observed very low prices for this electricity would probably be competed away in the future market, and it is only safe to assume what is feasible, that is the unit cost of non-combustible power.

For natural gas-based hydrogen, the system efficiency is assumed to be 70 percent and the non-energy costs to be €11 per GJ in the best case. In the worst case a system efficiency of 65 percent and non-energy costs of €15 is assumed. For electrolysis, the corresponding best-worst-case assumptions are 70/65 percent and €12/€16 per GJ. See Hansen (ibid.) for additional background information about these assumptions and the model used.

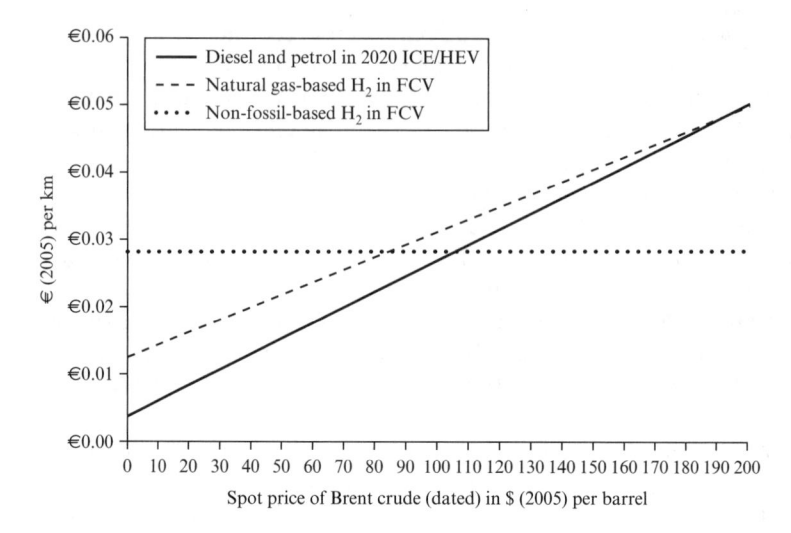

Figure 12.5 Per km fuel cost of ICE and FCEV technology with natural gas-based and non-fossil electricity-based hydrogen (best case, no taxes)

Based on these assumptions, the threshold prices can be identified as the intersections of the cost per km curves in Figure 12.5. The costs relating to conventional fuels are calculated as an average of petrol and diesel.

The fuel cost per km for hydrogen based on natural gas becomes competitive with that of conventional fuels at an international oil price of $299/bbl assuming best-case performance in hydrogen production (70 percent system efficiency and non-energy costs of €11/GJ H_2). Under worst-case performance (65 percent system efficiency and non-energy costs of €15/GJ H_2), hydrogen based on natural gas would not be competitive below $801/bbl. However, hydrogen based on non-combustible power becomes competitive at much lower oil prices: in the best case at $125/bbl and in the worst case at $202/bbl.

The comparison of per km costs in Figure 12.5 also shows that the order of competitiveness is very different for low and high oil prices. At oil prices below $95/bbl, the order of competitiveness is as follows: per km fuel costs are lowest for conventional fuels, highest for non-combustible-based hydrogen, and with natural gas-based hydrogen in the middle. This order of competitiveness is totally reversed beyond $299/bbl, but it begins at $95/bbl, where non-combustible-based hydrogen becomes competitive with natural gas-based hydrogen.

These tipping points are, of course, no more accurate than the efficiency and cost ratios assumed in the calculations, and different assumptions

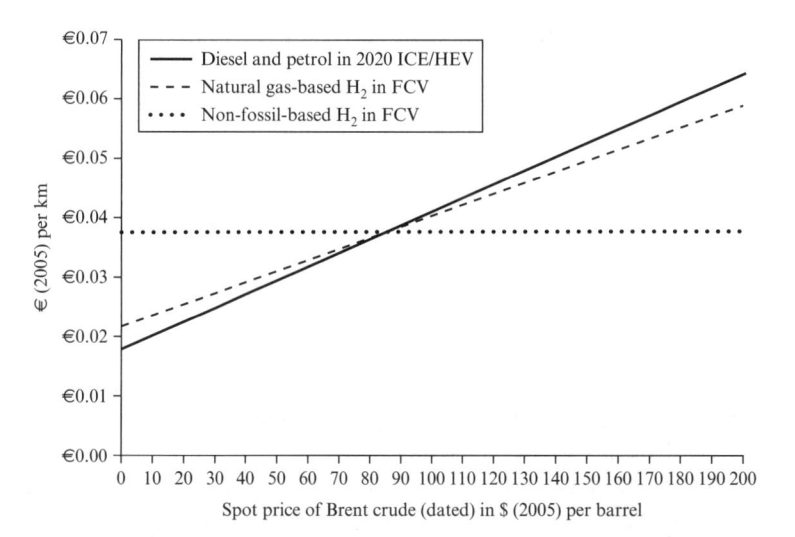

Figure 12.6 Per km fuel cost of ICE and FCEV technology with natural gas-based and non-fossil electricity-based hydrogen (best case, fuel tax €10/GJ on all fuels)

would yield different threshold values. The existence of a shift in the order of competitiveness is, however, robust to most of the likely assumptions. Moreover, fuel taxes lower the oil price at which this shift occurs.

Until now, fuel taxes have been excluded from the analysis, except for the minor cost of emission allowances to oil refining, expected to be mirrored by a tax on emissions from fuel production not included in the EU Emissions Trading System (EU ETS). Thus, the per km costs depicted in Figure 12.5 may be close to those paid by US consumers, but far from those paid by EU consumers, where high fuel taxes are applied. All the EU countries adhere to the minimum diesel tax of €8.3/GJ and petrol tax of €10.8/GJ and the EU-15 countries (except Spain) apply petrol tax rates from €16–26/GJ. High fuel taxes amplify the effect of the fuel efficiency on the cost per km and thus a cost advantage to the more efficient solution.

Figure 12.6 shows the cost per km when a €10/GJ fuel tax is applied equally to conventional and to hydrogen fuel. In addition to this tax, there is a CO_2 tax corresponding to €3/GJ on diesel, petrol and natural gas as well as a €1/GJ tax on the loss of energy in production of electrolytic hydrogen. A flat €15/GJ tax on all fuels without additional CO_2 and energy loss taxes would produce a similar outcome, but with correspondingly higher costs of all fuels.

Figure 12.6 shows that even a modest rise of the EU minimum fuel tax rate could suffice to make hydrogen competitive in all of the EU at an oil price around $90/bbl. Under these assumptions, it is not necessary to provide specific tax favors for hydrogen. In the worst-case scenario, however, hydrogen would not become competitive at oil prices below $150/bbl. This underlines the requirement for providing incentives to hydrogen producers to achieve the highest efficiency at the lowest infrastructure cost in hydrogen production and transport. Furthermore, the previous standard assertion that the primary energy basis of hydrogen in the beginning of the transition will be natural gas is only defendable at oil prices that are unrealistically low.

The conclusion is that it is not necessary to provide specific tax favors for hydrogen. A uniform fuel tax at a level like €15/GJ rewards the more efficient technologies. It materializes in a fuel cost advantage for the more efficient hydrogen and fuel cell concept.

Finally, the design of the minimum fuel taxes can be improved by applying more strongly the principle of uniform tax rates according to energy content, CO_2 emissions, and energy loss rather than different tax rates for different energy uses. In particular, it is important that the incentives to limit the energy loss in transformation are as strong as the incentives to limit the final use of the transport fuels. The extent to which the fuel cell and hydrogen technology in automotive use can contribute to cost-effective automotive transport depends crucially on the performance achieved in hydrogen production.

7 THE CONTRIBUTIONS OF HYDROGEN AND FUEL CELL TECHNOLOGY

How a transition from petrol and diesel to hydrogen as transport fuel will contribute to the goals of European energy and climate policies depends fundamentally on the primary energy basis of the hydrogen. The overall social benefit of using fuel cell and hydrogen technology in automotive transport is that it enables non-combustible electricity resources to serve as a primary energy basis for transport fuels.

In contrast to fossil energy, Europe is well endowed with renewable electricity sources, and the European energy and climate policy provides an economic framework for their rapid development in the 2010s and beyond 2020. Nuclear energy will play an increasing role in the 2020s in countries that have chosen to expand this energy source. An energy carrier, such as hydrogen, that enables electric power to fuel transport will increase the social benefit of developing the non-combustible electricity resources as well as of the use of the hydrogen technology. The same applies to batteries.

In many respects, we can expect the introduction of a large variety of BEVs, HEVs, and PHEVs announced by the car producers in 2009–11 to pave the way for a complete electrification of the car fleet. The limited energy density of batteries will, however, limit the demand for BEVs to the market segment for which the limited driving range per recharging (say, 150 km) does not represent a major disadvantage. This market segment will be expanded by HEVs, PHEVs, fast recharging, and battery replacement stations. However, an electrification of automotive transport beyond this segment is hardly possible without the fuel cell and hydrogen technology.

Europe faces an increasing competition for dwindling reserves of the oil on which almost all automotive transport is based. Replacing petrol and diesel by hydrogen based on natural gas would hardly be an escape from the problem. In the 2020s, when hydrogen is expected to be introduced, there will most likely be a similarly increasing competition for the natural gas reserves relevant to Europe. These reserves are likewise concentrated in a few countries. The natural gas price is likely to follow the oil price and it will not change the impact on the economy of a future oil price rise. The contribution to the curbing of emissions of CO_2 and local air pollutants would be marginal. On the other hand, under the assumptions of 'cheap oil' it would be the most economic solution. However, assuming that 'the era of cheap oil is over' it will be the more expensive solution.

The impact of adding CCS to the system has not been considered in detail, but it is safe to assume that it would further increase costs, but decrease CO_2 emissions. Switching to coal as a basis for hydrogen as transport fuel would probably improve the security of supply, but there is uncertainty about the competitiveness of hydrogen from coal produced in Europe versus hydrogen (or other transport fuels) from coal produced in the countries where the coal reserves are located. The considerable costs and energy losses of CCS processes adds to this uncertainty. Important reserves of lignite with low value as a fuel and a high environmental impact are available in some countries.

In contrast to these fossil fuel-based alternatives, shifting to European energy sources as the primary energy basis for transport would contribute significantly to all of the societal goals. A transition to indigenous energy sources would reduce the vulnerability of energy supply from the remaining and dwindling reserves of oil and natural gas. It would avoid the deterioration of the terms of trade that will result from the high prices of oil expected in the coming decades.

Replacing crude oil as the primary energy basis of transport by energy from these sources would reduce GHG emissions significantly. The European programs adopted to curb local emissions will not fully eliminate unacceptable health impairing and life shortening air pollution in the most polluted areas. Electrification of automotive transport will contribute crucially to achieving these goals.

NOTES

1. A full introduction to hydrogen and fuel cell technology is offered in Sorensen (2005).
2. See www.betterplace.com.
3. Hydrogen made liquid by freezing to −275°C.
4. An efficiency advantage of 66 percent equals an efficiency factor of 1.66, that is, that an FCEV runs 66 percent further than an ICE, given an equal energy content in the tank.
5. These databases are the leading sources of comparable data for present and future technology choices of automotive technology.
6. A car whose fuel consumption is 35 km/l has an efficiency advantage of 50 percent over one whose consumption is 23 km/l. Such levels of fuel efficiency could very well characterize the competition between FCEVs and other efficient vehicles in 2020.
7. The energy density of lignite is too low to warrant long transport.
8. Rising coal demand along with a rising oil demand from China and India also leads to simultaneously rising oil and coal prices at the world market. Rising coal prices at the world market may, on the other hand, make more European coal resources economically recoverable.
9. The drastic drop in the international oil price in the last months of 2008 is expected to be reversed when global economic growth takes off again.

REFERENCES

Argonne National Laboratory (2008): 'GREET Version 1.8b', Systems Assessment Section, Center for Transportation Research, Argonne, IL.
Alternative Fuels Contact Group (2003): 'Market Development of Alternative Fuels'.
California Environmental Protection Agency Air Resources Board (CARB) (2008): 'The Zero Emission Vehicle Program – 2008', Sacramento, CA.
Council of the European Union (2002): 'State aid to the coal industry', Council Regulation (EC) No. 1407/2002 on State Aid to the Coal Industry, Brussels, 23 July.
CUTE (2008): '"Clean Urban Transport for Europe" detailed summary of achievements', Ulm, Germany.
DOE (2007): 'Hydrogen, Fuel Cells, and Infrastructure Technologies Program. Multi-year research, development, and demonstration plan. Planned program activities for 2005–2015', 2007 revision, US Department of Energy, Washington, DC.
Edwards, R., J.G. Griesemann, J.F. Larivé and V. Mahieu (2007): 'Well-to-wheels analysis of future automotive fuels and powertrains in the European context', Version 2c, available at: http://ies.jrc.ec.europa.eu/wtw.html (accessed August 2009).
Energy Information Administration (EIA) (2008): *International Energy Outlook 2008*, US Department of Energy, Washington, DC.

Energy Watch Group (2007): 'Coal: Resources and future production', Berlin, Updated version: July 10, 2007.
European Commission (EC) (2007): 'Renewable Energy Road Map. Renewable Energies in the 21st Century: Building a More Sustainable Future', Brussels.
European Commission – DG COMP (2004): 'Energy Sector Inquiry', Brussels.
European Hydrogen and Fuel Cell Technology Platform (HFP) (2007): 'Implementation Plan – Status 2006', HFP, Brussels, March.
Fuel Cell Commercialization Conference of Japan (FCCJ) (2008): 'Commercialization of fuel cell vehicles and hydrogen stations to commence in 2015', Fuel Cell Commercialization Conference of Japan (FCCJ), Tokyo.
Gerling, P., H. Rempel, U. Schwarz-Schampera and T. Thielemann (2006): 'Reserven, Ressourcen und Verfügbarkeit von Energierohstoffen 2005' (Reserves, resources and availability of energy raw materials), Bundesanstalt für Geowissenschaften und Rohstoffe, Hanover.
Hansen, A.C. (2007a): 'Hydrogen and fuel cell technology in EU LDV transport: potential contribution to environmental goals', EECG Research Papers, Department of Environmental, Social, and Spatial Change (ENSPAC), Roskilde University, Roskilde, available at: http://rudar.ruc.dk/handle/1800/2434.
Hansen, A.C. (2007b): 'The international oil price and hydrogen competitiveness', EECG Research Papers, Department of Environmental, Social, and Spatial Change (ENSPAC), Roskilde University, Roskilde, available at: http://rudar.ruc.dk/handle/1800/2433.
Hansen, A.C. (2007c): 'The potential contribution of hydrogen to societal goals', EECG Research Papers, Department of Environmental, Social, and Spatial Change (ENSPAC), Roskilde University, Roskilde, available at: http://hdl.handle.net/1800/2979.
Hansen, A.C. (2007d): 'The supply security of hydrogen as transport fuel', EECG Research Papers, Department of Environmental, Social, and Spatial Change (ENSPAC), Roskilde University, Roskilde, available at: http://hdl.handle.net/1800/2978.
Hansen, A.C. (2007e): 'When will hydrogen become a competitive transport fuel?', EECG Research Papers, Department of Environmental, Social, and Spatial Change (ENSPAC), Roskilde University, Roskilde, available at: http://hdl.handle.net/1800/3011.
Hansen, A.C. (2010): 'Will hydrogen be competitive in Europe without tax favours?', *Energy Policy*, In Press, doi:10.1016/j.enpol.2009.03.035.
HyWays (2007): 'HyWays. The European Hydrogen Roadmap', FP6 Project, Commission of the European Communities.
IAEA (2008): *Uranium 2007: Resources, Production and Demand*, OECD Nuclear Energy Agency and the International Atomic Energy Agency, Vienna.
IEA (2005): *Prospects for Hydrogen and Fuel Cells*, Paris: OECD.
IEA (2007): *World Energy Outlook 2007*, Paris: OECD.
IEA (2008): *World Energy Outlook 2008*, Paris: OECD.
OECD (2008): *Deploying Renewables: Principles for Effective Policies*, Paris: OECD.
Ogden, J.M., R.H. Williams and E.D. Larson (2004): 'Societal lifecycle costs of cars with alternative fuels/engines', *Energy Policy*, Vol. 32, No. 1, pp. 7–27.
Panagiotidis, T. and E. Rutledge (2007): 'Oil and gas markets in the UK: evidence from a cointegrating approach', *Energy Economics*, Vol. 29, No. 2, pp. 329–47.
Sorensen, B. (2005): *Hydrogen and Fuel Cells*, Amsterdam: Elsevier.

US National Academy of Science (2004): *The Hydrogen Economy: Opportunities, Costs, Barriers, and R&D Needs*, Washington, DC: National Academies Press.
Villar, J.A. and F.L. Joutz (2006): 'The Relationship Between Crude Oil and Natural Gas Prices', Office of Oil and Gas, Energy Information Administration (EIA), Washington, DC.
WBCSD and IEA (2004): 'Sustainable Mobility Project Model', World Business Council for Sustainable Development and International Energy Agency, Paris.

13. R&D programs for hydrogen: US and EU

Steven Stoft and César Dopazo[1]

1 THE CASE FOR HYDROGEN

Hydrogen (H_2) is, among other things, an energy carrier very abundant in nature in combination with other chemical elements. Molecular H_2 can be synthesized by energy-intensive processes. H_2 must then be stored, distributed and finally utilized for energy generation. Internal combustion engines (ICEs), in the form of reciprocating machines or gas turbines, as well as electrochemical devices, known as fuel cells (FCs), can convert H_2 into mechanical energy and/or electricity.

Possible incentives for the development of hydrogen technologies are either related to the potential to store electricity from intermittent renewable energy sources or to provide an alternative fuel for transportation.

At present only three approaches seem plausible for powering cars and trucks without oil and without significant carbon dioxide (CO_2) emissions: (i) hydrogen, (ii) batteries, and (iii) biofuels. Each of these encompasses a number of distinct possibilities depending on the primary energy source. Hydrogen can be derived from natural gas, coal, fossil-generated electricity, or non-fossil electricity; if a fossil fuel is the energy source, it must be used with carbon sequestration to achieve significantly reduced CO_2 emissions. Batteries offer similar options. Biofuels are more restrictive, and only advanced biofuels, such as cellulosic ethanol, would have a large impact on the emissions problem

Of course, efficiency improvements to current internal combustion designs provide a partial fourth alternative, although there are physical limits. The US National Research Council (NRC) of the National Academies finds efficiency improvements more promising than hydrogen fuel cell vehicles (HFCVs) through about 2040. However, in combination with plug-in electric vehicles (PEVs) the possibilities are striking.[2]

Taking a longer view, the complete elimination of fossil fuel use is eventually inevitable. At that point, trucks and automobiles will be powered by electricity – from batteries or fuel cells – and by biofuels. Electricity is likely to be produced by wind turbines, solar energy and nuclear power. HFCVs will then require the production of hydrogen by electrolysis, or a high-temperature process with advanced nuclear reactors. Assuming 70 percent efficiency for electrolysis or a high-temperature process, and 70 percent efficiency for the fuel cell, the latter being a rather optimistic value, yields 49 percent efficiency for the conversion of bulk electricity to on-board electricity. Of course this ignores the need to pressurize and transport the hydrogen. By contrast, charging and discharging a battery is over 90 percent efficient. For this reason, the long-run dominance of HFCV technology over battery technology looks unlikely. However, this conclusion could be reversed by the development of cheap artificial photosynthesis for the production of hydrogen.

In the medium term, say between 2050 and 2075, hydrogen could possibly, although by no means certainly, dominate a combination of batteries, efficiency and biofuels.[3] It should, however, be remembered that, unlike cellulosic ethanol and engine efficiency improvements, HFCVs will not automatically do much to reduce CO_2 emissions. There must be an additional policy requiring carbon capture when hydrogen is produced from fossil fuels.

At present, there is no case for tipping the balance towards a hydrogen transition. As the NRC (2008, p. 104) explains, there is a 'downside risk of pushing HFCVs (or any other specific technologies) before they are really ready or if they turn out not to be the best option, which could be extremely expensive and disruptive'.

There is also no proof that hydrogen is the most likely long-run solution. However, there is an excellent case for the present R&D effort in hydrogen technology. There is a modest chance that it could provide an enormous payoff. For example, it is possible that batteries and cellulosic ethanol will remain too expensive to be practical, and that breakthroughs in fuel cell and hydrogen storage technologies will bring HFCVs to market sooner and more profitably than expected. And perhaps recent breakthroughs in artificial photosynthesis will eventually make solar hydrogen the fuel of choice.

Funding such high-risk advanced R&D is something the market does poorly. Because of this, the government should support hydrogen research, and quite likely at a higher level than the current one. That is the case for hydrogen. It is also the case for funding all of hydrogen's competitors, or as the NRC (2008b) says, taking a 'portfolio approach'.

2 R&D NEEDS

R&D opportunities appear along every stage of the H_2 chain. As already mentioned, if the hydrogen-containing molecule includes carbon as a component or the energy to be supplied is generated from fossil fuels or from a 'mix' including them, the H_2 production process emits CO_2, unless a carbon capture and storage (CCS) technology is incorporated; all these methods leave plenty of room for innovation and energy consumption minimization. Stationary and on-board storage demands high energy density materials. Distribution pipeline materials must be protected against the brittleness that H_2 can cause. The end use of H_2 in ICEs requires high-temperature process adaptation and materials, while its utilization in FCs needs important materials research in order to increase durability, robustness and significantly lower their costs.

Feeding FCs with conventional fuels such as coal, gasoil, gasoline, methanol or natural gas seems appropriate for stationary energy generation plants, to which a CCS technology can be added. For vehicle on-board FC applications, H_2 is the adequate energy carrier.

A hypothetical economy for transport, based solely on H_2, would require its massive production. Several thousand nuclear power plants or, alternatively, a gigantic penetration of renewable energies (in Europe) or possibly coal-based hydrogen synthesis plants (in the USA) would be necessary to supply the future fleet of vehicles. The H_2 storage and supply would demand rather expensive infrastructures for transport and distribution. On the other hand, from an efficiency standpoint, only the unmanageable electricity/energy supply (due to intermittency and unpredictability of wind and sun resources) should be used for H_2 generation, thus viewing H_2 as a storage method of electricity surplus. The previous remarks, added to the poor well-to-wheel efficiency as well as the elevated total (fuel production chain plus vehicle) emissions, as noticed by Wald (2004), would make FC, originally conceived for powering automobiles, a more suitable technology for stationary distributed combined heat and power (CHP) generation.

3 DETERMINANTS OF THE TRANSITION TO HYDROGEN

Most analyses of the hydrogen transition rest on what is referred to in this literature (for example, NRC, 2008b) as the 'chicken-and-egg problem' (which alludes to the popular puzzle: 'which came first the chicken or the egg?'). Applied to the hydrogen transition, this is interpreted to mean that

HFCVs must come before hydrogen fueling stations, and fueling stations must come before HFCVs.

This problem is, in fact, the central justification for a planned massive intervention by the US Department of Energy in the HFCV market. But as daunting as this problem sounds, there are many historical precedents for its solution by market forces and without government intervention. For example, gasoline cars and gas stations suffer from an identical problem, as do diesel trucks and diesel stations, and TVs and TV stations, and many other technology pairs.

So how does the market solve chicken-and-egg problems? The most important answer is niche markets.[4] In fact, the first use of hydrogen vehicles in several countries appear to be buses. Their obvious advantage is that city buses never take long trips and always return to the same point at night. A whole fleet needs only one fueling station and that station is well utilized. There are also many fleets of delivery vans, government cars and so on with similar attributes. But the point is not to solve the transformation problem, but rather to understand why the market is so much better at solving it than is the government. Governments can miss the most obvious niche markets, while those with money on the table are extremely creative at finding niches.[5]

The second most important answer to the problem is 'related markets', and NRC (2008, p. 14) has identified such a 'potential remedy to the "chicken-and-egg" problem of . . . investments in large-scale hydrogen production. . . . The flexibility of gasification systems to provide electric power as well as hydrogen can significantly reduce the financial risks associated with large-scale hydrogen production during the scale-up phase of HFCV commercialization'.

One likely scenario, without government subsidies, is that HFCV costs will come down gradually, lagging behind politically determined goals. As this happens, fuel cells, hydrogen storage tanks and HFCVs will find high-value niche markets, just as expensive solar panels and $98,000 plug-in electric cars have found high-value niche markets. This is the general path to market transformation. But if there is a sudden breakthrough, and HFCV becomes the clearly dominant technology, there will be a rush to avoid being left behind as too many companies compete to become the market leader in FCs, hydrogen storage tanks and HFCVs. The market problem will prove to be a bubble of the internet variety, with over- not underinvestment, followed by a shake-out. Of course, if the industry is asked how much subsidy is needed to kick-start the market, as good businessmen, they will be obliged to name the highest plausible number.

Another likely scenario, without subsidies, is that after a decade of slow progress on hydrogen, battery technology will surpass hydrogen technology fairly definitively, and only stationary FCs will remain in use. In this case, we shall possibly have saved the cost of subsidies and hastened the adoption of the superior technology by not getting in its way.

4 R&D EFFORTS IN THE US

4.1 Technology Goals

Most discussions of the federal hydrogen research initiative revolve around one or more of the DOE's technology goals (DOE, 2006). The US DOE, in conjunction with the auto industry, established the US FreedomCAR Fuel Partnership. This partnership set technology goals for 2015 that were considered sufficient to bring about the commercialization of HFCVs. The most important of these goals are listed in Table 13.1.

The first six of these goals, those concerning the cost of hydrogen and FCs, can best be understood by comparison with other methods of generating electricity. After all, a fuel cell is just an electric generator and the market provides many types of generators for comparison. For an FC meeting the DOE's 2015 assumptions, the variable cost of electric power would be $2/kg for hydrogen divided by 32.7 kWh per kg of hydrogen divided by 60 percent efficiency, or 10.2 ¢/kWh. If we assume a 100,000 mile life for the FC, and use the DOE's goal of 80 mpgge, and 32 kWh/gge, we find that the FC is expected to produce 41,250 kWh. At an FC cost of $2,400, this comes to a fixed cost of 6 ¢/kWh. This is fairly inexpensive for such a small-scale generator.

Such an FC would be most competitive with peaking generators, which are those with the lowest capital costs. According to the DOE (2008), the

Table 13.1 DOE's hydrogen technology goals for 2015

Technology	Goal for 2015
Hydrogen	$2/kg retail (by distributed methane reforming)*
Fuel cell	$30/kW at a volume of 500,000 units/year
Fuel cell	80 kW
Fuel cell	5,000 hrs life
Fuel cell	60% efficiency (80 mpg equivalent)
Fuel tank	$2/kWh (= $65.40/kg H_2)
Vehicle	300-mile range
Vehicle	80 miles/gasoline-gallon-equivalent (mpgge)

Sources: * DOE (2007, p. 19). Other goals are from NRC (2008b, p. S-4).

type of generator with the lowest capital cost is currently an advanced combustion turbine with a 230 MW capacity. The 'overnight' capital cost of such a generator is $450/kW, 15 times more than the DOE's FC.

In fact, the ORNL (2008) reports that large-scale hydrogen generation costs only between $1.00 and $1.50/kg. At $1.25/kg, the DOE's FCs installed at these plants would generate power for only a little more than a conventional peaking unit, which is quite surprising given that these FCs are so small and designed for the rigorous environment of driving.

4.2 How Much Progress?

With such optimistic goals, and no argument showing that they should be attainable by 2015, the only means of judging them is by the rate of technical progress. The DOE has issued annual progress reports from 2004 through 2007. These cover all aspects of hydrogen R&D, but one of the most closely watched is FC costs so it is informative to see what has been reported regarding this goal. The introductions to the four progress reports give the following values for FC cost per kW of capacity. All assume a factory production level of 500,000 units per year. In the early 1990s the cost was approximately $3,000/kW. In 2002, it was $275/kW, followed by $175 in 2004. In 2005 it was 'approaching $110', in 2006 it was $108, and in 2006 it was $107. The duplicate final date reflects the unusual fact that the November 2007 progress report gives a new, slightly lower value for 2006, rather than reporting the value for 2007, as would have been customary.

Adding to the puzzle of the missing 2007 value, the 2007 report contains the subcontractor reports which in the past have provided the DOE's cost estimates. The primary one, from TIAX LLC (Lasher et al., 2007), reports enormous cost reductions because of a reduced need for platinum. The other, from Argonne National Laboratory (Ahluwalia et al., 2007) reports the missing value. It reports $108 for 2006 and $67 for 2007. Apparently the DOE felt that the new results were too good, or too shaky, to report.

Two problems with FC cost evaluation make progress difficult to judge. First, the choice of characteristics of the evaluated FC changes frequently. Second, the estimates are based on very high production levels, even though no FCs of that year's assumed FC design may have ever been produced.

For example, the FC design in 2002 called for gasoline as a fuel and not hydrogen. The reported cost value has been partially adjusted for this change, but it does not appear that the $100/kW drop in cost from 2002 to 2004 was primarily due to technical progress. On the other hand, the assumed FC designs since 2004 appear to be increasing in performance.

The 2007 cost of $67/kW is based on a new design by the 3M Company, which utilizes 3M's nanostructured thin film (NSTF) catalyst support for the cathode (Ahluwalia et al., 2007; Lasher et al., 2007). The cathode uses the bulk of the platinum. NSTF (apparently a 'carbon fabric'), in conjunction with vacuum deposition of an iron–cobalt–carbon–nitrogen cathode catalyst followed by a heat treatment, has apparently been successful in cutting the platinum requirement by more than half while increasing performance (3M Company, 2007). The research team at 3M is also optimistic about production costs.

While the results sound promising, it appears difficult to estimate the per-unit cost of producing 500,000 units per year when not a single complete FC, never mind an 80 kW stack of them, has been produced. Although this estimation may be unusually heroic, which may explain why it was not reported by the DOE in its summary, all FC cost estimates have been based on volume extrapolations that reduce estimated costs by more than an order of magnitude, and sometimes by two orders of magnitude from the low-volume costs with which there is actual experience.

Progress on hydrogen storage is even more difficult to pin down. Kalhammer et al. (2007, p. 5) report: 'Unlike other major technologies being pursued in support of [zero-emission vehicles], hydrogen storage technologies have advanced relatively little in recent years'. The NRC's Case 1 scenario, which comes close to assuming that the DOE's goals will be met, assumes that hydrogen storage costs will be five times greater than the DOE's goal of $2/kWh in 2015 ($65/kg of hydrogen). But even this higher value is not an estimate of what will happen, but rather an optimistic possibility.

Because the DOE's on-board storage-cost targets appear to be so wide of the mark, the discussion of storage research goals is quite ambiguous. Apparently, there is a presumption that the goals will be widely missed and manufacturers will use rather costly high-pressure storage. Meanwhile, the research emphasis has shifted to a wide range of alternative approaches that are in the early stages of development. These are based on advanced (not yet fully developed) materials, and generally operate at fairly low pressures and at temperatures far above that of liquid hydrogen.

4.3 Research Funding and Subsidies

The NRC has compiled federal funding data for prior years, 2004–07, and has estimated likely necessary funding for future years, 2008–23. The two sets of estimates use different funding categories, but these have been aligned to the extent possible in Table 13.2. The NRC has also estimated that an additional $300 million per year will be needed for 2021–23, when

Table 13.2 DOE's R&D funding for hydrogen light-duty vehicles in selected years

	2004	2005	2006	2007	2011	2015	2020	2004–20
H_2 production & delivery	30	40	32	66	58	45	15	706
Production demonstration	–	–	–	–	17	50	0	223
Fuel cells and H_2 storage	53	68	59	90	115	110	110	1721
FC demonstration	–	–	–	–	50	20	10	355
Safety and codes	8	6	5	16	25	10	5	232
Nuclear H_2 production	6	9	24	19	–	–	–	58
Renewable H_2 production	–	–	–	–	30	30	30	404
System analysis	1	3	5	10	10	10	5	146
Technology validation	16	26	33	40	–	–	–	115
Science	0	29	33	50	60	60	60	892
Total	114	181	190	290	365	335	235	4852

Note: All figures in millions of 2005 dollars.

Source: Data from NRC (2008, Tables 13.1 and 13.2). The year 2011 is listed because the NRC estimates it to be the peak spending year.

it believes that HFCVs will break even and become self-supporting. This brings their estimated total R&D funding for 2008 through 2023 to $5 billion, although they say that more is likely to be needed, and that some funding is likely to continue after 2023.

In addition to federal spending on R&D, the NRC (2008) reports that over $2.5 billion was spent by the private sector through the end of 2006. Private spending was primarily by GM, Ford, Chevron/Texaco, United Technologies, General Electric and nine venture capital companies. The NRC also describes a survey conducted for the FC industry, which reports that combined federal and private R&D spending for the United States in 2005 was $320 million. Subtracting federal spending of $220 million leaves $100 million in privately funded research.[6] Because the response rate of the survey was only 37 percent, the NRC inflates this number to $700 million. Perhaps $270 million would be a better guess, and because of likely self-selection bias in the survey responses, this value must be considered highly uncertain. In any case, the NRC estimates that there will be $11 billion in privately funded R&D from 2008 through 2023. Anecdotal evidence, especially from GM and Toyota, indicates that interest in hydrogen may be waning as interest in plug-in hybrids accelerates.

Even if research funding achieves the goals set by the DOE, a widespread belief within the hydrogen-research community holds that the market will not adopt the new technology. Consequently significant effort is being expended to plan various ways to subsidize HFCVs and the fueling stations these will require. The authors of the NRC report judge that a 'realistic estimate' of government subsidies is the 'incremental cost of purchasing [producing] fuel-cell vehicles, plus about half the total cost of building and operating the [required] infrastructure' (2008, p. 98).

For Case 1, the NRC estimates the cost of these subsidies at $40 billion for the auto industry and $8 billion for the fueling industry. The ORNL (2008) propose three subsidy-policy scenarios to support Case 1.[7] These have a cumulative cost, through 2025 (at which time HFCVs are supposed to break even), of $8, $14, or $18 billion – considerably less than the NRC's $48 billion. No explanation of the consequences of the different subsidy levels is provided by the ORNL other than a calculation showing that with less subsidy and the same production level and prices, the auto industry will have lower profits. Apparently these three 'policy options' are all thought to be compatible with identical transitions to an HFCV future.

5 R&D EFFORTS IN THE EU

Germany is the most advanced member state within the EU in H_2 and FC technologies; this vantage position is a consequence of significant R&D spending for over 20 years. The National Organization for H_2 and FC Technology (NOW) was established in February 2008 as a component of the National H_2 and FC Technology Innovation Programme (NIP) in order to promote the development and commercialization of products and monitor the global program. A total of €1,000 million will be jointly spent by the federal government and German industry over the next 10 years for development and demonstration activities.

The EU has been modestly investing in H_2 and FC R&D. A fund of €8 million under the Second Framework Programme (FP2) for the 1986–90 period has been increased to €145 million and €320 under FP5 (1999–2002) and FP6 (2003–06), respectively. The last dedicated 19.3 percent of the budget to H_2 production, 8.1 percent to storage, 14.6 percent to basic research on FC, 19.3 percent to transport applications and 8.0 percent to stationary applications as depicted in Table 13.3. Several projects span

Table 13.3 R&D topics and projects under the European Commission 6th Framework Programme and 2015 EU targets

Technology	% € FP6	2015 targets	FP6 projects	Research
H_2 Production	19.3	2.5€/kgH_2	R: HYDROSOL-II, SOLREF; HYVOLUTION; SOLARH O: DYNAMIS, NEMESIS, HYTEC INNOHYP, GENHYPEM; HI2H2 DISTRIB. NATURALHY	Bioprocess Biomimetics Photolysis L&HT Electrol Th-Ch Cycles Materials/ Proc
H_2 Storage	8.1	CG: 0.025 kgH_2/l L:0.040 kgH_2/l	STORHY, COSY, HYCONES NESSHY, HYTRAIN	Comp/Cryog, Metal Hydrides Carbon Cones Nanoma- terials
H_2 Infrastructure			HyWays, Roads2HCom; DYNAMIS	
Bas. Research FC	14.6	2–3000 €/kW → 300 €/kW	PEM: FURIM, APOLLON-B, AUTOBRANE, CARISMA, FCANODE, SOFC: Real-SOFC, SOFC600 GENFC	PEM: Membram €, catalystno Pt, HTElec- trolymem SOFC: Material: bipolplates, ano cat, el. LT>600°C, manufact (screen print, tape cast)
Transport Appl.	19.3	Car: 40%/100€kW/ 5,000h Bus: 40%/100€kW/ 10,000 APU: 35%/500€kW/ 5000	HYFLEET-CUTE, HYCHAIN_ MINITRANS, ZERO_ REGIO, HYICE, HYTRAN, HYSYS, HOPE, FELICITAS, CELINA, MC_WAP	DEMOs
Stationary Appl.	8.0	Res: 40%, 6,000 €/ kW, >12,000h (10%) Ind: 40%, 15,000– 5,000€/kW, > 30,000h (10% degradation)	BICEPS, FlameSOFC, NextGenCell, Largesofc; MOREPOWER	DEMOs

over bioprocesses, biomimetics, photolysis, low- and high-temperature electrolysis as well as thermochemical cycles to produce H_2 with a target cost of €2.5–4.0/kg by 2015. The development of advanced materials (metal hydrides, carbon cones, nanomaterials) allowing the 2015 objectives of storing 0.025 kgH_2/l as a compressed gas or 0.040 kgH_2/l in the liquid state. Research on PEM (proton exchange membrane) fuel cells and SOFCs (solid oxide fuel cells) is aimed at reducing the cost by a factor of 10 by 2015 to figures of the order of 300 €/kW; materials for membranes, bipolar plates, anodes and cathodes, catalysts with no platinum, development of new manufacturing processes, and operation at lower temperatures in the case of SOFC are among the research topics covered. Some demonstration projects under the heading of transport applications aim to show the viability of increasing PEM efficiencies well above 40 percent and lifetimes of up to 5,000 hours for cars and 10,000 hours for buses. Projects related to stationary applications mostly concentrate on MCFCs (molten-carbonate fuel cells) and SOFCs and aim at demonstrating efficiencies above 40 percent with less than 10 percent degradation and costs in the range of 1,500–6,000€/kW as well as over 12,000 hours of operation for the residential sector and more than 30,000 hours for industrial uses.

A High Level Group (HLG) on H_2 and FC was created in 2003 by Ms Loyola de Palacio, EC Vice-President and Commissioner for Energy and Transportation, with the assignment of producing some recommendations on possible approaches to a hypothetical H_2 future economy. A report, 'Hydrogen Energy and Fuel Cells: A vision of our future' (HLG, 2003) proposed a roadmap and the creation of a European Hydrogen and Fuel Cell Technology Platform (HFP). The latter started operation in 2004, with a management structure comprising an Advisory Council, a member state Mirror Group and representatives from the EC. The Advisory Council, comprising major EU H_2 and FC industrial stakeholders and research community members, and its Executive Group constitute the governing board of the HFP. The interests of the member state are conveyed through the Mirror Group. Between 2005 and 2007 the HFP produced several key documents (Strategic Research Agenda, Deployment Strategy, Strategic Overview and Implementation Plan), apart from conducting several projects and initiatives, and holding an annual General Assembly. A reasonable 'Snapshot 2020' (Table 13.4) considers a realistic initial penetration of portable FCs, portable generators, stationary FCs and road transport FCs.

The interconnection of EU H_2 communities between 2015 and 2020 allowing travel from Madrid to Stockholm in an FC vehicle (HFCV) and the completion of the full H_2 infrastructures by 2050 are contemplated as ambitious targets. Four innovation and development actions

Table 13.4 'Snapshot 2020' with key assumptions on H_2 and FC applications for a 2020 scenario

	Portable FCs for handheld electronic devices	Portable generators & early markets	Stationary FCs CHP	Road transport
EU H_2/FC units sold per year projection 2020	~ 250 million	~ 100,000 per year (~ 1 GW$_e$)	100,000 to 200,000 per year (2–4 GW$_e$)	0.4–1.8 million
EU cumulative sales projections until 2020	n.a.	~ 600,000 (~ 6 GW$_e$)	400,000 to 800,000 (8–16 GW$_e$)	1–5 million
EU expected 2020 market status	Established	Established	Growth	Mass market roll-out
Average power FC system	15 W	10 kW	<100 kW (Micro HP) >100 kW (industrial CHP)	80 kW
FC system cost target	1–2 €/W	500 €/kW	2,000 €/kW (Micro) 1,000–1500 €/kW (industrial CHP)	< 100 €/kW (for 150,000 units per year)

on 'H_2 Vehicles and Refuelling Stations' (€2,661 million), 'Sustainable H_2 Production and Supply' (€759 million), 'FC for CHP and Power Generation' (€2,853 million) and 'FC for Early Markets' (€1,110 million) were established (the estimated budgetary needs between 2007 and 2015 appear within parentheses) and subdivided into well-defined tasks within the Implementation Plan.

The HFP proposed the creation of a joint technology initiative (JTI) to overcome the fragmentation of R&D activities by fostering the cooperation among industrial stakeholders; a consistent execution of the long-term strategy outlined at the Implementation Plan was aimed at ensuring well-defined R&D programs matching industrial needs. Some 58 companies from 15 different member states initially established in 2007 the Industry Grouping (IG) as a legal entity, leading way towards the JTI; over €10 million have been invested by the IG stakeholders for the JTI preparation.

The companies participating at the IG, both large corporations and small and medium-sized enterprises, represent 90 percent of the total industrial investment on H_2 and FCs and share 50 percent of the running cost of the JTI Program Office. By the end of 2007 the JTI proposal was adopted by the EC and the project FCHInStruct was launched as part of the JTI formal preparation; the latter was a Coordination and Support Action, co-funded by the EC and the IG. The creation of a Research Grouping (RG) was initiated in the second half of 2007; by January 2008, 49 participants from every member state had expressed interest in joining the RG to be integrated within the JTI. The JTI FCH received the approval of the Council and the European Parliament in spring 2008, and was formally launched in October 2008.

The JTI Governing Board comprise six members from the IG, five from the EC and one from the RG, and receive advice from a scientific committee and from the FCH States Representatives Group. The executive director head is the JTI Program Office; projects as well as coordination and cooperation with regional and international programs are the responsibility of this office. The FCH Joint Undertaking is the legal entity coordinating the use and efficient management of funds committed to the JTI; it was established under Article 171 of the EC Treaty and operates from 2008 to 2013, with a possible extension to 2017. The Stakeholders General Assembly of members from the IG, the EC, the RG, member states, regions, international organizations, non-governmental organizations and other industrial and research groups provide input to the JTI governance.

Some 50 percent of the JTI budget comes from industry, while 50 percent is provided by the EC, member states and the European regions. For the FP7 (2007–13) the EC has allocated €470 million, under Energy, Transport, Materials and Environment Programs; a matching budget of at least €470 million is supplied by the private sector. Additional contributions of about €200 million from non-EC public entities and private groups are expected. The innovation and development action, defined by the Implementation Plan with an overall budget of €7,383 million for the 2007–15 period, is streamlined and the different tasks prioritized and adapted to a reduced funding of about €1,100 million for the 2007–13 period.

6 COMPARISON BETWEEN EU (DG TREN AND DG RES) AND US DOE

As already mentioned, Germany is the leading EU member state in H_2 and FC R&D. H_2 funding reached its peak in 1991 and it has been reduced by a factor of 20 over 10 years. FC investment had its maximum in 1995 and declined by 30 percent until 1999. During the same period, the USDOE funding has more than doubled for both H_2 and FC starting from about $100 million/yr in 1992; that year the H_2 spending in the USA was more than twice that in Germany, while the FC investment in the USA was nearly 20 times that of Germany. That situation was even less favorable to Germany in 1996 with the previous figures being approximately 30 and 10, respectively.

In 2003 the EC HLG recognized the need for drastic action by the EU on H_2 and FC R&D in order to overcome its significant weakness in comparison with the USA and Japan. In some respects, the structural organization of the European JTI resembles that of the USDOE SECA (Solid State Energy Conversion Alliance), a public–private partnership, initiated in 1999, bringing together industry, research groups and government to foster the accelerated development of SOFC systems; however, the technical leadership of the USDOE in SECA seems, in principle, more significant than that of the EC, through DG TREN and DG RES. The JTI Implementation Plan bears some similarities to the Core Technology Program of SECA, supporting long-term research activities typically not prioritized by industry and with a dynamic annual peer review process.

It seems pertinent to remark that, while the operation of the USDOE could be compared with that of an orchestra, with well-trained musicians (15 national laboratories, nearly 300 university research groups and powerful companies) actively coordinated by a single conductor (the USDOE and its body of scientific advisors), the coordination of DG TREN and DG RES at the European member state level is hampered by 27 orchestras with 27 conductors, which insist on playing their own scores, with inefficient spending of funds and not profiting from synergies. The EC Joint Research Centre activities in support of EC policy measures are not comparable to those of the US national laboratories.

Research institutions, far from the standards of excellence, can be a part of joint European endeavors. The fragmented character of European R&D within several member states, with regional differential approaches not always well coordinated, compound the EC difficulties.

7 THE WAY FORWARD

The possibility of a future economy based on H_2 and FCs is both promising and uncertain. The fate of these technologies might well be conditioned by the decisions of China and India on their commercialization and deployment. However, due to low well-to-wheel efficiencies and to high fuel-chain plus vehicle emissions, H_2 and FCs, originally conceived for automobile propulsion, might be better suited for stationary CHP applications.

Regarding H_2 production, natural gas is a precious fuel and should not be used for the production of H_2, while its generation from coal should include CCS technologies. Massive nuclear and renewable power should be the carbon-free energy sources for possible H_2 large-scale production processes. A rational approach, contributing to alleviate the intermittent and unpredictable nature of some renewable energy resources, would produce H_2 from the unmanageable electricity surplus, storing it and, subsequently using it, either in ICEs or in FCs, for electricity generation in periods when there is no wind and/or sun. The use of ICE to burn H_2, as well as fossil fuels or biomass, with CCS, should be carefully evaluated in comparison with the alternative utilization of FCs. In any event, the development and demonstration of H_2 storage methods, as well as robust and inexpensive FCs seems to be a top priority.

Research on materials and processes for H_2 production and on materials and innovative manufacturing for FCs are fields of vital interest. Industrial development and demonstration projects should be actively launched in the near and medium terms, before medium-/long-term deployment, commercialization and market penetration can follow. Well-balanced funding strategies encompassing short-, medium- and long-term R&D activities should be undertaken by public–private partnerships. The allocated funds for H_2 and FCs should be necessary and sufficient, in competition with alternative/promising fuels and conversion technologies. High-quality management of programs and projects is essential and should include annual peer reviews and evaluations of performance and strategy; technical learning rates will equally depend on good funding, excellent R&D and quality management.

Temporary subsidies might be considered for penetration and deployment activities, while technology commercialization should be supported through rational and internationally homogeneous regulations, standards, procedures and licensing. However, it is likely that the market will find a cheaper path, through niche and related markets, to complete the transition to hydrogen – if that proves to be the right solution. And that is the real reason to let the private sector lead the market transition. In other words, the great advantage of relying on market incentives, such as a cap-and-trade system or a carbon tax, is that it will allow the market to determine the correct timing and the correct direction for the transition. This is what markets are good at and bureaucracies notoriously fail at. On the other hand, markets are poor at conducting advanced research, and governments have the required deep pockets and risk tolerance. It would be advisable to let both sides do what they do best.

8 CONCLUSION

Research on and development of hydrogen technologies are reasonable and necessary activities, in spite of uncertainties related to the future of a 'hydrogen economy' in competition with alternative solutions. However, it also entails important R&D spending. Therefore no single economy, either that of the US or that of the EU, can reasonably and profitably face the required efforts in isolation. Furthermore, experience shows that despite its faults, the US centralized research approach has been much more successful than the less-structured and fragmented EU pursuit.

In any event, the penetration of most hydrogen technologies is a long-term issue. A priori favoring of specific technologies for massive deployment can be an incorrect strategy and even detrimental to the long-term hydrogen future. Rather, proper economic public incentives should be established for different alternative competing technologies. Then entrepreneurs should look for the niche and related markets that could spawn earlier and wider applications of hydrogen.

NOTES

1. César Dopazo gratefully acknowledges the information on the Joint Technology Initiative and the FCHInStruct Support Action provided by Agustin Escardino and Andre Martin, members of the HFP Advisory Council and the Industry Grouping.
2. For instance, General Motors is currently hoping to begin selling PEVs with a 40-mile range in 2010. A 40-mile range is sufficient to cover 75 percent of the light-vehicle driving in the United States. Because the PEV is a hybrid, it will have increased efficiency; and, if used with a diesel engine, a doubling of efficiency should be easily achievable with present technology. This would result in roughly an 87 percent reduction in gasoline use. That is considerably better than what the NRC predicts could be achieved by HFCVs by 2050 under its optimistic Case 1.
3. The situation does not seem to be clearer in the short run, between now and 2050. For instance, the NRC concludes: 'advanced conventional vehicles and biofuels – have the potential to provide significant reductions in projected oil imports and CO_2 emissions [but] the deepest cuts in oil use and CO_2 emissions after about 2040 would come from hydrogen' (2008, p. 63). But this conclusion has been weakened by assuming that the future will bring only advanced conventional vehicles or biofuels, but not both. In fact the NRC reports (see ibid. Figures 6.29 and 6.30) what happens if both improvements are realized. In this case HFCVs do not provide deeper cuts in either oil or emissions until after 2050. This conclusion takes into account only what the NRC considers 'evolutionary' technology.
4. By way of contrast, the Oak Ridge National Laboratory's (ORNL, 2008) approach to the transition is to look for a way to jump-start the final mass market from day one. The ORNL suggest that because New York is compact it will be one of the easier mass markets to jump-start. This is probably correct, but it is unlikely to be the solution that the market would choose without heavy guidance. The trouble with trying to jump-start a metropolitan area is that most people who buy a car want the option of taking the occasional longer trip. Under the Gronich plan, New Yorkers would have to wait eight years to visit Washington DC and much longer to visit relatives in upstate New York.

5. Since the chicken-and-egg school of thought offers no detailed theory of why the market cannot self-start, let us consider their claims concerning the consequences of this problem. The ORNL (2008), in its Figure 16, shows that, without government support, the HFCV market would not achieve a 5 percent market share until 2048, while with $8–18 billion in subsidies it would achieve a 90 percent share by that date. Alternatively, this can be read as a 20-year delay if the government does not subsidize the transition. With $18 billion in subsidies, the industry experiences essentially no cost during the transition. Is it plausible that the need for an $18 billion investment would delay the industry for twenty years from moving to a new dominant technology?

Consider that in the third quarter of 2007, General Motors (GM) posted a $39 billion loss. Falling behind the market can be quite risky. With hydrogen technology already developed by 2015 and waiting only for a solution to the chicken-and-egg problem so that high-volume production levels can bring down the cost of HFCVs, every car company will live in fear that some other car company will move first and take over the new market. Once this market takes off, according to the ORNL, it will, without benefit of subsidies, grow from a 5 to a 55 percent share of new car sales in nine years. The companies left behind will suffer fates similar to GM's. That is why many are now researching hydrogen cars – they are not terribly optimistic, but they fear being left behind.
6. The $220 million figure includes $40 million not shown in Table 13.2 because it was earmarked by Congress for projects that were generally considered ineffective.
7. The NRC's Case 1, and the ORNL's (2008) Case 1 are both derived from Gronich (2007).

REFERENCES

3M Company (2007): 'Novel Approach to Non-Precious Metal Catalysts', Final Report, Prepared for the US Department of Energy, Washington, DC.

Ahluwalia, R.K., X. Wang and R. Kumar (2007): 'V.A.1 Fuel Cell Systems Analysis', Argonne National Laboratory, in DOE's 2007 Annual Progress Report.

DOE (2006): 'Hydrogen Posture Plan: An Integrated Research, Development, and Demonstration Plan', US Department of Energy, Washington, DC.

DOE (2007): 'Annual Progress Report, Hydrogen Production Sub-Program Overview', US Department of Energy, Washington, DC.

DOE (2008): 'Assumptions to the Annual Energy Outlook 2008', US Department of Energy, Washington, DC.

European Hydrogen and Fuel Cell Technology Platform (HFP) (2005–2007): 'Strategic Research Agenda', 'Deployment Strategy', 'Strategic Overview' and 'Implementation Plan', available at: http://ec.europa.eu/research/fch/index_en.cfm?pg=documents (accessed February 3, 2010).

Gronich, S. (2007): '2010–2025 Hydrogen Scenario Analysis', Presentation to the DOE Hydrogen and Fuel Cell Advisory Committee, Washington, DC, February 20.

Oak Ridge National Laboratory (ORNL) (2008): 'Analysis of the Transition to Hydrogen Fuel Cell Vehicles and the Potential Hydrogen Energy Infrastructure Requirements', Prepared by Oak Ridge National Laboratory, Oak Ridge, Tennessee, ORNL/TM-2008/30.

High Level Group (HLG) (2003): 'Hydrogen Energy and Fuel Cells: A Vision of Our Future', EC Report.

Kalhammer, F.R., B.M. Kopf, D. Swan, V.P. Roan and M.P. Walsh (2007):

'Status and prospects for Zero Emissions Vehicle Technology: Report of the ARB Independent Expert Panel', Sacramento, CA: State of California Air Resources Board.

Lasher, S., J. Sinha and Y. Yang (2007): 'V.A.5 Cost Analyses of Fuel Cell Stack/ Systems', TIAX LLC, in DOE's 2007 Annual Progress Report.

NRC (2008): *Transitions to Alternative Transportation Technologies – A Focus on Hydrogen*, National Research Council, Washington, DC: National Academies Press.

Wald, M.L. (2004): 'Questions about a hydrogen economy', *Scientific American*, May, pp. 40–47.

PART IV

Conclusions

14. EU energy security of supply: conclusions

Jean-Michel Glachant, François Lévêque and Pippo Ranci

1 INTRODUCTION

The chapters in this book have shown that energy security of supply is an extremely political matter. It involves consideration of international relations, geography and infrastructure control. Despite these realities economics has much to offer for improved policy making in the security of supply. In fact, economists can say a great deal about markets and policy. Markets (for example, for balancing, storage, reserves and access to cross-border capacity) as well as investment (which is guided by market prices) play a critical role in delivering security of supply, as does public policy (for example, for energy efficiency, storage obligations, and the fuel mix). Moreover, security of supply can give rise to 'free-riding', a phenomenon familiar to economists.

This final chapter presents the conclusions of the book and it corresponds to the scientific consensus of the CESSA project at large. While focusing on 'our' three sectors, natural gas, nuclear, and hydrogen, we are well aware of the fact that there are plenty of other important issues in energy supply security. Thus, most of our conclusions go beyond the more sectoral level.

The conclusions from the CESSA project are as follows:

1. Short- and long-term disequilibria in security of supply are different phenomena.
2. Solidarity between EU member states to cope with energy disruptions is necessary and even more so in the absence of satisfactory market liberalization and market design.
3. Information provision and sharing improve energy security for all.
4. Open access pan-European networks are the backbone of EU security of supply.

5. Public policies that are not specifically devoted to security of supply may be at least as critical in improving security of supply as ad hoc policies directed specifically at such objectives.
6. Markets are central to providing security of supply.
7. Security of supply is enhanced when markets are wider.
8. Public policies that facilitate market enlargement and competition improve security of supply.
9. Good competition policy enforcement helps security of supply.
10. Bad public policies, including security of supply policies, can severely damage long-term security of supply.

2 CONCLUSIONS OF THE CESSA PROJECT

1. *Short- and long-term disequilibria in security of supply are different phenomena*

They must be analyzed differently and may need to be addressed with different policy instruments.

Short-term disruptions in energy supply are transitory; they can be caused by a variety of factors (for example, equipment failure, human error, weather events, crime or accidents). Moreover, the provision of short-term security of supply may have to deal with the problem of free-riding. If one company (or country) invests in improving security, others can free-ride, deriving part of the benefit without paying for it. One measure to protect against such free-riding could be to charge appropriately for the protective service (for example, by short-term scarcity pricing of access to storage or reserves that others had secured on more favorable long-term contracts). In the absence of such cost-reflective charging, there is a risk of underprovision of security of supply services.

Long-term disequilibrium comes from a structural mismatch between supply and demand, such as that forecast by the IEA today in gas and oil. Long-term energy demand is especially difficult to predict with sufficient accuracy to assess future prices, whereas investments to respond to potential future shortages or high prices have to be decided now. Underprovision of substitute sources of energy or supply (for example, in nuclear power, LNG import facilities, or alternative pipeline routes) can therefore occur.

2. *Solidarity between EU member states to cope with energy disruptions is necessary and even more so in the absence of satisfactory market liberalization and market design*

Member states are, and will continue to be, key players in providing short-term security of supply within the EU. However, emergency access to gas reserves need to be guaranteed under well-defined criteria and on well-defined terms, otherwise it will be expected that they will be withheld to address national disruptions, with little credible improvement in EU security. A solidarity mechanism is required to ensure the necessary pooling of, and access to, resources that the whole EU requires. In electricity markets, regulatory barriers that prevent national and cross-border contracts from providing security of supply to national demand on equal terms should be removed. Otherwise each country will have to fully provide for its own security of supply. Under the right conditions, disruption in one place would be more quickly solved and would not contagiously spread elsewhere.

To date, the voluntary EU solidarity contract has failed to deliver satisfactory outcomes. It is the role of the EU institutions and rules to enforce cooperation whenever the common interest is larger than the sum of the member states' individual interests. The European Commission must propose and implement credible and robust solidarity mechanisms.

3. *Information provision and sharing improve energy security for all*

Information may be costly to produce (for example reliable energy and capacity forecasts) and information may deliver a strategic advantage to individual agents (for example, higher payback to a storage facility, higher scarcity prices for balancing services). It is therefore to be expected that there will be underprovision of some information as well as incentives not to communicate other information.

In order to improve responsiveness to disruptions and to provide resilience efficiently, EU energy security policy needs to mandate adequate information acquisition and dissemination. Transparency, open access, and real-time dissemination of information are key for short-term security of supply. It deals at least with a weekly statistics of gas storage in each country of the EU. Furthermore, explicit methodology and assumptions, scenario discussions, regionalization and collegialization of the forecasting studies, and an agreement on the data to be collected and made available (for example, storage capacity, import capacity, reserve margins, investment plans and so on) are vital mechanisms for long-term security of supply.

4. *Open access pan-European networks are the backbone of EU security of supply*

The electricity transmission grid and the gas transportation system contribute to both short- and long-term security of supply. They connect different sources of supply and therefore increase the availability of flexible and diverse resources that can be accessed to cope with a short-term disruption. They also ensure a better long-term match between energy demand and supply, because what consumers need and buy is *delivered* energy, the combination of energy and transportation services. They cannot use electricity or gas only provided at some distance to their home or business.

National network operators and regulators are responsible to their national consumers and governments. However, they frequently fail to deliver when cross-border links are at stake. At best, they may give equal priority to a pan-European investment as to a national investment. It is fair to say that there is a trans-European network policy that recognizes security of supply as one of its basic aims. However, its budget to invest in new links in gas and electricity networks is less than 1 percent of the annual spending of all national networks. More money to fund economic cross-border connections has to be found, as well as more incentives and duties for national operators and regulators to take pan-European investments projects seriously. This is where adequate information that allows external parties to assess the desirability of potential links is critical, as not all links are equally useful and the scarce funds should be allocated to those that deliver highest value.

Clear regulations that assign the responsibilities for the new investments, the participation of the stakeholders in the process and the allocation of cost to the network users are of essence in this respect. Moreover, policy makers have to deliver more efforts to fight against the NIMBY ('not in my back yard') syndrome that often obstructs the building of infrastructures that are so important to improve security of supply.

5. *Public policies that are not specifically devoted to security of supply may be at least as critical in improving security of supply as ad hoc policies directed specifically at such objectives*

Deciding on a mechanism for collective access to security reserves, building new grid connections and creating an information network, as mentioned above, are examples of specific policies directed at improving security of supply. It would be misleading to focus only on them, as security of supply is affected by other public policies.

Energy efficiency policy is a good example. While generally aimed at reducing CO_2 emissions and/or trade deficits, its impact on security of supply can be considerable. The more energy is saved, the more adequate will be the existing storage capacity as well as the margins of installed capacity over demand, and the lower will be the import dependence; reducing the addiction of the economy to energy is the most obvious way to reduce dependence on imports and the concomitant risk of supply security. However, it must be remembered that a more energy-efficient system may be less flexible: in case of emergency there will be fewer types of inefficient consumption that can be curtailed (for example, by reaction of consumers to higher prices or by compulsory rationing).

Technology policy is another example. Smarter networks, smarter devices that can reduce user demand at short notice, as well as diversity in power generation can do much to improve short- and long-term security of supply. Technological innovation and adoption in these areas can be facilitated by public policies.

A last example is given by foreign policy. As a truism, energy security depends on good relations with neighbors.

6. Markets are central to providing security of supply

The market is a mechanism that allocates scarce resources. When an energy disruption occurs, the market responds to the resulting short-term scarcity by rationing use through higher prices and moving flexible supplies to where they are most needed. The gas US market, for instance, succeeded in coping with the Katrina and Rita hurricanes that shut down 20 percent of the Gulf region capacity (amounting to 5 percent of national capacity) with quite modest price spikes and impacts on consumers.

As far as long-term security of supply is concerned, the market reacts to anticipated scarcity by incentivizing investments. Anticipated high prices trigger decisions to build new facilities.

Interestingly, when markets are opened to effective competition, there is frequently a rush of new projects, reducing future scarcity and lowering future prices. Thus in the EU gas sector we have recently observed a surge in investment in new LNG terminals and peak-shaving storage. The nuclear industry which also contributes to improving security of supply enjoys a revival, with market-based financing of new build now appearing as a viable business possibility.

7. Security of supply is enhanced when markets are wider

The larger the geographic extension of energy markets, the more they are able to absorb disruptions because more resources are available to damp price spikes. In addition to this insurance or resilience effect that improves short-term security of supply, wider markets also provide more diversity of primary fuel types and of geographic sources, and therefore ensure a better long-term security of supply.

Moreover, the wider the market, the higher the number of players, and therefore the more competitive the market. This point is important because uncompetitive markets allocate scarce resources inefficiently. For instance, if electricity companies cartelize they will restrict investment in new power plants to keep prices high. The cartel's interest is similar to a monopoly's interest, that is, to create, not reduce, scarcity.

8. *Public policies that facilitate market enlargement and competition improve security of supply*

Left to themselves, markets will expand and erode monopoly positions, but it can take time. This is especially true in electricity and gas where transportation costs can be high and innovation is slow. Public policies can speed up the process by decreasing barriers to trade and to entry. This is the reason why the EU internal energy market was created in 1996. Note that the opening of electricity and gas to competition also requires the creation of a complex series of related markets that must be designed. They bear esoteric names such as day-ahead markets, balancing markets, capacity payments markets, as well as institutions such as explicit auctions, and market coupling. Whenever they are badly conceived or implemented, trade is less efficient and usable cross-border interconnection capacity reduced. Competition is hindered and thus security of supply is reduced.

This is one area where the EU could make considerable progress at modest cost. The third package of directives related to the gas and electricity markets, once adopted, will contribute to such progress. As a consequence, a quick, full, and cooperative implementation of the new package would be an important step to improve the EU security of supply.

9. *Good competition policy enforcement helps security of supply*

Antitrust law, merger regulation and state aid control all contribute to protect competition in energy markets and therefore, as a side-effect, to the provision of security of supply in the EU. However, antitrust authorities are sometimes biased in reaching their decisions. They can put more weight on short-term losses than on long-term benefits, or on anticompetitive effects than pro-competitive effects, say of a merger or an agreement between firms, or on national rather than continental benefits.

Moreover, case law might not be sufficiently clear and decisions difficult to anticipate.

Gas purchasing long-term contracts is a topical illustration where antitrust biases and uncertainties may work against security of supply, for they can discourage investment.

10. *Bad public policies, including security of supply policies, can severely damage long-term security of supply*

Assuming that markets are competitive, very high prices signal unmanipulated scarcity and remunerate investments for extreme peaks of demand that may operate only a few days per year. High prices over a long period of time signal undercapacity in generation and attract new investments, as already mentioned above. But high prices hurt consumers. They may be quick to complain and induce policy makers to respond by capping prices on spot markets and/or maintaining low administered retail prices. Such short-term redistributive goals are achieved at the expense of future consumers, for they delay investment in new capacity, raising future prices and/or causing shortages and rationing.

Even policies aimed in principle at providing security of supply can have perverse effects. Consider the example of badly managed strategic reserves. Because high prices are politically unpopular, governments and/or their agencies can be tempted to open them in a period of high prices just to damp down markets that are handling temporary shortages satisfactorily. Such actions are counterproductive as releasing strategic reserves into the market lowers market prices and undermines the incentive for the market to invest to cope with future shortages. It reduces the capacity the market would have provided spontaneously and at lower cost, and may reduce future reserve capacity if the government cannot afford to take over this (unnecessary) responsibility.

These 10 CESSA conclusions may look unexciting to some, because they are based on economic arguments rather than identifying heroes and villains. Markets and public policies are both imperfect, and as usual economists recommend improving them both. They also argue that improving markets is normally cheaper and more direct than attempting to achieve the political consensus needed to impose change through policy. Improving markets is unfortunately less dramatic than making grand policy statements.

Name index

Adams, T.M. 194

Ball, J. 95
Barbir, F. 200
Barton, J.P. 192
Băsescu, T. 173
Bazilian, M. 130
Bossel, U. 215
Brunet, R. 243
Bucur, I. 175, 176
Bupp, I. 120

Castaneda, D.J. 34
Ceauşescu, N. 173, 174
Chan, C.C. 195
Chao, H.-P. 125, 126
Cheney, D. 61
Coase, R. 46, 50
Correlje, A. 68

de Esteban, F. 166–7
de Jong, J. 15
de Vries, L. 125
Derian, J.C. 120
D'haeseleer, W. 202
Diaconu, O. 174

Edwards, R. 254
Egging, R. 14
Esty, B. 118, 121

Finon, D. 136
Foraker, J. 35, 48, 51
Foss, M.M. 79
Fridley, D. 80

Gabriel, S. 14
Garcia Peña, F. 193
Garriba, S. 76
Gasmi, F. 15
Gerling, P. 258

Hallouche, H. 72
Hansen, A.C. 257–8, 262, 263, 267
Hartley, P. 92, 94, 95
Hashimito, K. 96
Hayes, M. 92, 94, 96, 97, 101
Hirschhausen, C. von 15
Holz, F. 14
Honnery, D. 205, 206, 207, 216
Hooley, R.W. 42
Hudson, G. 132

Iliescu, I. 173
Infield, D.G. 192
Ivy, J. 193

Jaffe, A. 94, 100
Jensen, J. 95
Johnson, A.M. 38
Joshi, S. 80
Joskow, P. 3
Jung, N. 80

Kahn, A.E. 43, 50
Kalhammer, F.R. 281
Kintner-Meyer, M. 195
Kirkilas, G. 180–1, 184
Kløverpris, J. 192

Lacy, B. 136
Leahu, C. 176
Lessard, D. 121
Levin, D. 213
Levy, B. 120
Lewis, S.W. 96, 102
Luostarinen, K. 141

MacKerron, G. 109
Mares, D.R. 97, 98, 103
Martin, P. 77
Medlock, K.B. 92, 94, 95
Miller, R. 121

Mitchell, J.V. 67
Moriarty, P. 205, 206, 207, 216
Moynihan, D. 42

Nauduzas, V. 181
Nelson, J.R. 50
Neuhoff, K. 125
Ni, M. 207, 210
Nicholas, M. 225
Nuttall, W.J. 158

Olcott, M.B. 94
Olson, W.P. 45

Piebalg, A. 181
Pouret, L. 158

Roques, F. 130, 158
Rothwell, G. 164

Sanders, M.E. 40
Sandulescu, A. 175, 177–8
Schenk, N. 216
Shepherd, R. 95
Sherif, S. 202, 216
Simpson, A. 192
Skinner, R. 62–3

Smeers, Y. 14–15
Soligo, R. 100
Spiller, P. 120
Staropoli, C. 136
Stern, J. 63, 64, 73, 75, 84, 94

Tarjanne, R. 141
Taylor, S. 155, 162
Thomas, S. 122
Toenjes, C. 15
Tolnay, A. 173
Troxel, C.E. 37–8
Turnock, D. 173–4

van der Linde, C. 68
Veziroglu, T.N. 200
Victor, D.G. 92, 93, 94, 96, 97, 98, 101–2, 102, 103
Victor, N.B. 94, 97, 103
Vilemas, J. 181, 184–5
Voets, P. 205, 206
von der Mehden, F. 96, 102
Voorspools, K. 202

Wong, Y.S. 195

Zhuang, J. 14

Subject index

accidents, nuclear 163, 164, 169
accounting regulation 45, 53
adminstrative procedures 33, 41–2
Africa 68
   North 64–7, 71, 82–3
   West 67, 71, 83
agricultural wastes 207
Algeria 65, 66, 71, 101
   GECF 72, 73
   Skikda liquefaction plant 74
anaerobic bacteria 213–14
Angola 67, 71
AREVA 122, 124, 129, 140–1, 143, 144
Argentina 97–8
Arun, Indonesia 96, 102
Asia 12–13
   Central 69
   see also under individual countries
Atlantic Basin
   emerging LNG market in 76–9
   GECF 72, 73
Atomic Energy Canada Ltd (AECL) 174, 176
Azerbaijan 70

Baker Institute World Gas Trade Model (BIWGTM) 92–103
Baltic states 172, 178–85
battery electric vehicles (BEVs) 249–50, 271, 275–6
Belarus 59, 62, 74, 81
biofuels 192, 260, 275–6
biological processes 207
   hydrogen production 212–14
biological water-gas shift reaction 213
biomass 192, 207–8, 215, 216
   hydrogen production 210–11, 212–14, 235
biophotolysis 212–13

bipolar electrolyser 208
Blue Banana region 243–5
Bolivia 97
Brazil 97, 103
British Energy 146, 152, 155, 159, 161, 164
   financial crisis 155, 156–8, 162
   risk management strategies 162, 163
   share price 156, 157
British Gas 10
Bulgaria 177
buses, hydrogen 255, 278
byproduct hydrogen 234, 235–6, 253

CAFE policies 263–4
California
   energy crisis of 2000–2001 25, 26–8, 51
   zero emission programme 251
Canada 76–7
CANDU-6 nuclear technology 174–5
capacity credit 202
carbon capture and storage (CCS) technology 19, 111, 193, 259, 263, 271
carbon dioxide emissions 84–5, 199
   reduction target 261
   see also greenhouse gas emissions
carbon dioxide price risk 124–5, 143–4
cartels 300
   gas 4, 72–3, 100–101, 257
cash-flow analysis 241–3
Caspian region 70–71, 82
catastrophe risk 163, 164
Central Asia 69
Central Europe 11, 18, 110, 171
Centrica 10
Cernavoda power plant complex 174, 175–6
Chernobyl disaster 164, 179

Chicago cold snap 25, 26, 27
Chile 97–8
China 12
  LNG imports 79–81
Cities Alliance 40
climate change 4, 167
  see also carbon dioxide emissions; greenhouse gas emissions
climate policy 109–10
  hydrogen transition and 248–74
  natural gas and 6, 19
coal 85, 110–11, 160
  hydrogen production from 193, 196, 234
    security of supply to Europe and 256, 257, 258–9, 271
collaboration between gas exporters 73, 84
  see also Gas Exporting Countries Forum (GECF)
combined-cycle gas turbine (CCGT) plants 121, 124, 131
combustible hydrogen resources 252–3
combustion 207
'Commodities Clause' 33, 43–4
commodity markets 160–1
common carriage (TPA) 33, 34–7
Commonwealth of Independent States (CIS) 59, 69
competition
  natural gas 6
    global competition for LNG supplies 76–81
    US and European gas pipelines 21–55
  and security of supply 6
    policy enforcement 296, 300–1
    public policies facilitating competition 296, 300
compressed gaseous hydrogen 236, 237–8, 239
concentrated-user scenario 225, 230, 231, 233, 241
concentrating solar thermal systems 203–4
consortia 122, 126, 128–9
  of consumers 139–42, 148–9
  of producers 137, 146, 148
Constellation Energy 139

construction risk 118, 121–3, 137, 140–1, 144, 148
consumers' cooperative 139–42, 148–9
continental Europe 11–12
contracts 99
  gas pipelines and 21, 23, 24, 48–9, 99, 301
  nuclear power 161–2
    compatibility with financing arrangements 127–35
    contractual arrangements and new nuclear investment 117–54
    risk management 161–2
contractual congestion 51
cooperative, consumers' 139–42, 148–9
corn 160
corporate finance 112, 118–19, 127, 129–32, 141, 144, 147, 148
costs
  cost of capital 132–5, 147–9
  cost efficiency of FCEVs 266–70
  of hydrogen 239–41
  sharing in a consortium 146
credit rating 133–4

damless hydropower 205
dark fermentation 213–14
debt finance 128
decentralized markets 136–42, 148
delivered gas era 24–5
demand
  hydrogen demand rollout 229–30, 231
  natural gas 5
    development in Russia 63–4
    uncertainty 4
Department of Energy (DOE) (USA)
  hydrogen technology goals 279–80
  progress reports 280
  R&D funding 281–2
    comparison with the EU 287–8
derivatives markets 160, 161
direct biophotolysis 212–13
distributed-user scenario 225, 233, 241
diversification
  of gas supply 8, 17–18, 63, 101–2
  of portfolios 126–7, 130
dual hydrogen supply 239

Eastern Europe 11, 18, 110, 171
  nuclear investment 145, 146
eco-efficiency 261–6
economic risk 111–12
Electricité de France (EDF) 122, 143–4, 146
electricity 298
  competition between hydrogen and 215
  markets
    and gas 103
    meaning of a liberalized market 159–61
    regulation 111
  from renewables 192, 199–200, 202
electricity companies 134–5
electricity forward agreement (EFA) market 161
electricity prices
  and British Energy's financial crisis 157–8
  Lithuania 181–2
electrolysers 208–10, 216
electrolysis 193, 208–10, 235
electro-motor 249
ENEL 144, 145
Energy Charter Treaty 11, 60, 81
energy crops 207
energy efficiency
  hydrogen transport 253–5
  policy 298–9
energy mix 114–15
energy policy, EU 178, 248–74
energy storage *see* storage of energy
Energy Watch Group 258
enlargement of markets 296, 300
environment
  eco-efficiency and transition to hydrogen 261–6
  impact of hydrogen production from renewables 214–15
  see also climate change
E.ON 144, 145, 146
Equatorial Guinea 67, 71
equity investment 128–9, 132–3
Estonia 178, 184
European Atomic Energy Community (Euratom) 166–7
European Commission 8, 40–1, 44–6, 60, 114–15

European Defence Community 166
European Economic Community (EEC) 166
European Emissions Trading Scheme 109
European gas pipeline network 21–2, 23, 24, 29, 30–2, 48–50
European Hydrogen and Fuel Cell Technology Platform (HFP) 255, 285–6
European spatial structure 243–5
'European Strategy for Sustainable, Competitive and Secure Energy' (Green Paper) 8, 60, 114–15
European Union (EU) 3
  Air Strategy and Maximum Climate Action 264–5
  energy policy goals 178
  EU–Russia Energy Dialogue 17, 60–1, 81
  integrated energy and climate policy 248
  transition to hydrogen and 248–74
  member states
    diversity of opinions on nuclear power 110
    need for solidarity between 295, 297
    visions of hydrogen sources 226–9
  nuclear energy *see* nuclear energy
  R&D efforts in hydrogen 283–7
    comparison with USA 287–8
  Second Gas Directive 11, 35–6, 39, 40, 42, 45

facility incidents 73–4
Federal Energy Regulatory Commission (FERC) 21, 24, 41, 42, 43, 44–5, 46–7, 49
federal guarantees 123, 137–9
Federal Trade Commission (FTC) 37
Fennovoima 142, 152
finance
  gas pipelines 24–5, 42
  new nuclear build 112, 117–54
    case studies 135–49
    compatibility with contractual arrangements 127–35

corporate finance 112, 118–19, 127, 129–32, 141, 144, 147, 148
  hybrid finance 118, 127, 131–2
  impact on cost of capital 132–5
  project finance 112, 118, 127, 129–32, 138–9, 147, 148
  Romania 176
financial crisis 4
financial leverage 158–9
Finland 169
  Okiluoto III project 140–2, 148
first-of-a-kind (FOAK) plants 112, 121
fixed-price contracts 122
Flamanville 3 reactor project 143–4, 148
fleet cash-flow graphs 241–2, 243
foreign policy 299
forestry wastes 207
fossil fuels 191
  Europe's security of supply 256–9
  hydrogen production from 192–3
  reserve depletion 199
  see also coal; gas, natural; oil
France 243
  nuclear power 169–70
    Flamanville 3 reactor project 143–4, 148
free-riding 295, 296
fuel cash-flow graphs 241–3
fuel cell electric vehicles (FCEVs) 195, 196, 221, 222–3, 249–51, 275–6
  cost efficiency 266–70
  eco-efficiency 261–6
  energy efficiency 253–4
fuel cells 194, 195
  costs 280–1
fuel mix 167
fuel taxes 269–70
fuelling stations, hydrogen 224–5, 230–2, 233, 236, 240
funding for hydrogen R&D 289
  USA 281–3
    comparison with the EU 287–8
fusion, nuclear 113, 196

gas, natural 3–20, 160, 271, 298
  cartel 4, 72–3, 100–1, 257
  CCGT plants 121, 124, 131
  declining production 58–9, 82
  geopolitics *see* geopolitics
  hubs 10, 11
  hydrogen production from 234
    security of supply to Europe and 256, 257–8
  import dependence 56, 58–9, 73–5
  Lithuania 183
  mixtures with hydrogen 216–17
  pipelines *see* pipelines
  regulation *see* regulation
  Romania 177
  security incidents 73–4
Gas Exporting Countries Forum (GECF) 4, 72–3, 100–101, 257
GasAndes 97–8, 99
GasBol pipeline 97
gasification 211, 234, 235
Gazprom 59, 61, 63, 64, 73, 75, 99
gearing ratio 128, 132, 150
'Generation 3' nuclear technologies 122
geopolitics 4, 15, 16–17, 91–105
  new security environment for gas 56–90
  policy recommendations 17–18
  see also global gas market
Georgia 69, 171
geothermal energy 206–7, 215
Germany 283, 287–8
global gas market 92–103
  emergence 92, 93–5
  government roles 93, 95–100
  and supply security 93, 100–2
  uncertainty and the shift to gas 93, 102–3
governments
  emergency financial support 155, 157
  federal support for nuclear investment 123, 137–9
  role and global gas market 93, 95–100
  subsidies 123, 283, 289
greenhouse gas emissions 178
  carbon dioxide *see* carbon dioxide emissions
  hydrogen transition and reduction of 261–3, 265–6
  renewables and 214–15
guaranteed lifetime load factor 124

Subject index

heat, from hydrogen 223
heavy water 174–5
Henry Hub 25, 26, 28
Hepburn Amendment 34–5, 43
High Level Group (HLG) on hydrogen and fuel cells 285
high-temperature electrolysis systems 209–10
holding companies 37–8
hurricane season of 2005 25, 28, 51
hybrid electric vehicles (HEVs) 249, 271
hybrid financing approaches 118, 127, 131–2
hydroelectricity 205–6, 215, 216
hydrogen 191–8
   competition between electricity and 215
   early transit road network 230, 231
   early-user centres 224–5, 230, 231
   end uses 194–5
      deployment of end-use applications 222–4
   EU member state visions of hydrogen sources 226–9
   infrastructure 193–4, 216–17
      build-up in Europe 221–47
   production 192–3
      primary energy basis 252–3
      production mix 232–6
      way forward 289
   R&D 195–6, 275–92
   from renewables *see* renewables
   security of supply 255–61
   transition
      determinants of 277–9
      and EU energy and climate policy goals 248–74
   transport *see* fuel cell electric vehicles (FCEVs); transport
hydrogenase 212
hydropower 204–6, 215, 260
HyWays project 221–2, 224, 254

Ignalina nuclear reactor 179–85
import dependence
   gas 56, 58–9
      security and 73–5
   uranium 167
incremental pricing 46, 47, 53

India 12
   LNG imports 70, 79–81
indirect biophotolysis 213
Indonesia 13, 78, 84, 96, 102
Industry Grouping (IG) 286–7
infrastructure
   failure 81–2
   hydrogen 193–4, 216–17, 221–47
   investment
      gas 8, 91–2, 103
      hydrogen 239–41
institutions
   gas pipelines
      central role 22–3
      institutional divide in US and European pipeline networks 29–47
   institutional arrangements and nuclear new build 136–49
integrated gasification combined cycle (IGCC) 211
integrated global gas market *see* global gas market
Interconnector pipeline 29, 31, 74
intermittency 192, 199–200, 202
internal combustion vehicles (ICVs) 250, 267, 268
International Energy Agency (IEA) 58, 61–2, 63, 64–7, 92, 121, 222, 223
Interstate Commerce Commission (ICC) 34–5
investment
   gas security investments in liberalized markets 75–6
   infrastructure
      gas 8, 91–2, 103
      hydrogen 239–41
   in new gas fields 63
   nuclear new build 117–54
investor confidence 102
investor–owners 33, 34
Iran 65, 66, 69–70, 72, 81, 94
   LNG project 70
Iraq 70
Italy 73, 76, 96–7, 99, 167–8, 232

Japan 12, 91, 96, 251
Joint Technology Initiative for Fuel Cells and Hydrogen (JTI) 250, 286–7, 288

Kazakhstan 70
Korea, North 173
Korea, South 12, 91
Krishna Godavari Basin 80

Latvia 178, 184
lead-times for nuclear build 121
learning costs 122–3, 137–8
liberalized markets
  contractual and financing
    arrangements for new nuclear
    investment 117–54
  electricity markets 159–61
  gas 10–11
  nuclear power in the UK 155–65
Libya 65, 66, 71
Lietuvos Energija 182, 183–4
life-cycle assessment (LCA) 214–15
liquefaction of biomass 207
liquefied natural gas (LNG) 11, 12,
    16–17, 75, 91, 103, 257
  Asian countries 12–13
    China and India 79–81
  Atlantic Basin 76–9
  GECF 72, 73
  global competition for LNG
    supplies 57, 76–81
  government role in global gas
    market 96
  West African exports 67, 83
liquid hydrogen 194
  truck deliveries 236, 237, 238–9
Lisbon Agenda 4
Lisbon Treaty 166
Lithuania
  National Energy Strategy 180,
    181–3, 185
  nuclear energy 172, 178–85
  treaty of accession to the EU
    179–80
loan guarantees 138–9
loan instruments 42
local politics 102–3
local pollutants 264–5, 266
local-use scenarios 229–30
long-distance road scenario 229–30
long-term contracts
  gas pipelines 21, 23, 36–7, 47,
    301
  nuclear energy 125, 138

Magnox Electric 156
Malaysia 13
market risks 118, 124–7, 138, 141,
    143–4, 148
markets
  commodity markets 160–1
  hydrogen and 278, 291
  R&D 289
  natural gas 6, 9
    integrated global gas market *see*
      global gas market
    interconnectedness 8, 18
    modelling 14–15
    new market structures 95–100
    regional approaches to security
      9–13
  niche 278
  nuclear power and liberalized
    markets
    contractual and financing
      arrangements for new
      investment 117–54
    UK 155–65
  related 278
  and security of supply
    centrality of markets 296, 299
    public policies facilitating
      enlargement of markets 296,
      300
    wider markets 296, 299–300
merchant model for nuclear new build
    136–9, 148
mergers 300–1
Mexico 76–7
Middle East 13, 64–7, 67, 68, 69–71,
    82, 94
mineriads 173
minimum efficient scale (MES) of
    capacity 161
Mobil Corporation 96
Moldova 69
monopolies
  natural 22, 50
  public utilities 23, 49
MOREHyS model 226
municipal wastes 207

Nabucco pipeline 70
National Power 160
natural gas *see* gas, natural

natural monopolies 22, 50
Netherlands, the 58, 185–6
network effects 196
networks, open-access pan-European 295, 298
new build nuclear power 112
   contractual and financing arrangements 117–54
niche markets 278
Nigeria 67, 68, 71
NIMBY ('not in my back yard') syndrome 298
nitrate oxides 263
non-combustible hydrogen resources 252–3
Nordpool 186
Nordstream pipeline 62
Norpipe 29, 30
North Africa 64–7, 71, 82–3
North American gas market 76–7
Norway 58, 85
NRG Energy STP 137–9
Nuclear Electric 156
nuclear energy 109–16, 200
   characteristics 158–9
   context 110–11
   in the EU 166–88
      current situation 167–8
      diversity of member state opinions 110
      in the EU-15 168–70
      fifty-year history 166–8
      new member states 170–86
      nuclear renaissance 109–10
   European Commission 2006 Green Paper 114–15
   fuel security and the nuclear fuel cycle 112–13
   future nuclear energy systems 113
   hydrogen production from 193, 234, 235
      EU security of supply and 259
   new build
      contractual and financing arrangements 117–54
      and the supply chain 112
   nuclear plant ownership and credit rating 133–4
   regulation 111–12, 120–1
   UK 112, 121, 145, 146, 155–65, 169

nuclear fuel cycle 112–13
nuclear fusion 113, 196
nuclear weapons 114, 175
Nuclearelectrica 174, 176

OECD 67, 68–9
oil 25, 26, 101, 160, 191, 240, 248–9, 266–7
   prices 268, 269, 270
   security of supply to Europe 256, 257–8
   US oil pipeline system 38, 48
Okiluoto III project 140–2, 148
onsite hydrogen production technology 235, 236
open-access pan-European networks 295, 298
operating leverage 158–9
operating risk 118, 123–4
Orenburg pipeline system 29
overlapping jurisdictions 39–41
ownership separation 18, 33, 43–4
Oxford Institute for Energy Studies (OIES) 58

Pacific Basin 78, 79
Pakistan 80, 81
pan-European networks, open-access 295, 298
particulate matter (PM) 263–5
penetration scenarios for hydrogen vehicles 222, 223, 229
performance risk 118, 123–4, 137, 140–1
Petrobras 97
photo-electrochemical decomposition of water 204
photofermentation 213
photoheterotrophic bacteria 213
photovoltaic (PV) cells 202–3, 214
pipelines
   gas 7, 36–7
      China 80, 81
      India 80–1
      Middle East/Caspian region 70–1
      North Africa 71
      regulation 15, 15–16, 18, 21–55
   hydrogen 194, 236, 237, 239
plug-in-hybrid electric vehicles (PHEVs) 195, 200, 249, 250, 271

Poland 97, 99, 103, 182, 184, 185
policy
   climate policy *see* climate policy
   energy policy 178, 248–74
   roadmaps 115
   and security of supply 296, 298–9
      bad policy and the long term 296, 301
political risk 118, 120–1, 140, 143
portfolio bidding 126–7, 130
portfolio diversification 126–7, 130
PowerGen 160
price risk 161–2
Priority Interconnection Plan (PIP) 18
private carriers 33, 34–7
private commercial players 98–100
privately funded research 282–3
privatization 33
production tax credit (PTC) 123, 137–8
project financing 112, 118, 127, 129–32, 138–9, 147, 148
project management 122
property rights 33, 46–7
property value 33, 41–2
proton exchange membrane (PEM) fuel cells 208–9, 222
public utility holding companies 37–8
public utility monopolies 23, 49
Putin government 60, 68–9
PVC 160
PVO 140
pyrolysis 210–11

Qatar 65, 66, 69, 72, 96

reaction turbines 205
RECLUS 243
regulation
   gas 15, 15–16, 18, 21–55
      institutional divide in US and European pipeline networks 29–47
   nuclear power 111–12, 120–1
regulatory risk 118, 120–1, 138, 140, 143
relearning costs 122–3, 137–8
renewables 109–10, 178, 192
   energy sources 201–8
   EU security of supply 259–61
   hydrogen from 199–220, 234, 235, 252–3
   environmental impact 214–15
   production technologies 208–14
   storage problem 195, 200
rents, regulation of 33, 46–7
reprocessing of nuclear fuel 113
research and development (R&D) into hydrogen 195–6, 275–92
   comparison of EU and USA 287–8
   EU 283–7
   needs 277
   USA 279–83
Research Grouping (RG) 287
resilience 8–9
resource nationalism 67–8
restructuring 3–4
risk management 161
   strategies for nuclear power 161–2, 163
risk premium 147
risks
   in new nuclear investment 117–18
      mitigating or shifting away from investors 119–27, 149
   nuclear power and 162–4
   see also under individual types of risk
Romag-Prod facility 174–5
Romania 172–8
Rome Treaties 166
Rough storage facility 74, 75
Russia
   Energy Strategy 63
   gas exports 11, 12, 16, 67, 68, 68–9, 101, 171
   crisis with Ukraine 56, 59, 60, 74, 171
   EU–Russia Energy Dialogue 17, 60–1, 81
   GECF 4, 72, 100–1, 257
   global gas market 94
   LNG 77–8
   long term 82
   pipeline to Poland 97, 99
   short term 81
   supplies after Ukraine crisis 59–64
   Lithuania and energy dependence on 180–1
RWE 144, 145, 146

safety regulation 111, 120–1
Saudi Arabia 66–7, 94
Scottish Nuclear 156
Scully Capital 131–2
seasonal storage 7
secondary trading 9, 46, 47
Securities and Exchange Commission (SEC) 37, 39
segmenting of capacity 46
SEMO 186
Shell International 68
short term
 disequilibria 295, 296
 and gas supply security 5, 7
  new security environment for gas 81–2
silicon wafers 203
single regulatory authority 33, 39–41
Sixth Framework Programme (FP6) 283–5
'Snapshot 2020' 286, 287
Societatea Nationala Nuclearelectrica (SNN) 176
solar energy 202–4, 214, 234, 235, 260
solid polymer electrolyte (SPE) electrolyser 208–9
Solid State Energy Conversion Alliance (SECA) 288
solidarity between EU member states 295, 297
South-Eastern Europe 18
South Stream Gas Pipeline 62
South Texas Project (STP) 136–9, 148
Soviet Union 178–9
 see also Russia
Spain 96–7
special purpose entities (SPEs) 127, 130–1
Standard Oil Company 34–5
storage of energy
 hydrogen 191–2, 200, 223–4, 267, 281
 natural gas 7, 18
  investment in 75–6
  problem and renewables 195, 200
 strategic reserves, management of 301
subsidies
 hydrogen R&D 283, 289
 nuclear new build 123

Sucursala Cercetari Nucleare (SCN) 174
Suez-Electrabel 144, 145
sunk costs 99
sustainability 6
Sweden 169, 182, 184

TAP pipeline project 80
tariff regulation/administration 33, 41–2
technology policy 299
thermal dissociation of water 193, 211
thermochemical processes 207
 hydrogen production 210–11
third party access (TPA) obligations 33, 34–7
3M Company 281
tidal power 205
tidal stream power 205
Trans Europa Naturgas Pipeline (TENP) 29
Transelectrica 174
Trans-European Network (TEN) 18
Transgas pipeline system 29, 30, 98, 99
transit incidents 73–4
Trans-Mediterranean (Transmed) Pipeline 29, 30, 74, 96–7
transparency 8–9
 information about gas pipelines 33, 44–6
transport
 gas see pipelines
 of hydrogen 193–4, 236–9, 246
 using hydrogen as fuel 191, 192, 194–5, 222–3
  case for 275–6
  hydrogen transition and goals of EU energy and climate policy 248–74
  infrastructure analysis 224–46
  see also fuel cell electric vehicles (FCEVs)
Trinidad 77, 99
truck transport of hydrogen 236, 237–9
Turkmenistan 70, 80
turnkey contracts 122–3, 129, 140–1

Ukraine 62, 69, 81
 gas crisis with Russia 56, 59, 60, 74, 171

unconventional gas reserves 77
unipolar electrolyser 208
United Kingdom (UK)
  Energy Act 1983 164
  and EU energy policy goals 178
  gas 5
    declining production 58
    pipeline system 49
    restructured markets 10–11
    security incidents 74
  Nuclear Installations Act 1965 164
  nuclear power 112, 169
    investment 121, 145, 146
    liberalized market 155–65
United States of America (USA)
  Administrative Procedures Act 1946 41–2
  Department of Energy *see* Department of Energy (DOE)
  development of transport policy 250–1
  Energy Policy Act 2005 (EPACT) 121, 123, 138
  gas 5, 94
    CCGT plant bankruptcies 131
    competition with EU 11
    growth of the pipeline network 23–5
    independence of gas and oil markets 25, 26
    institutional divide between US and European pipeline networks 29–47
    LNG 76–8
    market and stress 25–8
    MENA gas exports 66
    regulation 16, 21–55
    restructured markets 10
    secondary trading 9, 46, 47
  Holding Company Act 1935 37–8, 39

hydrogen R&D efforts 279–83
  comparison with the EU 287–8
Interstate Commerce Act 1887 36
  Hepburn Amendment 1906 34–5, 43
Natural Gas Act 1938 40
nuclear power 159, 164
  new build investment 136–9, 148
Price-Anderson Act 1957 164
uranium 259
Uranium National Company (UNC) 174

Venezuela 72
vertical integration
  gas pipelines 23, 24, 38–9
  nuclear new build 117, 126–7, 132, 142–6, 148
volatile organic compounds 263

waste, nuclear 112–13, 120, 158–9, 169, 175
water
  electrolytic hydrogen production 193, 208–10, 235
  photo-electrochemical decomposition of 204
  thermal dissociation of 193, 211
water-gas shift reaction, biological 213
water turbines 205, 206
waterwheels 205
wave power 205
weighted average cost of capital (WACC) 133, 134
West Africa 67, 71, 83
wind energy 201–2, 214, 216, 234, 235, 260, 261
World Gas Model (WGM) 14

Yamal Peninsula 63

zero emission program 251